In Search of Cell History

In Search of Cell History

The Evolution of Life's Building Blocks

FRANKLIN M. HAROLD

THE UNIVERSITY OF CHICAGO PRESS   CHICAGO AND LONDON

Franklin M. Harold is professor emeritus of biochemistry at Colorado State University and affiliate professor of microbiology at the University of Washington.

The University of Chicago Press, Chicago 60637
The University of Chicago Press, Ltd., London
© 2014 by The University of Chicago
All rights reserved. Published 2014.
Printed in the United States of America

23 22 21 20 19 18 17 16 15 14   1 2 3 4 5

ISBN-13: 978-0-226-17414-3 (cloth)
ISBN-13: 978-0-226-17428-0 (paper)
ISBN-13: 978-0-226-17431-0 (e-book)
DOI: 10.7208/chicago/9780226174310.001.0001

Library of Congress Cataloging-in-Publication Data

Harold, Franklin M., author.
    In search of cell history : the evolution of life's building blocks / Franklin M. Harold
        pages cm
    Includes bibliographical references and index.
    ISBN 978-0-226-17414-3 (cloth : alkaline paper)—ISBN 978-0-226-17428-0
(paperback : alkaline paper)—ISBN 978-0-226-17431-0 (e-book)   1. Cells—
Evolution—Popular works. 2. Cytology—Popular works. 3. Life (Biology)—Popular
works   I. Title.
    QH582.4.H37 2014
    571.6—dc23

                                                                    2014006515

♾ This paper meets the requirements of ANSI/NISO Z39.48–1992 (Permanence of Paper).

Contents

Preface   ix

Acknowledgments   xiii

CHAPTER 1.   Cells, Genes, and Evolution: On the Nature
and Workings of Life   1

CHAPTER 2.   The Tree of Life: Universal Phylogeny
and Its Discontents   18

CHAPTER 3.   A World Mostly Made Up of Microbes: Bacteria, Archaea,
and Eukarya   35

CHAPTER 4.   The Deep Roots of Cellular Life: The Common Ancestry
of Living Things   51

CHAPTER 5.   The Perplexing Chronicles of Bioenergetics:
Making a Living, Now and in the Past   68

CHAPTER 6.   Life's Devices: On the Evolution of Prokaryotic Cells
and Their Parts   86

CHAPTER 7.   Emergence of the Eukaryotes: The Second Mystery in
Cell Evolution   104

CHAPTER 8.   Symbionts into Organelles: Mitochondria, Plastids,
and Their Kin   126

CHAPTER 9.   Reading the Rocks: What We Can Infer
from Geology   145

CHAPTER 10. Ultimate Riddle: Origin of Cellular Life   164

CHAPTER 11. The Crooked Paths of Cell Evolution:
             Cell Evolution Is Special   190

CHAPTER 12. Summing Up: Journey without Maps   215

Notes   231

Glossary   249

References   261

Index   289

Preface

Biology presents us with innumerable puzzles and also a handful of mysteries, questions that point beyond the conceptual framework of our own day. None is more fraught than the genesis of life, that unique state of matter in which the dust of the universe attains complexity, autonomy, purpose, and the capacity to reflect on its own nature. How living things came to be is arguably the most momentous issue in biology today; the object of this book is to consider how far we have come in the quest for rational understanding of our deepest roots.

Life is so familiar and ubiquitous that it is easy to forget how astonishing it is, and how sharply living things differ from those that are not alive. Living things draw matter and energy to themselves, maintain their identity, reproduce their own kind and evolve over time. Nothing else in the known universe has this capacity. Living things are made up of lifeless chemicals; their composition, and everything they do, is consistent with the laws of physics and chemistry. And yet there is nothing in those laws that would lead one to expect a universe that harbors life. At the heart of the mystery lurk cells, the elementary units of life and the smallest entities that display all its characteristics. Every living thing is made up of cells, either one cell or many, and every cell is itself a highly integrated ensemble of millions of molecules structured in space. Truly, "the cell is the microcosm of life, and in its origin, nature and continuity resides the entire problem of biology."[1] When we inquire into the genesis of life, the primary objective must be to understand the origin of cells and of cellular organization.

The history of life unfolds over a time span so vast it boggles the imagination. I, for one, can draw no meaning from a million years, let alone a billion, and prefer a geographical metric. Let 1 millimeter, the thickness of a dime, stand for 1 year. Then 1 meter makes a millennium, 1 kilometer

1 million years, and the age of the earth (about 4.5 billion years) spans 4,500 kilometers, a little more than the distance between Miami and Seattle. As you fly northwest across the United States, armed with the latest in spyware, you may spot the first signs of life over Tennessee. Unmistakable bacterial cells appear over Kansas, eukaryotic ones over Nebraska. Western Montana features the Cambrian explosion, mammals make their debut near Spokane; and the plane is preparing to land before the first hominids rise up on 2 feet, just 4 or 5 kilometers short of SeaTac Airport. That huge span of time between Memphis and Missoula, 3,400 to 600 million years ago, is the era of cell evolution; and it remains today as thinly charted as the West was when Lewis and Clark set out on their Voyage of Discovery.

The evolution of life as displayed in museums of natural history turns on forms and functions whose scale matches our own: skulls, wings and fins, leaf imprints, ammonites and the puzzling creatures of the Burgess Shale. But all these are latecomers. For more than three-quarters of life's history, all the life that lived consisted of single-celled microorganisms invisible to the naked eye. By the time multicellular organisms appear in the geological record some six hundred million years ago, the evolution of cells themselves had largely run its course. To be sure, cell evolution continues even today, with the proliferation and diversification of bacteria and protists and the elaboration of the specialized cells of animals and plants. But the core of the subject is the genesis of the basic cell types, and their spatial and functional organization. Everything we know suggests that between Memphis and Missoula, cells arose from the primordial slime; acquired the familiar complement of parts and functions including ribosomes, cell walls, flagella, protein synthesis, and photosynthesis; and generated the molecules that make up those structures and underpin their operations. Cell evolution is, first and foremost, about how cells came to be; and as we pursue this inquiry, we are bound to achieve better insight into the nature of life itself.

The genesis of cell organization is a historical subject, for which the sources of information are quite limited. In principle, the most direct is the testimony of the rocks. Fossils have been invaluable in tracing the evolution of animals and plants, but are much less enlightening about unicellular microbes. Even so, geology supplies a chronological framework, a window onto the environments in which life first took hold, and an invaluable corrective to exuberant speculation. Synthetic biology, the ongoing effort to construct living cells in the laboratory, may in the future offer useful insights but has thus far generated more heat and smoke than light. The most productive mine of information by far has been the biology of living organisms, genomics in particular. In the 1960s it was first realized that the sequences of proteins and nucleic acids hold a record of the evolution of those molecules, and by implication of the organisms that house them. Forty years on, we are well lost in that Library of Babel: "We have now come to rely on gene phylogenies to recreate evolution's pattern, and to see genomes as surrogates for organisms, as well as chroniclers of the process of evolution: we have reduced evolution to (phylo)genetics and genomics."[2] Anyone who reflects on evolution must draw heavily on genomic data, and so will I. But the interpretation of the genomic record is anything but straightforward, and very few inferences enjoy general assent. Besides, it turned out that the history of cells is not identical with the history of their genes, and there is reason to believe that not all the information that specifies cellular architecture is carried in the genome. In the end there is no safety in sequences, or anywhere else.

Cell evolution is a quirky subject, out on the margins of serious science. The traces of events that occurred in the remote past are commonly faint and ambiguous. Interpretations rely heavily on extrapolation and conjecture; opinions are passionately held, but rest as much on conviction and rhetoric as on objective facts. The standard method of science, the formulation and testing of falsifiable hypotheses, can seldom be applied. But those who love science for its big questions and windy spaces find the lure of cell evolution irresistible: here, more than anywhere else, we touch the outermost limits of what we know, perhaps the limits of what we can know.

This book is a sequel to my previous one, *The Way of the Cell*,[3] which sought an answer to the question most famously posed by the physicist Erwin Schrödinger, What is Life? That book touched on many evolutionary matters, but it was not until I began to compose its successor that I realized just how explosively the literature has grown over the past decade. I have tried to ground the present volume squarely in the current literature; my object is not to promote my own interpretations (which continue to evolve) but to render a fair and comprehensible account of the ideas and controversies that keep the field in turmoil. This has entailed many personal choices among competing claims, which explains the frequent appearance of the first-person singular in these pages. To hold the length within reason, it was necessary to assume that the reader has a basic knowledge of cellular and molecular biology, but I have made every effort to keep the text accessible and technicalities to a minimum. Readers who wish to delve deeper will find numerous portals among the references, largely selected from publications over the past decade; responses to the literature extend through 2012.

I came to cell evolution through the back door by way of a long engagement with the spatial organization of biochemical processes, first in bacterial energy transduction and later in cellular morphogenesis. For that reason I look on cell evolution from the perspective of a physiologist, as the progressive emergence of spatially ordered systems of molecules that possess the extraordinary capacity to make themselves. This book was written in the belief that one can learn something about biological organization from the history of cells, and come to better understand cell history by examining it through the prism of organization. Inevitably, many matters discussed in this book fall outside the range of my professional expertise. This is a familiar predicament that confronts anyone who would write of science for a wider audience; and so I take comfort from the apologia of Thomas Sprat, who set out to write a history of the Royal Society of London three hundred years ago.

> Perhaps this task which I have propos'd to my self will incur the censure of many judicious men, who may think it an over-hasty and presumptuous attempt . . . Although I come to the performance of this work with much less Deliberation and Ability than the Weightiness of it requires; yet, I trust, that the Greatness of the Design itself, on which I am to speak . . . will serve to make something for my Excuse.[4]

Acknowledgments

The object of science is to make sense of the world, but our immediate product is commonly information, reams and reams of it, far more than any one person can assimilate. So I am indebted first and foremost to the scholars of science, those who set in order the raw output of research in the form of reviews and commentaries. This book could never have been written without them. Many such contributions are cited in the text, but formal acknowledgment fails to do justice to mentors, colleagues, and friends, past and present, who sharpened my understanding of evolution through their writings or by instruction, correspondence, and conversation. Ford Doolittle, Stephen Jay Gould, Arthur Koch, Nick Lane, Lynn Margulis, William Martin, Daniel McShea, Peter Mitchell, Harold Morowitz, Norman Pace, Moselio Schaechter and Roger Stanier, thank you all. The alphabet dictates that Carl Woese come last on this list, but in my debt he ranks first, for it is his vision of cell evolution that set the course for this journey. Special thanks are also due to Loren Eiseley—naturalist, philosopher, and poet—who taught me at a critical juncture that the true use of science is to make the world intelligible and renewed my commitment to my profession.

Colleagues and friends in Seattle and elsewhere reviewed portions of this book during its gestation. John Leigh and Moselio Schaechter commented on individual chapters; Eric Fourmentin, Ruth Harold, Nick Lane, Daniel McShea, Diana Sheiness, and James Staley loyally stayed the entire course. They, as well as two anonymous reviewers, spotted more ambiguities, misstatements, omissions, and outright errors than I care to admit; and I have corrected those as best I could. In the end, the opinions and attitudes expressed here are my own and no one else should be held responsible.

I write by hand, and so I am much indebted to Jennifer Chubb for turning stacks of scribbled yellow paper into neat typescript. Thanks are due also to colleagues who allowed me to adapt diagrams from their work or supplied original photomicrographs, and to the journals in which those contributions were first published. Kate Sweeney, medical and scientific illustrator at the University of Washington, and artists at Visual Health Solutions in Fort Collins, Colorado, drew the illustrations. I am doubly grateful to Jennifer Chubb, Diana Sheiness, and to my ever-helpful editor Christopher Chung for keeping this project on track when I was sidelined by illness while preparing the manuscript for publication.

Finally my thoughts turn to my wife Ruth, who aided and abetted this project from the beginning and cheerfully put up with an oft-distracted husband. In the lab, wandering the globe, and throughout life's journey, I could not have wished for better company.

CHAPTER ONE

Cells, Genes, and Evolution

On the Nature and Workings of Life

Men talk much of matter and energy, of the struggle for existence that molds the shape of life. These things exist, it is true; but more delicate, elusive, quicker than the fins in water, is that mysterious principle known as "organization," which leaves all other mysteries concerned with life stale and insignificant by comparison. For that without organization life does not persist is obvious. Yet this organization itself is not strictly the product of life, nor of selection.
— Loren Eiseley, *The Immense Journey*

The Doctrine of the Cell
Microbes Come on Stage
Prokaryotes and Eukaryotes
Molecular Systems of Daunting Complexity
Genes Rule
Cell Heredity
Cell Evolution: What Nobody Is Sure About

Unlike most husbands, I quite enjoy doing the dishes; it seems to satisfy a need to impose order on my corner of the world, at least for a little while. I begin with the sink piled helter-skelter with soiled plates, cups, and cutlery; twenty minutes later all are clean and neatly ranged in the rack, the large ones in the rear and the small ones in front. I look upon my work and see that it is good, and I have no doubt that the same need to find order in the universe motivates much of science.

Biologists encounter the tension between order and randomness every day, for living things differ from nonliving ones most strikingly in their degree of order. When used in its technical sense, the term "order" refers to regularity, predictability, and conformance to law. Wallpaper is ordered, repeating a particular pattern over and again. The deep blue bird that occasionally visits my garden is called a Steller's jay. It furnishes a spectacular example of order, for that label immediately implies a host of regular and predictable features: forms and colors, a set of anatomical and biochemical characteristics, even a pattern of behavior. Regularity and predictability are also found in the nonliving world (the solar system comes to mind), but not to the same degree. Besides, the order that living things display is of a special kind, commonly termed "organization": that jay's patterns of order have purpose or function. This feature is seen only in living things and their artifacts, such as airplanes, spider webs, or the shells constructed by testate amoebas. Nevertheless, their intricate organization notwithstanding, living things are creatures of contingency. They are not manifestations of physical laws as the solar system is, nor were they designed for a purpose like the airplane, but rather they evolved by the interplay of random variation and natural selection. Living things conform to the laws of physics and chemistry but are not fully explained by those laws; and their existence could not be deduced from physics and chemistry. Despite their familiarity and ubiquity, living things are truly strange objects.

The Doctrine of the Cell

One of the earliest and most profound statements about biological organization is the cell theory, rightly acclaimed in every textbook as a cornerstone of biological science. The theory asserts that all the infinite diversity of living things is constructed on a single architectural plan: every organism is made up of cells, consisting either of a single cell or of a society of many cells. Cells are the atoms of life, and life is what cells do.

The idea of the cell emerged gradually over a period of two centuries.[1] Robert Hooke coined the term in 1665, when he examined thin slices of cork through a microscope and saw a pattern of rectangular boxes that reminded him of monks' cells (we now know that he was not looking at cells in the modern sense, but at their empty rigid walls). As microscopes grew sharper and more powerful others reported similar patterns, and the suspicion grew that cells might be a general feature of living things. It fell to a pair of German scientists, the botanist Matthias Schleiden and the physiologist Theodor Schwann, to articulate the consensus that was waiting to be born (1838 and 1839). Their "cell doctrine" stated that the tissues of plants and animals were not homogenous wholes, but rather composed of innumerable tiny individual cells. Each cell consists of a droplet of jelly, later called protoplasm, enclosing a dense central kernel, or nucleus. And they explicitly recognized that each cell is itself a complex and organized structure and the seat of the organism's vital activities.

Two decades later the prominent pathologist and physician Rudolf Virchow took the next step. Schleiden held the position that cells formed by aggregation of protoplasm around the nucleus. Virchow knew better: in his textbook of pathology, published in 1858, he insisted that every cell originates by division from a pre-existing cell. His famous aphorism, *Omnis cellula e cellula* ("every cell from a previous cell") remains another landmark of biological science. Cellular organization has passed continuously from the dawn of life to the present day. Organization is sometimes transmitted by division, sometimes by the fusion of gametes, but it never arises *de novo*. With the rise of molecular science, Virchow's law has been marginalized, but it continues to hold and is central to the present book.

Microbes Come on Stage

The pioneers of the cell doctrine thought entirely in terms of higher organisms, the multicellular animals and plants; of the microbial world the early nineteenth century knew very little. Microscopic organisms had been seen and described by Anton van Leeuwenhoek (1632—1723), merchant and civic official of Delft in Holland and a lens grinder of extraordinary skill. With the aid of what was, in effect, a powerful magnifying glass, Leeuwenhoek observed spermatozoa and red blood cells, capillary vessels, all the major kinds of algae, protozoa, and yeast, and even some bacteria. He reported his discoveries in a stream of letters (in Dutch) addressed to England's Royal Society, which duly translated and published them. But the pace of discovery slackened after Leeuwenhoek's death, largely for technical reasons, and quickened only with the advent of more advanced microscopes.

The novel and sometimes peculiar creatures thus revealed posed problems for biologists, whose interests centered on taxonomy. Scientific tradition reaching clear back to Aristotle recognized two kingdoms of living organisms, animals and plants. Some of the microorganisms could be shoehorned into one or the other kingdom, but that procedure grew increasingly unsatisfactory as microscopists discovered tiny organisms that were neither plants nor animals yet had qualities of both; the alga *Euglena*, for example, which is both green and motile. There was much argument over whether such organisms should be considered unicellular or noncellular, but the instruments that revealed the internal structure of protozoa and algae also documented their essential affinity with the cells of higher organisms. In 1866, Ernst Haeckel, Darwin's champion in Germany and one of the most prominent scientists of his day, published a universal classification with three, rather than two, major categories: animals, plants and protists. His kingdom Protista included a grab bag of "lower" creatures: the protozoa, unicellular green algae, fungi, diatoms, and much else besides, including the bacteria. The latter were set apart in a subgroup of their own, the Monera. Among the protists, Haeckel was convinced, would be found not only the ancestors of plants and animals, but also descendants of the primordial organisms with which life began.

The bacteria never nested comfortably among the other protists. The cells of protozoa, algae, and also fungi were organized along the same lines as those of plants and animals, with a nucleus that divides by mitosis, a bounding membrane and various internal organelles and inclusions. Bacterial cells were much smaller and lacked a nucleus, they did not divide by mitosis, and even seemed to do without heredity. As early as 1938, a formal proposal was made to remove bacteria from the protistan realm and assign them a kingdom of their own. Bacteria were clearly fundamentally unlike other cells, but the nature of the difference remained undefined for another quarter of a century.

Prokaryotes and Eukaryotes

By the middle of the twentieth century, microbiology was becoming a hotbed of intense research. Bacteria had been well recognized as a large and diverse group of organisms, the agents of human diseases and industrial processes as well as the grand nutrient cycles, and probably the oldest forms of life. Moreover, thanks to their small size and relatively simple organization, bacteria had become the beacon that would illuminate all of cell physiology, biochemistry and genetics. *Escherichia coli* reigned as everyone's favorite model organism. Yet those microbiologists whose interests ran toward natural history often felt frustrated by their inability to achieve an objective classification of the bacteria, or even to define the relationship of bacteria to the rest of the living world.

The question of the essential nature of bacteria was tackled by two of the most respected microbiologists of the time, Roger Stanier (1916–1982) and C. B. van Niel (1897–1985), in a magisterial paper entitled "The Concept of a Bacterium" (1962)[2] By then, thanks chiefly to the perfection of the electron microscope, enough had been learned about the ultrastructure of bacteria to differentiate them unambiguously from other kinds of cells. Stanier himself was especially influenced by his mounting interest in the cyanobacteria, photosynthetic organisms familiar to everyone as the green scum that forms on the surface of stagnant ponds. The "blue-green algae" were traditionally considered to be simple plants and studied by botanists; yet their fine structure clearly ranked them with the bacteria and demanded that they be reclassified. To give substance to the concept of bacteria as a separate, kingdom-level class of organisms, Stanier and Van Niel adopted and promoted terminology that had been mooted by the French protozoologist Édouard Chatton thirty years before: eukaryotes and prokaryotes, cells endowed with a true nucleus and cells without.

The tale of how this fundamental distinction came to be naturalized in biology is curious, and illuminates how science actually works. Several generations of students have been taught to credit Chatton with its discovery. But when the historian Jan Sapp reexamined the matter,[3] he found that Chatton himself had made little of the distinction between prokaryotes and eukaryotes; to him these were just convenient labels, not an insight into the nature of things. It was really Stanier and Van Niel who drew the bright line across biology, separating two modes of biological organization.

The cells of plants and animals, including our own, display the eukaryotic mode. The same is true of fungi and of protists (fig. 1.1). All possess a membrane-bound nucleus that contains chromosomes and divides by mitosis. They contain organelles such as mitochondria, golgi, and (in photosynthetic organisms) plastids. An intricate system of internal membranes pervades the cytoplasm, and a cytoskeleton can often be made out. They also have cilia and flagella, organs of motility built around microtubules (I like Lynn Margulis's term "undulipodia" to designate these structures and shall use it hereafter). Prokaryotic cells are much smaller and simpler in structure. They lack a nuclear membrane, chromosomes, and mitosis; there are no organelles and (usually) no internal membranes. Their flagella differ from undulipodia in structure and operation, and their cell walls are chemically different from those of eukaryotes. Stanier and Van Niel defined prokaryotes more in terms of what they lack than by any

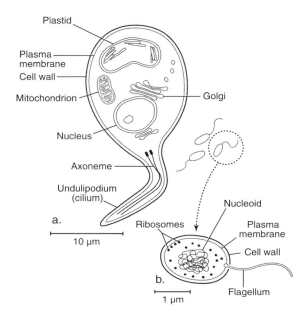

FIGURE I.I. Eukaryotes (a) and prokaryotes (b). Schematic sketches of generalized cells; note the disparity in size and architectural complexity.

positive attributes, but those differences sufficed. "The distinctive property of bacteria and blue-green algae is the procaryotic nature of their
cells. It is on this basis that they can be clearly segregated from all other
protists (namely, other algae, protozoa, and fungi), which have eucary-
otic cells."[4] Until very recently, few would have questioned the judgment
that "this basic divergence in cellular structure . . . represents the greatest
single evolutionary discontinuity to be found in the present day world."[5]

The division into two kinds of organisms, eukaryotes and prokaryotes,
was quickly and enthusiastically accepted and was soon reflected in the
large-scale classification of living things. In 1969, when the ecologist Robert Whittaker revised the scheme he had formulated a decade earlier, he
divided up the world among five kingdoms—four eukaryotic and one prokaryotic.[6] Animals, plants, fungi and protists (or protoctists, in Margulis's
terminology) compose eukaryotes. All the bacteria—and only bacteria—
were placed in the kingdom Monera. The scheme omitted viruses, which
are not made of cells. Five kingdoms were never intended as a statement
about who begat whom, but came to be taken as such: the presumption

was, and commonly still is, that prokaryotes preceded eukaryotes and were the latter's evolutionary precursors.

Five kingdoms proved to be a practical system for putting in order the overwhelming diversity of life on earth; writers of textbooks, obliged to survey the landscape and make it comprehensible to students, found it indispensable. But it must be said that as a guide to the evolution of life, the scheme is profoundly misleading. It puts all five kingdoms on an equal footing, implying that the difference between prokaryotes and eukaryotes is of the same kind and magnitude as that between animals and plants. It gives no clue to the gulf of time that separates the familiar multicellular creatures from the microbial world. And there is surely something lacking in a scheme that covers all of life but has no place for viruses. It is true that viruses are not cellular in nature and are obligatory parasites upon other organisms; whether viruses should be considered living depends on your definition of life.[7] But they are made of the same kinds of molecules as true living things, reproduce with heredity, and evolve all too quickly; they are linked to the great tree of life and must eventually be represented there. Just how drastically our perceptions had to change only became clear with the development of novel, molecular methods to assess relationships among living things and explore the patterns of descent.

But before we go there, let us pause and consider what the concept of the cell means today. A century ago, it meant first and foremost a structural motif: a droplet of cytoplasm, a nucleus, perhaps some organelles, and a plasma membrane to separate inside from outside. The majority of living things is, indeed, constructed upon this plan; but we know so many exceptions and variations that its universality is no longer obvious. The cells of eukaryotes and prokaryotes differ profoundly in size and organization, and each comes in many versions. Discrete nuclei and organelles are common but not indispensable. And many highly successful organisms live as collectives in which numerous nuclei inhabit a common cytoplasm (fungi, oömycetes, many algae, also muscles). A membrane that segregates the cytoplasm from the rest of the world still seems an absolute requirement (to me, at least), but even this has been called into question.[8]

What, then, is the cell that we should praise it as a cornerstone of biological science? Perhaps we should look beyond the architectural commonalities and emphasize that cells are the minimal units capable of underpinning the phenomenon of life. Life, at least the kind we know, is unavoidably complicated. To metabolize, grow, reproduce, and evolve requires a system made up of thousands of collaborating molecules, spatially organized within a sheltering boundary. Discrete units, the gametes, are necessary at least once in a life cycle, to evade the inevitable ravages of time and start afresh with a clean slate. Single cells are the smallest units that meet these requirements, but more elaborate variations are possible. Nature is endlessly flexible and will endorse whatever works in any particular milieu.

Molecular Systems of Daunting Complexity

When examined under the light microscope, most bacterial cells present no more than tiny plain blobs, commonly cylindrical or spheroidal. Modern staining technology, and even more so the electron microscope, reveal structure and order within: flagella and other appendages, a multilayered envelope, various granules, a nucleoid, and, in some instances, internal membranes or a cytoskeleton. When set beside eukaryotic cells with their intricate internal architecture, prokaryotic ones do appear simple; but that illusion quickly vanishes when one turns from the cellular level to the molecular.

Take *E. coli*, that workhorse of biochemistry and molecular biology, and of all living creatures the one best understood. A single cell takes the form of a short rod, a cylinder some 2 micrometers long and 0.8 wide, with rounded caps. Under optimal conditions, 20 minutes suffice for each cell to elongate, divide, and produce 2 where there had been 1 before. But what a prodigious task this is! In that brief span of time the original cell will have produced some 2 million protein molecules, potentially of 4,000 different kinds; some 22 million lipid molecules, composing 60 varieties; 200,000 molecules of various RNAs; and nearly 1,000 species of small organic substances, some 50 million molecules in all. It will also have duplicated two unique giant molecules. One is the circular, double-stranded DNA helix, consisting of about 4.6 million nucleotide pairs; were it uncoiled, it would stretch for 1,600 micrometers. The other is the peptidoglycan layer of the cell wall, composed of some 2 million repeating units cross-linked into a huge bag-shaped molecule that encases the whole cell. All these are crammed and folded into a volume of about 1 cubic micrometer, a minute capsule filled with a concentrated gel whose properties bear little resemblance to the dilute solutions that laboratory scientists prefer.[9]

Bacterial cells supply instructive, and relatively simple, subjects for reflection on the nature and reach of biological order. The molecules of

have taken to it without reservations. We have for many decades focused attention on the relationship between genes and traits, which in microorganisms is commonly direct and straightforward. A cell that carries a mutational defect in any of the genes that specify the pathway of arginine biosynthesis will require exogenous arginine for growth. A bacterium that has acquired a plasmid that contains a gene for penicillinase becomes resistant to the antibiotic. We have learned that much of the molecular architecture of cells is quite directly specified by their genes: the primary sequences of all proteins and RNAs, many of the agents that regulate gene expression and the sites at which they act, even the localization of membrane proteins and whether they are to be secreted or inserted into the membrane. Numerous organelles, including ribosomes, microtubules, actin filaments, and many others arise by the self-assembly of preformed components without input of additional information or even energy; the structure and operation of these molecular machines must in some sense be implicit in the genetic instructions that specify their components.

Recent discoveries have tended to bolster the genocentric view of life. One is the recognition that, on the millennial timescale, genes are readily transferred across taxonomic boundaries, carrying traits into new settings and opening evolutionary opportunities. At the risk of some exaggeration, one can argue that all the genes in the prokaryotic pool make up a single common market, and organisms represent nothing more than the samples that proved successful (see chapter 2). And now come Craig Venter and his colleagues, reporting the transplantation of the entire naked genome from one species of *Mycoplasma* to another, turning the recipient species into the donor by both genetic and phenotypic criteria.[15] One can quibble over the details of these remarkable experiments, but they certainly reinforce the perception that a cell is the product of instructions encoded in its genes, and nothing more. It's not even so very complicated: a set of about three hundred genes suffices to specify the essential functions of a minimal cell.

No one denies the wealth of understanding that flowed from the transformation of biology into a molecular science; but as is commonly the way with revolutions, this one also entailed losses as well as gains. Chief among the losses is the appreciation of organisms, and of all the higher levels of order that makes them tick. Misgivings surface as soon as we focus on those quintessential qualities of living things: their growth, reproduction, and morphology. Is it true that, as François Jacob famously put it, "The whole plan of growth, the whole series of operations to be carried out, the order and the site of synthesis and their coordination are all written down in the genetic message"?[16] If so, can we specify how a linear genome on the nanometer scale determines the structure of a three-dimensional organism on a scale at least three orders of magnitude larger? Is form specified by a particular subset of the genes? Does the phrase "self-assembly" supply a satisfactory image of biological growth, as Roy Britten has argued?[17] Is it true that, when we grind living matter into a homogenate, nothing is lost that is irretrievable in principle? If something essential has been destroyed, what is it? Variation and selection at the level of genes is central to evolution; do they make up the whole story, or do more cellular modes of heredity contribute to the evolutionary play? I am but one of a cohort of skeptics who suspect that the gene-centered view of life is incomplete and fundamentally misleading; and in the following section we shall spell out the deficiencies and ask what should take its place.

Cell Heredity

The patron saint of those who harbor misgivings about the gene-centered view of life is surely Rudolf Virchow, who proclaimed 150 years ago that every cell must come from a preexisting cell. Virchow's dictum continues to hold, even in this era of genome transplantation, because the DNA that Lartigue and colleagues transplanted was copied from natural DNA and slipped into the matrix of a recipient cell.[18] Over the years a small school of contrarians and dissenters has articulated an alternative to the conventional wisdom that puts the cell in the center, rather than its genome.[19] Genes remain important, of course, as the key repository of information in an integrated multimolecular system, but they are no longer perceived as the central directing agency that makes the cell and rules its operations. There are no experiments that can prove one view true and the other false, and there will never be. Both are aspects of a rounder reality, and depending on the specific problem one stance or the other will turn out to be the more illuminating.

Good reasons to doubt that genes occupy the summit of a linear chain of command come from within genetics itself. Over the past two decades it has become abundantly clear that it is not the genes that decide which protein is to be made under what circumstances. Thanks to alternative splicing, even the protein's primary sequence may not be determined solely by its cognate gene. Instead, all these turn out to be functions of the regulatory dynamics of the cellular system.[20] Let me here pursue a different line of argument, which stems from our increasing knowledge of cell growth and morphogenesis in both prokaryotes and eukaryotes. There is not space here to do the subject justice,[21] but even a capsule summary demonstrates that the way cells make cells does not fit smoothly into a framework in which genes issue all the orders.

Briefly, the spatial organization of cells, including the arrangement of cytoplasmic elements and the cell's overall form, is not explicitly mandated by the genome. Genes do specify the primary sequences of all macromolecules (with some ambiguity, see above), and portions of those sequences commonly play a role in the localization of the product. But genes do not encode dimensions or architecture on the cellular scale. Those arise "epigenetically" from the interactions among numerous individual gene products. Many of those interactions, perhaps most, fall under the heading of self-organization, either by simple molecular self-assembly or by dynamic self-construction. Now, when this self-organizing chemistry takes place inside a living cell, it comes under constraint and guidance from the cellular system as a whole. New gene products, and also the output of biosynthetic processes, are never altogether free to diffuse at random; they are released into a milieu that is already spatially structured, and they find their place in that pattern under the influence of the existing order. These influences take diverse forms: cellular compartments, spatial markers and vectors, heritable membranes, the tracks and channels of intracellular transport, gradients and fields that supply positional information, and an array of physical forces including turgor pressure. The cytoskeleton is the chief instrument for imposing order on the construction site, and is itself a product of its own operations. Every cell is continuous with its progenitor, not only genetically but also structurally; and the new cell will in turn serve as a source of configurational instructions for its own daughters. Growing cells are not organized by genomes alone; they model themselves upon themselves, and therefore the smallest entity that can be truly said to organize itself is a whole cell. It follows that when biochemists grind cells into a pulp, something irretrievable is indeed lost: the organization that makes the cellular system work.

So does it still make sense to insist that all the essential information required to construct a cell resides in its genes? In my opinion, the answer is clearly no. With regard to cell growth and morphogenesis, it seems plain that a conceptual framework centered wholly on genes leaves out too much, particularly the spatial organization that is essential to many cellular operations. Because cellular structures are often continuous from one generation to the next, cell organization can be transmitted in ways that do not involve genes. Mechanisms of this kind have been known for fifty years, especially in eukaryotes, and are designated structural (or cell) heredity.[22] Their meaning has been much misunderstood (by myself, among others). Structural heredity is not an alternative mechanism for the transmission of specific characters; that's what genes are for. Rather, cell heredity is a consequence of the continuity of cell architecture, and complementary to the work of the genes. At the risk of some oversimplification, I argue that all instructions that can be expressed in linear, digital form end up in genes. Those concerned with three-dimensional order, such as topology and polarity, are carried by cell heredity. As the details of the genome-transplantation experiments are worked out, I expect them to prove entirely consistent with the preceding formulation.[23]

Take cell forms and organization. Genes specify all cellular proteins, including those that generate morphology (the cytoskeleton in particular), and in that indirect manner architecture comes under the genes' writ. Innumerable mutants of both prokaryotic and eukaryotic cells testify to the role of genes in shaping cells. But there is no way to predict a cell's form from its genome alone, because cellular mechanisms are required to carry the patterns of spatial order from one generation to the next. Among the agents of continuity the most prominent are membranes, including the plasma membrane and those of eukaryotic organelles such as the Golgi, endoplasmic reticulum, mitochondria, and plastids.[24] Membranes, it appears, are never constructed *de novo* but arise by the growth of a preexisting membrane. Like DNA, they have been passed continuously down the generations, ever since the dawn of cellular life; and that alone suffices to explain why every cell must come from a preexisting cell. I take the continuity of membranes as a broad hint that genes were expressed in a cellular context from the very beginning, and that structural heredity reaches equally far back in time.

What is at issue here goes far beyond figuring out how cells grow and divide. The deeper questions probe the essence of organisms, and their place in the hierarchy of nature. Cells do not simply express the instructions encoded in their genes but embody an integrated system of which genes are a part. That viewpoint has been articulated many times over the years but remains very much a minority view. A charming recent formulation comes from the physiologist Dennis Noble, contrasting his interpretation of the place of the genes with that of his Oxford colleague Richard Dawkins; Dawkins' selfish-gene phraseology is listed first below, followed by Noble's, in italics:

Now they swarm in huge colonies, safe inside gigantic
Now they are trapped in huge colonies, locked inside highly
lumbering robots, sealed off from the outside world,
intelligent beings, moulded by the outside world,
communicating with it by tortuous indirect routes,
communicating with it by complex processes,
manipulating it by remote control.
through which blindly, as if by magic, function emerges.
They are in you and me;
They are in you and me;
they created us, body and mind;
we are the system that allows their code to be read;
and their preservation is the ultimate rationale for our existence.
and their preservation is totally dependent on the joy
we experience in reproducing ourselves.
We are the ultimate rationale for their existence.[25]

The wide-open question is how cell heredity plays out in cell evolution. Does cell heredity generate a pool of variations upon which selection can act? That probably happens sometimes, as Conrad Waddington first recognized sixty years ago, but I suspect that is not its real significance. Variation and selection is the business of genes; cell heredity supplies the continuity of organization which underlies that business and makes it possible. If genes are the pieces in the great game of chance and necessity, cell heredity keeps the board.

Cell Evolution: What Nobody Is Sure About

There really are only two possible explanations for the existence of life on earth: an act of creation by some higher power or evolution by natural causes from the raw materials of the cosmos. Darwin convincingly marshaled the evidence that life has changed and evolved over time; but his leap of genius was to apprehend how evolution works.[26] Living things arose by "descent with modification," and the underlying mechanism is the interplay of heredity, variation, and natural selection. His theory became the cornerstone of biology because it makes the subject intelligible. Broadly speaking, evolution explains the pattern of nature: why cats resemble tigers but are quite unlike birds, how organisms adapt to their environment, how the living world became so diverse, and why lineages eventually go extinct. The important corollary, that all living things are members of a single extended family and ultimately share a common ancestry, has been validated many times and at all levels of the biological hierarchy. It is manifest in the pattern of anatomical homologies, such as those that link the bones of animal limbs to those of birds' wings; it is reinforced by the pervasive unity of biochemical processes; and loudly proclaimed by the universality of the genetic code. Moreover, evolution is not some special attribute of our particular world. As the late John Maynard Smith observed a quarter of a century ago, "Given a population of entities having the properties of multiplication, variation and heredity, and given that some of the variation affects the success of these entities in surviving and multiplying, then that population will evolve; that is, the nature of its constituent entities will change in time."[27]

The general case for evolution as the sculptor of living forms and functions has been made many times by some of our most articulate writers.[28] Anyone who lets himself be guided by reason will find the argument overwhelming, and I do not propose to repeat it here. Like the great majority of my colleagues, I start from the premise that both the origin of life and its subsequent expansion should be, and eventually will be, wholly accounted for by natural processes operating within the framework of chemistry and physics. Specifically, I look to a generalized theory of evolution to supply an intelligible and convincing account of the beginning of life and its early history. But this is no more than a statement of principle; much flesh must be put on those bones before genesis feels fully at home in the natural sciences, and subsequent chapters will strive to do that.

Within the sprawling domain of evolution studies, cell evolution makes up a special province for two reasons. First, natural selection can only act upon entities endowed with the capacity for multiplication, heredity, and variation; cells of some sort must have come first. In fact, I take the origin of cells to be synonymous with the origin of life. Second, cell evolution takes us clear outside and beyond standard Darwinian thought. Even the simplest cells are full-fledged organisms made up of millions of molecules, arrayed in functional order and acting in a purposeful manner to harvest energy, maintain their integrity and reproduce their own kind. How can such dynamic Gordian knots have arisen of their own accord from what Loren Eiseley called the "stolid realm of rock and soil and mineral"?[29] And having done so, how and why did they diversify into the major cell types that dominate the living world today?

The evolution of cells took place so long ago that all but the faintest traces of how it came about have been erased. We will probably never know exactly how cellular life came to be. What we can hope to do is to discern the forces that drove the emergence and transformations of early life, and to identify some landmarks in that history. In order to help readers keep their bearings during the immense journey on which we are about to embark, let me close this chapter with a short list of key questions to which we seek answers.

How many kinds of cellular design does our world hold, and how are they related to each other?

Ever since Darwin, biologists have envisaged the history of life as a great tree. Is that metaphor accurate, or has it outlived its usefulness?

What are viruses, and how do they relate to cellular organisms?

Can one construct a timeline for the origin and early history of life?

Do all living things share a common ancestor, and what was the nature of that entity?

Why are eukaryotic cells so much more complex than prokaryotic ones, and how did they arise?

Has functional, adaptive organization increased over time, and if so, why?

Is there an alternative way to generate functional organization that does not depend on heredity, variation, and selection?

How, where, and when did life emerge from the lifeless world of physics and chemistry?

Can a generalized theory of evolution account for the origin of life?

Is the history of life a succession of contingent events, or does it have direction and meaning?

I do not promise to answer the questions, only to engage with them. But by the end of this book, readers should have a clear sense why the answers have proven so elusive and for so long.

CHAPTER TWO

The Tree of Life

Universal Phylogeny and Its Discontents

Humility is not a frame of mind conducive to the advancement of learning. — Sir Peter
Medawar

A Molecular Chronometer
Microbes Traffic in Genes
Still Seeking the True Tree
Timeframe Wanted
A Tree to Hug Lightly

We make sense of the world by sorting the endless multiplicity of objects and events into categories that suit our own purpose. As a rank amateur, I find good use for field books that arrange wildflowers by color but lay no claim to deeper meaning. Even the hierarchical classification of all living things devised by Linnaeus at the end of the eighteenth century, which puts every organism into a bin nested within a larger bin—species, genus, family, and so on, up to phylum and at last kingdom—was intended for practical purposes, not as a statement about the nature of life. Only with Darwin did the utilitarian craft of taxonomy acquire a higher goal: to devise a natural classification of all living things, one that corresponds to their phylogeny, or line of descent. And it was Darwin who bestowed upon this field the compelling image of a tree:

> The affinities of all the beings of the same class have sometimes been represented by a great tree. I believe this simile largely speaks the truth. The green and budding twigs may represent existing species; and those produced during each former year may represent the long succession of extinct species. . . . The

limbs divided into great branches, and these into lesser and lesser branches, were themselves once, when the tree was small, budding twigs; and this connection of the former and present buds by ramifying branches may well represent the classification of all extinct and living species in groups subordinate to groups.[1]

The quest for a universal phylogeny has again moved to the forefront of biological science for it furnishes the foundations of any inquiry into the history of life.

A Molecular Chronometer

A century and a half after publication of *The Origin of Species*, the history of life on earth continues to produce surprises and to generate controversy. In both respects, the most exhilarating development is surely the emergence, over the past thirty years, of a unified genealogical tree covering all the kingdoms of nature, just as Darwin had envisaged. The tree is unfinished; many of its features remain uncertain, unknown or in dispute; the very existence of a tree of life has been questioned. Nevertheless, it's a safe bet that future versions of biological history will not spring forth altogether afresh, but rather descend with modifications from the current universal tree. Perhaps the best measure of the conceptual significance of phylogeny is the fact that, more than twenty years after it was formulated, the tree is still under assault.

By the time Stanier and Van Niel divided the living world into prokaryotes and eukaryotes, the phylogeny of plants and animals had been mostly worked out; so had the conventional understanding of both the course of evolution and its underlying mechanisms, drawn entirely from work with higher organisms and their rich fossil record. By contrast, the phylogeny of microorganisms remained virtually unknown, with that of prokaryotes an area of particular darkness. Morphology, physiology, and biochemistry furnished no convincing clues to the evolutionary history of prokaryotes, and bacteriologists had effectively conceded defeat. The solution to this and other problems came from advances in technology, specifically, the advent of molecular sequences in the 1960s, with sweeping consequences for all of evolutionary thought.

Very briefly, proteins and nucleic acids are large linear molecules made up of monomeric units (amino acids and nucleotides, respectively) strung together in series, like letters in a very long word. The sequence of monomers is a characteristic and nearly invariant feature of any given molecule, but it changes on the evolutionary timescale as a result of mutations, errors in replication, and rearrangements. The hemoglobin of a horse is nearly identical to that of a donkey, differing in just one or two positions; pig hemoglobin differs more extensively, and whale hemoglobin even more so. The differences can be expressed numerically and furnish a measure of the evolutionary distances among the organisms from which the hemoglobin came. As methods were developed to determine sequences quickly and cheaply, they unlocked a huge trove of evolutionary information and brought within sight the realization of Darwin's dream of a universal tree of life.

As is so often the case in science, the breakthrough came at the hands of an outsider. Carl Woese (1928–2012), a biophysicist by training and an early master of molecular biology, was interested in the translation of genetic information into protein structure. That is the business of ribosomes, tiny molecular machines that zip the amino acids into the finished protein. Ribosomes consist of 2 subunits, a larger and a smaller, each of which is made up of some 24 proteins and a long strand of RNA (as a rule, a short extra strand of RNA is present as well). In the early 1970s, Woese conceived the idea of using those RNA molecules to assess phylogenetic relationships among the organisms from which they came. At that time, methods to determine the sequence of nucleotides had not yet been developed, so Woese devised an indirect and exceedingly laborious method to achieve his purpose. RNA isolated from the smaller subunit (known in the trade as 16S RNA, or SSU rRNA), some 1,500 nucleotides long, was carefully digested with T1 ribonuclease to generate specific fragments up to 20 nucleotides in length. These could be separated by electrophoresis and sequenced, to produce an "oligonucleotide catalog" characteristic of the RNA under examination. In closely related organisms, most of the oligonucleotides should be identical or nearly so; the more distant the relationship, the greater the differences in their oligonucleotide patterns, which can then be fitted onto trees by statistical procedures. For the first time, a procedure was at hand to assess relationships among organisms independently of phenotypic characters such as form and function, and to do so in a quantitative manner.

There was nothing casual or accidental about the choice of ribosomal RNA as a "molecular chronometer." Ribosomes are constituents of every cell and organism; they are plentiful (a cell of *E. coli* harbors some twenty thousand of them) and perform the same essential function in all. RNA is a major constituent, making ribosomal RNAs the most abundant and accessible of all the species of macromolecules. Moreover, RNA is one component of an intricate machine; its function (now known to be catalysis of peptide bond formation) depends on interaction with many of the ribosomal proteins. RNA evolution should be highly constrained, very slow, and confined to segments of the molecule not critically involved in its function. Ribosomal RNA should therefore evolve in a steady, clocklike fashion, well suited to tracing ancient divergences; it proved to be a timepiece to track the ages.[2]

Woese and his colleagues began by preparing and comparing oligonucleotide catalogs from the more familiar bacteria. Some of their patterns confirmed the taxonomic validity of traditional groupings, such as Cyanobacteria and Spirochaetes; many others, however, turned out to be collections of unrelated organisms. And several, including the methanogenic bacteria and the salt-loving halobacteria, proved to be entirely different beasts: their oligonucleotide catalogs did not group with those of the other bacteria at all. The central conclusion was startling: the kingdom Monera, comprising all the prokaryotes, was heterogeneous, conflating two distinct kinds of organisms. On one side were the eubacteria, all the familiar denizens of soil, clinic, and laboratory. On the other, a small coterie of organisms found in extreme environments: hot, acidic, or hypersaline, which differed from ordinary bacteria in many aspects of their structure and physiology. George Fox and Carl Woese proposed to call these archaebacteria, for they seemed well suited to the environments thought to prevail on the early earth.[3]

Oligonucleotide catalogs have long since been superseded by methods to sequence ribosomal RNAs and their cognate genes directly and to extract maximal phylogenetic information from comparisons among the sequences. The principles are simple, the practice far otherwise. Ribosomal RNAs differ not only in sequence but also in length and secondary structure; for example, in eukaryotes the small subunit RNA is longer and is designated 18S. The rate of RNA evolution is not steady, nor is it uniform along the length of the molecule. Meaningful comparisons demand careful alignment, and must perforce rely on a battery of statistical algorithms, each with its own merits and limitations. These technical matters are beyond the scope of this book (and of its author's competence); treatises on genomics supply an introduction.[4] For the purpose of tracing cell history, what matters is that the conclusions have been solidly confirmed and reinforced. Small-subunit RNA sequences proved to be a practical way to assign organisms, from bacteria to tree frogs, each to a proper taxon. Within a decade, the accumulation of ribosomal RNA sequences allowed Woese to construct the first full and rational phylogeny of the prokaryotes.[5] Supplemented by biochemical and structural criteria, supplied most notably by the laboratories of Wolfram Zillig and Otto Kandler in Germany, the sequences reinforced Woese's leap of intuition that eubacteria and archaebacteria are fundamentally distinct kinds of creatures, as different from each other as each is from eukaryotes. In fact, judging by RNA sequences (and by many other indicators; see chapter 3), archaebacteria appeared to be more closely related to the eukaryotes than either is to eubacteria.

In 1990 Carl Woese, Otto Kandler, and Mark Wheelis formalized the idea that had been incubating for a decade: that the traditional classification of living things had to be abandoned, and replaced with a natural one corresponding to their phylogeny. The latter, in turn, could be inferred from molecular criteria, specifically ribosomal RNA sequences.[6] Figure 2.1 depicts not five kingdoms but three great stems designated domains: Bacteria, Archaea, and Eukarya. Each consists of numerous divisions, or phyla, which can be resolved by judicious comparison of those sequences. The universal tree could also be rooted, at least tentatively:[7] the first node, representing the most recent common ancestor of all living things, puts the Bacteria on one side and Archaea and Eukarya as "sister" domains on the other. Note that the universal tree envisages the Eukarya as an independent cellular lineage that reaches almost as far back in time as the divergence of the prokaryotic domains. The most portentous assertion of the universal tree is that the living world is made up mostly of microbes, which hold most of its diversity and made most of the evolutionary running. The familiar kingdoms of higher organisms, the animals, plants and fungi, are no more than twigs on the huge and ancient stem of the Eukarya.

Such a drastic overhaul of the entire biological landscape was bound to raise hackles, and so it did. The late Ernst Mayr, dean of American evolutionists, spoke for many organismal biologists when he protested against the creation of a separate domain for the archaebacteria.[8] Morphologically and in most physiological respects, archaebacteria are scarcely distinguishable from other bacteria; differences are apparent only at the molecular level, and even those are of a sort likely to be appreciated only by specialists. It made no sense to honor them with a separate domain, a taxon equivalent to that which holds all the eukaryotes from *Paramecium*

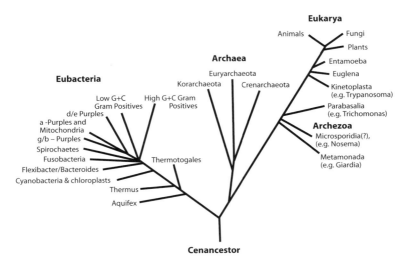

FIGURE 2.1. The canonical universal tree of life inferred from ribosomal RNA sequences. The lengths of the lines are proportional to evolutionary distance, not to time. (From Brown and Doolittle 1997, with permission of the author and the American Society for Microbiology.)

to the Blue Whale. According to Mayr, archaebacteria are indeed curious and divergent, but they are bacteria, prokaryotes: the living world features only two proper domains, prokaryotes and eukaryotes.[9] But Woese and his colleagues would not be put down. In a sharp and combative rejoinder centering on the very purpose and meaning of biological classification, they argued that what was at issue here was not taxonomic niceties but the evolutionary history of all cellular life.[10] That history can only be inferred from its molecular traces, and it clearly displays not two stems but three. The argument has not entirely died down. Thomas Cavalier-Smith, whose views will be considered in chapter 4, continues to maintain much the same position as Mayr's, and I suspect that many zoologists and botanists quietly do likewise. But in the opinion of most microbiologists, Woese carried the day. The universal tree with its three ancestral stems has become a staple of textbooks, and few would deny it the status of one of the premier intellectual achievements of twentieth-century science.

Happily, science remains the endless frontier, and the universal phylogenetic tree is not an endpoint so much as a new beginning. It proclaims not a truth graven in stone but a bold and far-reaching theory about the early evolution of cellular life. The late Lynn White, a distinguished historian of science, understood just what scientists mean by truth: "It is not a citadel of certainty to be defended against error; it is a shady spot where one eats lunch before tramping on."[11] Much of this book revolves around issues linked to the universal phylogenetic tree: the nature of Archaea, the origin of the Eukarya, the last universal common ancestor, the time frame of cell evolution, and just what the metaphor of the tree of life means today. But before we pick up packs and trekking poles, a few words must be said about words.

The terms "Bacteria," "Archaea" and "Eukarya" (or "Eucarya") have become standard in the molecular and microbiological literature, and will be used hereafter in capitals. The traditional term "bacteria" (lower case) continues to be useful, and will occasionally make an appearance as the informal equivalent of "prokaryote." The latter has itself become the subject of considerable dispute,[12] which seems to me blown quite out of proportion. From the standpoint of cell organization, the domains Bacteria and Archaea share many features of structure and physiology that set them apart from the Eukarya; not to mention the latter's uncertain origins and unique evolutionary potential. Prokaryotes do not constitute a clade, and are not a valid taxonomic unit; but the term does designate a grade of cellular organization visibly simpler than that of the eukaryotes, and will be freely employed throughout this book in that restricted sense.

Microbes Traffic in Genes

If the universal tree of life correctly depicts the course of descent from ancestral forms to modern ones, one would expect that same pattern to be recovered whenever phylogenetic trees are constructed from sequences of any class of widely distributed macromolecules. Quite a few cell proteins meet this criterion, and some do indeed fall into three coherent families. RNA polymerases are one example, the proton-translocating ATPases (ATP synthase) another. Judging by sequence similarity, the ATPases of Bacteria make up one cluster (which includes the ATPases of mitochondria and plastids), those of Archaea another and those of Eukarya a third; the latter resembles the archaeal proteins more closely than the bacterial, just as their ribosomal RNAs do. But many other proteins fail to conform. For example, sequences of several of the enzymes that activate amino acids for protein synthesis group the Archaea with the Bacteria and put Eukarya on a separate branch, and so do many enzymes of the metabolic pathways.[13]

At first, this discovery raised doubts about the validity of the phylogenetic tree inferred from ribosomal RNA. But those protein trees disagree, not only with the RNA tree but with each other; they fail to support a clear alternative. The interpretation generally accepted today preserves the universal tree but invokes a startling expansion of the modes of inheritance: the familiar pattern of vertical inheritance, from parents to offspring, must be supplemented by occasional acquisition of genes from the outside, across the boundaries of species, phyla, even domains. Lateral (or horizontal) gene transfer is defined as "the movement of genetic information between two distantly related genomes (i.e., not by sexual recombination within one species), resulting in a genome that contains an expressed, functional gene from a foreign source."[14] Lateral gene transfer is almost unheard of among animals and plants but turned out to be common among prokaryotes.

Lateral gene transfer did not come altogether as a surprise: the phenomenon had been discovered in the laboratory half a century earlier. Bacteria exchange genes by a kind of sexual congress, but they can also pick up naked DNA from the environment or acquire foreign genes by infection with viruses or plasmids. This, in fact, is how resistance to antibiotics spreads so quickly from one bacterial species to another. A gene so acquired will occasionally be incorporated into the genome of its new host; and if it proves beneficial (or at least harmless), it may become fixed in the population and gradually modified ("ameliorated") to work more harmoniously with the resident genes. What was unexpected was the prevalence of lateral gene transfer in microbial evolution, especially in prokaryotes. Over three billion years, most if not all prokaryotic genes seem to have undergone replacement; most commonly by a gene from a related species, but sometimes by one from another phylum, even another domain. At the risk of some exaggeration, one can even regard all the prokaryotes as members of a free-trade zone that share a common pool of genes on a millennial timescale.[15] Lateral gene transfer has also been documented among eukaryotic microbes but is less conspicuous than among prokaryotes.[16]

How can one tell whether a given gene is "native" to the genome in which it resides or was imported from somewhere else? The most frequent transfer from a closely related species may indeed be undetectable, but genes transferred from some distant organism often retain traces of their original home; and now that whole genomes tumble like ninepins before banks of automated sequencing machines, those traces can be read.

A newly arrived gene will likely differ from neighbors long in residence in details of composition, such as the frequency of guanine and cytosine nucleotides. It may be marked as foreign by the absence of homologs in the genomes of related species. Most convincingly, when the phylogenetic tree of some class of proteins puts one of them in an incongruous position, there is often good reason to suspect that lateral gene transfer is to blame.

Examples have been noted and debated ever since the advent of genome sequencing.[17] Thus, J. G. Lawrence and H. Ochman inferred that 18 percent of the *E. coli* genome had been transferred there from elsewhere since *E. coli* diverged from *Salmonella* about a hundred million years ago; that fraction includes the very determinants which identify *E. coli*. In the genomes of the Archaea *Methanococcus jannaschi* and *Methanosarcina mazei* about one-third of the genes appear to be of Bacterial rather than Archaeal provenance. By the same token, thermophilic Bacteria of the phylum Thermotoga have acquired many genes from Archaea. According to a recent study by Nelson-Sathi and colleagues, the salt-loving and oxygen-respiring Haloarchaea are derived from anaerobic methanogens but acquired about a thousand genes from Bacteria.[18] Dozens of metabolic enzymes appear to have been trafficked, and so sometimes have systems of considerable complexity including ATP synthases and elements of the photosynthetic apparatus and respiratory chain. Bacteriorhodopsin, a unique light-driven proton pump, was first found in an Archaeal phylum, the Haloarchaea, but it also occurs in many Bacteria of the open ocean, probably as the result of lateral gene transfer. The rate and extent of such transfer are hard to measure. Genes do not flit lightly between one genome and another, particularly when donor and recipient are phylogenetically distant; barriers to integration and retention ensure that vertical transmission will always be the overwhelming favorite. All the same given endless time, huge populations and recurrent opportunity, improbable transfers become inevitable.

Not all genes are equally likely to undergo lateral transfer. Those whose product is part of some intricate and integrated machine, such as a ribosome or an ion-translocating ATPase, will have been honed for precise fit to other members of the complex; they are far more likely to be inherited vertically, and will seldom be displaced by an adventitious substitute (though they may on occasion be trafficked as a unit). Genes that encode metabolic enzymes or transport carriers are another matter: these operate more or less as free-standing modules and can therefore be more readily gained, lost, or exchanged. For the purpose of using genes to reconstruct the phylogeny of organisms, it is helpful to distinguish between "informational" genes and "operational" ones.[19] The former are more likely to be transmitted vertically and to retain a phylogenetic signal. By contrast, operational genes that encode most housekeeping enzymes are too flighty to be useful for phylogeny over a time span of millions of years. Ribosomal RNAs are prime examples of the informational class. They represent the organism's "genetic core," and phylogenetic trees drawn from their sequences are probably as close as we can come to the genealogy of the organisms themselves.[20]

How large is this genetic core, made up of genes that are invariably transmitted vertically from parent to progeny, and whose lineage may be tracked by ribosomal RNAs? There has been heated debate about the extent of lateral gene transfer, the proportion of the genome that is never trafficked, and about the degree to which those promiscuous goings-on erode the family tree. For some scholars, such as Charles Kurland, lateral gene transfer is real but relatively uncommon;[21] the barriers to the acquisition of foreign genes and particularly to their retention, are high enough to assure that genes pass vertically almost all the time. Lateral gene transfer is a nuisance but not so prevalent as to seriously confound the lines of descent. Ford Doolittle and Eric Bapteste represent the opposite camp: on the evolutionary timescale lateral gene transfer is rampant among prokaryotes, and no gene can be assumed to be immune.[22] From their perspective, transfer is so prevalent as to call into question the very existence of a phylogenetic tree, whose lower limbs must be represented as a web (fig. 2.2). The emerging consensus is that the core of genes that are always transmitted vertically, and seldom or never laterally, cannot be large—perhaps one or two hundred out of the three thousand to five thousand that make up a standard prokaryotic genome. In addition to ribosomal RNAs it includes many ribosomal proteins, various ATPases, RNA polymerase and others. Even these stalwarts must not be taken for granted: lateral transfer of core genes seems to be very rare but exceptions have been claimed, and it is quite conceivable that no unbreachable core exists. Lateral gene transfer is less common among eukaryotic microbes, but even here examples are coming to light.[23] And lateral gene transfer, in the special form of endosymbiosis, played a major role in the genesis of the eukaryotic cell itself. Could it be that the venerable tree of life is but a figment of our need for order, a construct that we impose on nature rather than a reality that we find there?

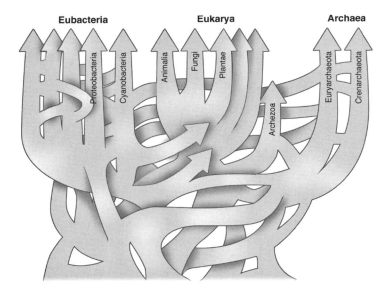

FIGURE 2.2. Tree and web. The early history of life is represented as an era of rampant gene exchange, out of which emerged the three major domains. The proposed subdomain Archezoa has since been withdrawn. (From Doolittle 1999, with permission of the author.)

Still Seeking the True Tree

The discovery of lateral gene transfer has not halted the quest for a universal tree of life on which every species has its rightful place, but it is forcing us to reconsider both the means and the goal. Most biologists, following in Darwin's footsteps, take it for granted that a single rooted and dichotomously branching tree best represents the history of life; that the task of phylogeny is to reconstruct this tree; and that, for major groups of organisms, this has been accomplished. For higher organisms—the animals, plants and even fungi—this attitude is well founded, but it becomes dubious when extended to prokaryotes.[24] Lateral gene transfer refutes the premise that genes always go with the organisms in which they are found, and therefore casts doubt upon the use of genes as proxies for organisms. In reality, trees based on sequences are gene trees, and in principle every gene has its own history. Moreover, the greater the extent of lateral gene transfer, the more the tree of life turns into a web. There is good reason to suspect that lateral gene transfer played an even greater role in the earliest stages of cell evolution. What, then, is the meaning of the universal tree when vertical descent and lateral transfer mingle on every branch?

One of the few certainties in cell evolution is that every cell that ever lived descended by division from a previous cell; sometimes fusion of cells enters into it, as in the case of sexual reproduction, but no cell has ever arisen *de novo* (this generalization may have to be qualified if the practitioners of synthetic biology have their way). If we had a continuous record of all divisions since genesis, and the software to analyze it, we could reconstruct a universal genealogy, a tree of cells. No such record exists, of course, nor is there a practical substitute; cell membranes and other features of structural organization leave no permanent trace. The promise of molecular phylogeny was based on the assumption that genes serve as reliable proxies for the organisms in which they are found. Lateral gene transfer undermines that hope and teaches that there is no certainty in sequences or anywhere else. Nevertheless, as long as they are handled with care and open-minded skepticism, molecular sequences remain the most powerful tools in the box.

For most biologists, ribosomal RNAs serve as the best approximation to a tracer for the organismal lineage, and we are indebted to Norman Pace for a masterly survey of where this approach presently stands and what has been learned from it.[25] The databases currently hold more than four hundred thousand small-subunit rRNA sequences (and a smaller number of large-subunit sequences), including many from organisms that have not been grown in culture; some may never have been seen and are known only from environmental samples. An invaluable gift of RNA-based phylogeny is to have granted us glimpses of that cryptic majority, well over 90 percent of all organisms, and thus vastly expanded the biological universe. Ribosomal RNA sequences unequivocally define three domains and suggest an outline of their evolutionary histories (fig 2.3). Briefly, the domain Bacteria is composed of more than seventy divisions, commonly referred to as phyla. Various relationships among the phyla have been proposed, but SSU rRNA sequences do not reliably resolve the order of branching. That may also be why all the phyla appear to emerge simultaneously at the start of the Bacterial radiation. By contrast, Archaea split early into two distinct phyla, the Euryarchaeota and the Crenarchaeota, each of which diversified into many classes. The Eukarya are different again. Judged by their nuclear lineage, Eukarya are one of life's primary lines of descent, present almost from the beginning. Their RNA sequences also suggest that cellular complexity emerged progressively over time. The deepest branches represent anaerobic protists devoid of mitochondria, many of them parasitic. More advanced protists appear later, and the eukaryotic lineage culminates with its latest twigs, the fungi,

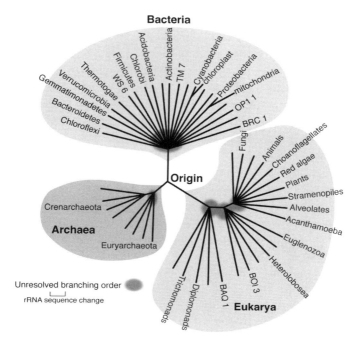

FIGURE 2.3. A current molecular tree of life, based on comparison of ribosomal RNA sequences. For clarity, only a few of the many known lines of descent are shown. (From Pace 2009b, with permission of the author and the American Society for Microbiology.)

plants and animals. The root of the tree puts Bacteria on one side and Archaea and Eukarya as sister lineages on the other. In outline, and also in most details, the current ribosomal tree accords very well with that envisaged by Woese, Kandler, and Wheelis twenty-five years earlier and stands as a tribute to their sagacity.

So can one really take a single molecular species as an adequate tracer for the history of cellular life? Doubts have proliferated along with the accumulation of sequences. They include the failure of the SSU rRNA tree to take account of the widespread lateral transfer of genes and also several statistical problems that can distort the length of branches or cause them to be misplaced. The position and nature of the tree's root are particularly questionable. There is even legitimate doubt whether the early history of life can be recovered at all, given that over the stupefying span of four billion years most mutable sites will have suffered multiple changes that can no longer be discerned. An appealing way to counter some of these uncertainties is to ground phylogeny, not in a single molecular species but in the genome as a whole, or at least in that fraction which has homologs all across the board. With more than a thousand fully sequenced genomes now at hand, genome phylogeny has become a practical proposition; a variety of methods has been devised to that end, and generated a small forest of higher-order trees.[26] They tend to differ from the SSU rRNA tree in details, particularly in the order of branches, but most of them recover its central feature: three distinct domains, two prokaryotic and one eukaryotic, with the Archaea usually more closely allied to Eukarya than to Bacteria. The consistency of this pattern underpins the confident assertion by Berend Snel and his colleagues that "genome evolution is largely a matter of vertical transmission."[27] That proclamation was promptly challenged by a chorus of critics,[28] who point out that many of those trees are based on a minute fraction of the total gene complement ("the tree of 1 percent"; W. Martin); that the great majority of cellular genes have undergone transfer; and that the genomic history of the Eukarya clearly has the topology of a ring rather than a tree. The argument has reached an impasse, and seems unlikely to be resolved by genomics alone.

Timeframe Wanted

For those whose interests center on the history of cellular life, the most glaring shortcoming of ribosomal RNA as a molecular chronometer is that it measures, not time elapsed, but "evolutionary distance." These distances are proportional to the number of base changes required to transform one RNA sequence into another, and there is no straightforward way to translate distance into years. The reason is that the rate of RNA transformation is not constant: it varies from one lineage to another, even from one segment of the sequence to another. One can read off the arrangement of branches, but not its chronology.

Proteins keep somewhat better time. Proteins evolve by the occasional substitution of one amino acid for another, and most replacements have little effect on that protein's function. Only a few critical amino acids are strongly conserved by natural selection, the remainder vary more or less at random, and in animals they do so at an average rate of one substitution every four million years. One can therefore use the number of amino acid replacements in, say, hemoglobins of mice and humans to estimate when they last shared a common ancestor (twenty-seven replacements, or about a hundred and ten million years ago). On balance, dates estimated by molecular phylogeny are in good agreement with those deduced from the fossil record. Unfortunately, no single class of proteins reaches as far back in time as the nodes of the universal phylogenetic tree.

Russell Doolittle and his colleagues refused to be dismayed. In the 1990s they probed deep time by pooling more than 300 amino acid sequences from 64 different enzymes shared by representatives of all three domains and then calibrated the collective tree by reference to the fossil record of animals.[29] They concluded that protists began to diversify almost 1,450 million years ago; Bacteria diversified earlier, 2,100 to 2,500 million years ago; eukaryotes diverged from Archaea during that same window, 2,100 to 2,500 million years ago; while Archaea split from Bacteria as long ago as 3,200 to 3,800 million years. The last common ancestor of all living things would have flourished earlier still.

One should certainly not take this as gospel truth, but as rough estimates go these have held up well. Remember that all the ancient dates are obtained by extrapolation from the ages of fossils barely one-fifth as old. By the same general procedure, Douzery and colleagues arrived at divergence times for the eukaryotic kingdoms between 950 and 1260 million years ago, in good agreement with other recent studies (see chapter 8).[30] The prokaryotes will have diversified earlier and are correspondingly harder to locate in time. Blair Hedges and colleagues put the radiation of Bacteria between 2,500 and 3,200 million years ago (cyanobacteria come out around 2,300 million years ago, in neat agreement with current geological data); Archaea diversified rather earlier, 3,100 to 4,100 million years ago.[31] These dates have been widely cited, but the statistical methods employed have come in for a withering critique,[32] leaving one with little confidence in their accuracy. Molecular phylogeny in itself is probably not sufficient to date events that took place billions of years ago, but taken in conjunction with geological findings these rough estimates are not bad at all.

A Tree to Hug Lightly

Like the flag with fifteen stars that flew over Baltimore's Fort McHenry that sultry summer night in 1814, at the end of the bombardment the tree is still there. It is there because in order to make sense of the world, we need a framework on which to hang the observations, and the conventional rRNA tree serves that practical purpose admirably, better than any alternative presently on hand. It is also there because, in the opinion of many biological scientists, the tree captures deep truths about the evolution of cellular life; and I share that opinion. But a tree is not an adequate representation of that history, because tree and web are complementary aspects of a multifaceted reality whose lineaments fade as we descend ever deeper into the distant past.

Two principal claims of the universal tree of life as sketched in figures 2.1 and 2.3 are that the living world is naturally divided into three major stems, each of which diversified to produce the organisms that we see around us; and that the genetic core, a subset of genes that are never (or at least seldom) transferred, tracks the continuity of cell history. I believe these claims to be correct and have made them the bedrock upon which this book stands. I take the three domains to be a real feature of the world because they are recovered by most genome trees, and also because each domain displays substantial internal cohesion as judged by cell structure and physiology. We shall explore both unity and diversity in chapter 3. I also take the three domains to be extremely ancient, relics of early phases in the evolution of cellular life (chapters 4 to 6). However, the relationship among the three kinds of cellular design, according to Woese, remains murky, and particularly so the origin of the Eukarya. Even before the formulation of the universal tree, it had become clear that both mitochondria and plastids are descendants of Bacteria that had established a symbiotic relationship with a precursor of today's eukaryotic cells; many of the symbionts' genes wound up in the host's genome. It follows that eukaryotic cells are fundamentally chimeric, cobbled together from at least two kinds of cells and genomes, and a tree of life drawn from the genetic core represents at most a partial ancestry of the nuclear lineage (fig. 2.2). We shall venture into this jungle in chapters 7 and 8.

So the metaphor of the tree of life still "largely speaks the truth," but must not be taken literally. The topology of the tree is more banyan (or even coral) than pine, and the predictive power of that phylogeny is limited and not uniform across the biological spectrum. In the case of higher organisms, whose genes pass reliably from parent to offspring, knowing an organism's place on the tree allows one to predict most general features of morphology, physiology and development. You can conjure up an entire animal from a single fossil bone, as the French zoologist Georges Cuvier is supposed to have done. This is far less true for prokaryotes since, thanks to lateral gene transfer, many metabolic characters have become scattered about, within and even between domains. Thermophilic Bacteria have acquired useful genes from Archaea living in similar habitats. Genes involved in pathogenicity are mobile, and therefore the intelligence that I harbor *E. coli* in my gut is not sufficient to tell me whether my guests are harmless commensals or life-threatening invaders. Lateral gene transfer and its grander cousin symbiotic gene transfer conflict in principle with the description of genealogy as a tree, and have historically been regarded as marginal contributors to the evolutionary show; yet both are major participants in cell evolution, most prominently in the origin of the eukaryotic cell. Lateral gene transfer has enormously accelerated and enriched the process of cell evolution; but at the same time, it slowly and relentlessly erases the record of original ancestry, and it is conceivable that the prehistory of cellular life has long since been irretrievably lost. The assertion, that the three domains are relics of divergences that occurred two or even three billion years ago, is a plausible interpretation of what we know, but will probably always remain in the realm of hypothesis rather than demonstrable fact.

A century ago, physicists perceived the world afresh when they accepted the counterintuitive idea that light has the qualities of both waves and particles. Biologists are just beginning to work out the evolutionary implications of the discovery that heredity is conducted both vertically and horizontally. The living world is simultaneously treelike and weblike, and in its earliest phases it may not be possible to tell the difference.

CHAPTER THREE

A World Mostly Made Up of Microbes

Bacteria, Archaea, and Eukarya

The heat of history has melted and folded [organisms] into one another's crevices, in unpredictable outcrops and striations. — Neal Ascherson, *Black Sea*[1]

Life Comes in Three Flavors
Unexpected Outcrops
The Enigmatic Viruses

The universal tree of life proved to be a door that opens onto an unfamiliar landscape, full of surprises. In that fresh yet unfathomably ancient world, all living things are divided into three grand domains, not two or five. The phenomena to be explained are first and foremost the genesis of cells and their parts, not forelimbs and feathers. And the course of evolution has been enriched, accelerated, and also obscured by the exchange of genes across all borders. It is a world made up largely of microorganisms. Their aggregate mass equals that of all other creatures taken together, and their numbers dwarf the competition.[2] Microbial habitats range from boiling hot springs to the frozen deserts of Antarctica and reach downward into the rocks far below the surface crust. The greater part of biological diversity—and the most profound divisions—are found in the microbial world; and for three-quarters of biological history, microbes had the stage to themselves. It all takes some getting used to but makes for a much more spacious view of evolution. In this chapter we turn from the genes, which dominate reflection on the tree of life, to the organisms that inhabit it.

The transition is considerable and calls for a shift in the frame of reference. A century ago, organisms were biologists' chief objects of study.

Eclipsed in the twentieth century by their molecular constituents, organisms have come to be seen as little more than epiphenomena of their genes, and relegated to the backwaters. Some have even questioned whether biology still needs the concept of organisms! To my mind, this is reductionism gone berserk. The first thing one notices about the living world is that the phenomenon of life comes dispensed in the form of organisms.[3] And it is organisms, not molecules or genes, which reproduce and become the targets of natural selection. "The organism is at least as fundamental to biology as cities or firms should be to economics, or molecules to chemistry."[4] Nevertheless, and despite their ubiquity and visibility, it is not easy to pin down what makes an organism, and there is no agreed-on definition. Part of the problem is that living things vary in the degree to which integration of their parts prevails over conflict between them: "organismality" is a continuum, not an all-or-none category. Still, in principle, it is not hard to spot the key feature: organisms act in a purposeful, goal-directed manner, and their parts have functions. "We suggest that the essence of organismality lies in this shared purpose; the parts work together for the integrated whole, with high cooperation and low conflict."[5] From this viewpoint, organisms are not only the units of selection, they are the units of adaptation. Organisms are indeed central to what Richard Dawkins called the "greatest show on earth": their nature and origin are what we strive to explain.

Life Comes in Three Flavors

Eukarya

How many kinds of organisms does the world hold? Of biology's many recent accomplishments, few have so captured the imagination as the discovery that the short answer is three: three themes of cellular organization. One kind is represented by the Eukarya, whose distinctive nature has been recognized for more than a century while their origin remains one of the deep mysteries of cell evolution. The other two, Bacteria and Archaea, can be broadly assigned to the prokaryotic grade of order. They share many features of structure and physiology but display different modes of cellular design that stand forth clearly when the organisms are viewed from the molecular perspective. Additional designs may exist (or may have existed in the past), but despite assiduous and continuing search no others have been discovered. In this chapter I try to highlight the general

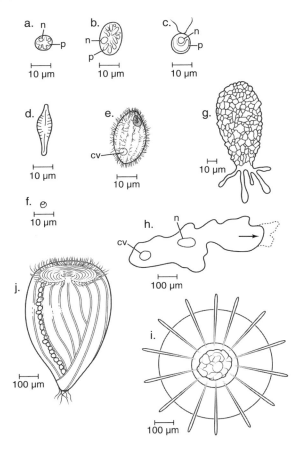

FIGURE 3.1. A sampler of unicellular eukaryotes; note the huge range of forms and sizes. (a) The red alga *Porphyridium*; (b) *Glaucocystis*, a glaucophyte alga; (c) the green alga *Chlamydomonas*; (d) a generic diatom; (e) the ciliate *Tetrahymena*; (f) a picoeukaryotic red alga; (g) *Difflugia*, a test-building amoeba; (h) *Amoeba proteus* on the prowl; (i) a generic radiolarian; (j) the giant ciliate *Stentor*, easily visible to the naked eye. (After Sleigh 1989; Lee 1989.)

features of these designs, purposely setting aside the numerous variations and exceptions.

It is the eukaryotic cells of protists, fungi, plants, and animals that historically have defined what the term "cell" means. Most are relatively large, ten micrometers or more in length, some quite visible to the naked eye. Eukaryotic cells come in a huge range of forms and sizes (fig. 3.1) but always contain a prominent nucleus with multiple chromosomes bounded by a membrane studded with characteristic pores. Cell reproduction is a performance that features chromosome duplication, partition of the products by the stately minuet of mitosis, and finally division of the cytoplasm and cell separation. Energy production is housed in discrete organelles: respiration and oxidative phosphorylation in mitochondria, photosynthesis in plastids. An intricate network of membranes pervades the cytoplasm, creating multiple intracellular compartments. Mobile vesicles link one chamber to another and mediate both the secretion of cell products and the uptake of particulate matter from the environment ("phagocytosis"). Mechanical integration of cellular space is one function of the cytoskeleton, an interconnected scaffold of filaments composed of tubulin, actin, and many ancillary proteins. The cytoskeleton also supplies tracks for the transport of cargo within the cytoplasm and underpins movement of the cell as a whole. Swimming is powered by undulipodia,[6] large and very complex appendages that row or beat; they have a characteristic structure built of microtubules, two central ones and nine doublets around the periphery. Polysaccharide cell walls are common, usually made of either chitin or cellulose. Genomes are large, typically ten thousand protein-coding genes or more, but those genes are apt to be islands in a sea of noncoding DNA whose functions are still being worked out. Besides, most of the coding sequences are interrupted by noncoding stretches called introns that have no obvious function. Finally, eukaryotic cells often harbor endosymbionts, foreign cells that have taken up residence within the cytoplasm of their host. Every one of these general features comes with variations and exceptions, but the pattern is consistent and unmistakable.

In contrast to their sophisticated architecture, the metabolic patterns of eukaryotic cells are comparatively simple. Fungi, many protists, and all animal cells rely on organic matter from their environment, either in the form of small soluble molecules or of solid particles; many protists and some animal cells prey upon smaller organisms. Nutrients are broken down to a handful of central metabolites by anaerobic reactions in the cytoplasm, and then (usually) oxidized by respiration with molecular oxygen as the oxidant. The alternative "plant" lifestyle depends on photosynthesis for the production of both cell constituents and energy. Photosynthesis is based on chlorophyll to absorb light, and on the light-activated splitting of water to generate reducing power (with oxygen as a byproduct). We shall see later that both organelles of energy production, mitochondria and plastids, are descended from bacterial endosymbionts. Unlike many of the Bacteria and Archaea, eukaryotes never use inorganic substrates for energy generation, and there is no evidence that they did so in the past.

About 1.5 million species of Eukarya have been described and named, less than a quarter of the number thought to exist on earth today. Protists alone make up more than 250,000 known species. The diversity of forms and habits among these "lower eukaryotes" is astonishing and has fueled a century of disputation about their phylogeny. Nevertheless, both molecular and structural criteria document that all eukaryotes belong to a single clade and shared a common ancestry in the remote past. We humans tend to focus on the distant descendants of those single-celled protists, the "higher" eukaryotes: fungi, plants, and animals. It is surely remarkable that eukaryotic cells, and they alone, had whatever it took to evolve huge multicellular creatures, differentiated yet fully integrated, even displaying various degrees of intelligence. Can anyone doubt that the world revolves around us? Yet from the perspective of the universal tree of life, higher organisms appear as recent and somewhat specialized offshoots, whose significance in the scheme of things has yet to become apparent.

Bacteria

The two prokaryotic domains, Bacteria and Archaea, are populated almost entirely by minute cells, one to five micrometers in length, whose spare and streamlined organization contrasts with that of the larger and more elaborate Eukarya (fig. 3.2a). The differences are conspicuous and important. Prokaryotic nuclei are not bounded by a membrane, nor do they contain discrete chromosomes. Instead, genes are arrayed on a long molecule of DNA, usually circular rather than linear, which is attached to the plasma membrane. Genes are transcribed and immediately translated within the same cytoplasmic compartment. By contrast, in eukaryotic cells genes are transcribed in the nucleus, and messenger RNA must be translocated into the cytoplasm prior to translation. Another significant difference is that prokaryotic genomes are made up almost entirely of coding sequences, with minimal noncoding DNA within or between genes. Energy transduction makes a third major difference: it is localized to the plasma membrane rather than to specialized organelles, and is effected by a current of protons (or of Na^+) across that membrane. Our conception of prokaryotic cell organization was drawn very largely from research with organisms now classified as Bacteria; no one anticipated that it would come in two versions.

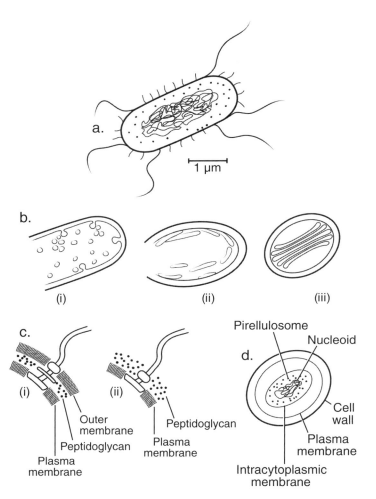

FIGURE 3.2. Membrane organization in prokaryotic cells. (a) The basic bacterial cell. Note the nucleoid devoid of a bounding membrane, the cytoplasm filled with ribosomes, and flagella emerging from the cell envelope. (b) Many Bacteria feature intracellular membranes, characteristically extensions of the plasma membrane. The patterns shown here represent two photosynthetic Bacteria, *Rhodobacter* (i) and the cyanobacterium *Synechococcus* (ii), as well as the nitrite-producing *Nitrosococcus* (iii). (c) Diderms and monoderms. Diderm (gram-negative) Bacteria (i) are surrounded by two lipid-bilayer membranes, the plasma membrane and an outer membrane, with a thin peptidoglycan layer in between. Monoderm (gram-positive) Bacteria (ii) have only the plasma membrane enclosed in a thick peptidoglycan wall. Bushings pass the flagella across the layers of the envelope. (d) In planctomycetes, a unique set of internal membranes surrounding the nucleoid suggests analogies to the eukaryotic nucleus. (Generalized sketch after Fuerst and Sagulenko 2011, with permission.)

Turning first to the Bacterial version, much of cellular physiology con-
verges on the cell envelope, which is made up of several successive layers.
Innermost is the plasma membrane, a bilayer of phospholipids studded
with diverse proteins that mediate the transport of nutrients and ions into
the cell, and of waste products out. Respiration, photosynthesis, and the
generation of ATP are localized to the plasma membrane and its extensions, which sometimes make up an extensive array of intracellular membranes (fig. 3.2b). Next comes (usually) a cell wall, strong and fairly stiff
but porous, composed of polysaccharides cross-linked into a single large
sacculus. Its functions are primarily mechanical; were it not for the wall,
cells would burst due to the osmotic pressure of the cytoplasm. The physical properties of the wall are primarily due to a particular class of polysaccharides called peptidoglycan, which is characteristic of Bacteria and
never found in Eukarya or Archaea (biology is beset with quibbles: some
Bacteria lack peptidoglycan and have a proteinaceous surface layer instead, as do most Archaea; Archaea never possess peptidoglycan but some
Euryarchaeota do have a related polysaccharide). Finally, many Bacteria
feature an outer lipid membrane that shields the cell from environmental
insults, while allowing ions and small metabolites to pass through specialized pores (fig 3.2c).

Bacterial morphology is largely determined by the cell wall, whose
construction is guided by a cytoskeleton. This is a recent discovery which
overturned the long-held belief that prokaryotes have no cytoskeleton,
and much effort is presently focused on working out the cytoskeleton's
composition and operations; suffice it to note here that the chief proteins
are distant relatives of the tubulin and actin of eukaryotic cells. We shall
return to the cytoskeleton in chapter 6.

Many Bacteria are motile thanks to an array of mechanisms that support swimming, gliding, or creeping. By far the best understood is swimming effected by flagella, highly distinctive organelles utterly unlike the
undulipodia of eukaryotes in structure and operation. Bacterial flagella
are made up of the protein flagellin, powered by a rotary motor set into
the plasma membrane, and driven by the circulation of protons or of
sodium ions.

Bacteria excel at biochemistry, having discovered almost all the known
ways of utilizing environmental sources of chemical energy (methanogenesis, unique to the Archaea, is the chief exception). Among the familiar
pathways is the utilization of carbohydrates, amino acids, and carboxylic
acids, either by anaerobic fermentation or by oxidation, with oxygen,

sulfate, or nitrate as the oxidant. Some Bacteria live entirely by inorganic reactions, including the oxidation of hydrogen gas or the reduction of ferric iron. Photosynthesis based on chlorophyll is the main alternative economic basis, one confined to the Bacteria alone. It comes in two versions: one in which growth requires an external source of reducing power, such as H_2S, and another in which water serves as the reductant and oxygen is formed as a byproduct. The latter is confined to the cyanobacteria, one of which subsequently became the precursor of plastids and donated the capacity for photosynthesis to the Eukarya. An altogether different mechanism of photosynthesis, with bacteriorhodopsin as the light-harvesting pigment, occurs sporadically in both prokayotic domains. It's a good rule of thumb that if a reaction generates enough energy to live by, there will be a prokaryotic species to exploit it.

The phylogeny of Bacteria drawn from ribosomal RNA sequences shows all the phyla emerging in a tight cluster more or less at once (see fig. 2.3). But the structure of the cell envelope suggests the existence of at least two subdomains, traditionally designated gram-negative and gram-positive (after a staining procedure devised by the Danish microbiologist Christian Gram in the nineteenth century). In gram-negative Bacteria, such as *E. coli*, the envelope consists of two concentric phospholipid membranes, the plasma membrane and an outer membrane, with a thin peptidoglycan wall in between. Gram-positive Bacteria, such as *Bacillus subtilis*, have only a single phospholipid membrane, the plasma membrane, with a thick peptidoglycan wall on the outside (fig. 3.2c). Since what matters here is the structure of the cell envelope, rather than the staining reaction, per se (which is sometimes ambiguous), I shall adopt the apt terminology proposed by Radhey Gupta (discussed further in chapter 4), and refer to these two groups as diderms and monoderms, respectively. The division is likely to be an ancient one and features prominently in several accounts of cell evolution that will be considered below. Despite the existence of these two subdomains, all Bacteria appear to share a common ancestry; Arthur Koch has suggested that what first defined the domain Bacteria was the invention of the peptidoglycan wall, which allowed early cells to maintain their integrity in the face the cytoplasm's osmotic pressure.[7]

Archaea

Prokaryotes that inhabit extreme environments or make a living in peculiar ways such as the reduction of CO_2 to methane were known long before the advent of modern phylogeny. It occurred to no one that these might be anything more than specialized bacteria. Indeed, from the viewpoint of cell structure and operations, the similarities between Archaea and Bacteria seem to outweigh their differences. Both domains feature small, walled cells whose cytoplasm is filled with ribosomes of the same general size, composition, and function. Archaeal genes are arrayed on a long molecule of DNA, usually circular and attached to the plasma membrane; they are organized into operons, often in the same order as in the Bacterial equivalent. The lipoidal plasma membrane is studded with transport proteins of the kind found in Bacteria; they mediate the transport of metabolites and ions and carry out chemiosmotic energy transduction. The structure of the Archaean envelope resembles that of monoderm Bacteria, consisting of the plasma membrane and a single layer of wall made up of either polysaccharide or S-proteins; there is no outer lipid membrane. Many pathways of central metabolism are the same in both domains, or at least similar. Yet the Archaea are phylogenetically a well-defined group. Most Archaeal genes have their closest homologs among the Archaea, and a subset of critical molecules bears witness to the ancient separation between the two prokaryotic domains. Any account of cell evolution must take notice of both the commonalities and the differences.[8]

Until recently Archaea were associated with unusual and extreme environments, and organisms extracted from boiling hot springs, brine ponds, and acid mine effluents have captured the microbiological imagination. But the Archaeal domain is almost as diverse as the Bacterial one and harbors a variety of habits and forms. The SSU rRNA tree (see fig. 2.3) shows at least two prominent stems, which can be thought of as subdomains but are traditionally designated phyla. The Euryarchaeota (from the Greek *euryos,* for diversity) house the salt-loving Haloarchaea, several wall-less thermophiles, and all the species of methanogens, a metabolic capacity found only among the Archaea. The Crenarchaeota (from the Greek *crenos,* for spring or origin) boast most of the extreme thermophiles. But this same phylum also makes a major contribution to the ecology of the open ocean; one representative, as yet uncultured and unnamed, is thought to make up some 20 percent of the planktonic picoflora, which would make it one of the most abundant cell types on earth. A third phylum, designated Thaumarchaeota, has been proposed to accommodate a cluster of lineages that includes ammonia oxidizers and various thermophiles.[9] The deep branch designated Korarchaeota, known only from ribosomal RNA sequences recovered from hot springs,[10] may represent yet another phylum. The origin and significance of these deep divisions within the Archaea is quite unknown, and so is the origin of the domain as a whole.

Woese envisaged the ancestral archaeon as an anaerobic thermophile that lived by the oxidation of sulfur. The domain may be rooted in adaptation to a thermophilic lifestyle, but David Valentine has recently made a strong case for the proposition that Archaea evolved to thrive under conditions of energy stress.[11]

The features that distinguish Archaea from Bacteria naturally command the most attention, for the claim that these represent distinct cell designs turns on those differences. The roster is substantial, chiefly functional structures and molecules whose counterparts occur in both domains but in significantly different versions. Some pertain to the expression or replication of genetic information, others to membrane architecture or energy transduction. Notably, in most instances the Archaeal version resembles that of Eukarya more closely than the Bacterial one. Here is a short list.[12]

i   Signature sequence differences in all ribosomal RNAs, which first indicated the existence of two radically different kinds of prokaryotic cells.

ii   RNA polymerase, subunit composition and sequences; the Archaeal version has been described as a stripped-down version of the Eukaryal one.

iii   Ribosomes. Archaea are resistant to most of the antibiotics that inhibit protein synthesis in Bacteria, but are susceptible to those that affect Eukarya. Ribosomal proteins and the genes that specify them typically resemble the eukaryotic ones.

iv   Transcription factors of Archaea resemble those of Eukarya, not Bacteria.

v   DNA polymerase is another important instance in which the Archaeal enzyme looks like a simplified version of the Eukaryal one, but is quite unlike that of Bacteria.

vi   In Euryarchaeota, but not in Crenarchaeota, DNA is associated with histones homologous to those of Eukarya. Histones are never found in Bacteria.

vii   Motility in Archaea, as in Bacteria, commonly depends on rotary flagella. However, while in Bacteria an ion current supplies the power, Archaea use ATP. Moreover, Archaeal flagella are thinner than those of Bacteria, composed of altogether different proteins, and grow from the base rather than the tip.

viii   The ion-translocating ATPases of Archaeal plasma membranes, designated A_1A_0, are homologous to the F_1F_0–ATPases of the Bacterial plasma membrane, but quite distantly so. A closer resemblance links the Archaeal ATPases to those of eukaryotic vacuoles.

ix   Archaeal plasma membranes are made up of isoprenoid lipids ether-linked to glycerol-1-phosphate, whereas Bacteria make theirs of fatty acids ester-linked to glycerol-3-phosphate. Remarkably, in this instance Eukarya group with the Bacteria, not the Archaea.

xi   The apparatus of cell division in Archaea is quite unlike that of Bacteria.

In spite of various exceptions, most of them readily explained by lateral gene transfer, the differences are clear and not in dispute. The question is what they tell us about the evolution of microbial cells. As I read the evidence, Bacteria and Archaea are obviously different versions of a "prokaryotic" pattern of organization. Common features include a unitary compartment that houses both transcription and translation, a streamlined genome and energy transduction by means of ion currents across the plasma membrane. But the differences between Archaea and Bacteria are profound and consistent and appear to represent a deep and very ancient bifurcation of cell evolution. In fact, the separation of Bacteria from Archaea represents the most ancient event in cell history for which we have solid evidence; it is also the most consequential, for the rise of the eukaryotes hinged on it.

Unexpected Outcrops

Scientists are forever torn between the innumerable particulars and the few generalities, between the urge to uncover all the facts of nature and the obligation to put them in order. At the deepest level of molecular biology, unity prevails over diversity: ATP, DNA, RNA, and the canonical set of twenty amino acids are more or less universal. But once we reach the level of systems, every general statement must immediately be qualified.

Take cell size. The great majority of prokaryotes fall into the range of one to five micrometers, but there are spectacular exceptions: *Epulopiscium fishelsoni,* an inhabitant of the gut of certain fishes, attains a length of seven hundred micrometers and is visible to the naked eye; another such giant among the Bacteria is *Thiomargarita namibiensis,* found in large numbers off the coast of Namibia, which oxidizes H_2S. Both are consortia of thousands of nucleoids enclosed within a single plasma membrane. At the other extreme some microbiologists claim to find nanobacteria, less than one-tenth of a micrometer in diameter and barely within the theoretical limits for a viable cell.[13] Eukaryotic cells also range upward and downward from the canonical ten micrometers. Axons in the giraffe's neck must extend for several meters to reach from the cell body to their

target. At the same time, the oceans swarm with picoeukaryotes as little as
one to two micrometers long but still of standard eukaryotic organization,
complete with a mitochondrion, a plastid, and an endomembrane system.
The reason is, of course, that living things were not designed to a plan or
purpose, but evolved by variation and selection within limits that are little
known but certainly very broad. Irrepressible diversity is to be expected;
but when variations seem to breach the categories we have established to
make sense of the living world it is time to take notice.

One of the verities of cell biology is that there are no intermediates between prokaryotes and eukaryotes. That is the generalization challenged
by planctomycetes, and by some verrucomicrobia, phyla of Bacteria se-
curely anchored there by ribosomal RNA sequences, whose structural
organization has much in common with that of eukaryotes.[14] The cell envelope lacks peptidoglycan; the cell wall is proteinaceous, and beneath lies
a plasma membrane that appears to be the locus of energy transduction.
So far, so prokaryotic; but the cytoplasm is divided into internal compartments bounded by lipid bilayer membranes (fig. 3.2d). An inner region
called the paryphoplasm, devoid of ribosomes, is separated by a membrane
from the ribosome-rich pirellulosome, which also holds the nucleoid. In
one planctomycete, *Gemmata obscuriglobus,* the nucleoid is further defined by a double set of closely apposed membranes, as in eukaryotes. The
cells contain clathrin-like proteins, and there are hints of structures akin
to nuclear membranes pores. Recent studies have even documented the
uptake of proteins by a process akin to endocytosis.[15] Some planctomycetes synthesize sterols and contain tubulins that are clearly albeit distantly
related to those of eukaryotes.

There are, as always, several ways to envisage the relationship between
the prokaryotic planctomycetes and the much grander Eukarya.[16] Might
planctomycetes be intermediates in the transition from prokaryotes to
eukaryotes? No, the genomic evidence excludes that hypothesis. Instead,
one can argue that the two lines evolved separately but convergently; the
features they share are curious analogies but are not homologous. Lateral
gene transfer may have played a role, particularly in shuttling tubulins.
And then there is the intriguing possibility that planctomycetes, which by
some phylogenies represent an early branch of the Bacterial stem, retain
cellular characteristics and their cognate genes from a common universal
ancestor that was structurally more elaborate than streamlined contemporary Bacteria (chapter 4). Complex internal organization is not confined
to the Bacteria, but has also been noted in some Archaea such as *Ignicoc-*

cus hospitalis. Here is a window that looks out upon the beginnings of cellular evolution from an unexpected angle.

The Enigmatic Viruses

Of all the dubious entities that microbiologists encounter, viruses are the most abundant and familiar yet also the most perplexing. Viruses are everywhere, in astronomical numbers. Estuarine waters swarm with them, some 10^7 per milliliter, or about 10 virus particles for every cell. Their abundance in marine mud can be 10 times as much again.[17] The oceans are thought to contain of the order of 4×10^{30} viruses, equivalent in mass to 75 million blue whales. If all these particles, each 100 nanometers long, were lined up end to end they would span 10 million light-years, about 100 times the width of our galaxy. This makes viruses the most abundant biological objects on earth, comparable to all prokaryotes taken together. Viruses are a major cause of mortality, not only among plants and animals but for bacteria as well; there is probably no kind of life that evades predation by viruses. They are major players on the ecological stage, and by shuttling genes between cells they are prominent agents of evolution. Yet because they lack ribosomes, viruses have no home on the tree of life, and that is disquieting.

Are viruses alive? The answer depends not on the facts of nature but on one's definition of life. Consider just two prominent ones. According to the late Lynn Margulis, promoting the position of the Chilean biologists F. G. Varela and H. R. Maturana, living things are autopoietic systems—they make themselves. John Maynard Smith points in a different direction when he argues that life can be defined by the possession of the properties required for evolution by natural selection: multiplication, heredity, and variation. By the latter definition viruses are alive, by the former they are not. But rigorous definitions cannot do justice to objects that are by nature ambiguous. All viruses are obligatory parasites, utterly dependent for their reproduction on a host cell whose biochemical machinery they commandeer. So viruses do not make themselves; they are made. Simple ones can even be made by chemists, who have learned to combine the capsid proteins of the viral shell with the cognate genome to reconstitute infectious virus particles. Yet viruses obviously bear a deep-seated relationship to cellular life. They are constructed of the same kinds of molecules, produced by the same procedures; they display biological kinds of heredity and variation, and engage intimately with cellular life. Alive or not, viruses are part of the biological universe and they straddle the divide between the living and nonliving spheres.

The origin of viruses has baffled biologists ever since their discovery more than a century ago. Three general ideas have structured the discussion: that viruses are relics of the origin of life, older even than cells; that they are descended from cells by the kind of reductive evolution that gave rise to other parasites; and that they began as genetic material that escaped from cellular control, clothed itself in capsid proteins and learned to reproduce by infection. None of these views has many adherents today. As obligatory parasites of cells, viruses can scarcely be more ancient than cells; unlike other parasites, viruses display no relics of any cellular ancestry; and the notion of "escape" is devoid of content in the absence of detail. The advent of viral genomics has refreshed a conversation gone stale and reoriented it utterly. Viruses, it appears, may be products of an evolutionary history linked to that of cellular life and of comparable antiquity, but maintaining a separate and distinct identity.[18]

There are now more than 250 complete viral genomes on record, more than sufficient to document a pattern of branching descent from a common ancestor such as the history of cellular life displays. But that is apparently not the way the virosphere came into existence: the evolution of viruses is not a single story line but a tangled web of overlapping lineages that have freely exchanged genes (in fact, the idea of lateral gene transfer was naturalized in virology long before it became fashionable in cell evolution). Lineages are poorly defined, but five great clusters have been recognized: positive-strand RNA viruses, retroid viruses, small DNA viruses, tailed bacteriophages, and nucleocytoplasmic large DNA viruses. They appear to be of separate origins and evolved by a combination of both vertical inheritance and lateral gene exchange, generating enormous diversity. Environmental sampling keeps turning up novel sequences, suggesting that the viral genome space is huge. Perhaps the most striking discovery is that viral genes have very little homology to cellular ones. Most viral genes have no cellular homologs, and in the case of viruses of Crenarchaeota that proportion rises to over 90 percent. Viruses, it seems clear, are not descended from LUCA, the last universal common ancestor of cellular life (or else, if they did, variation has erased all traces of their ancestry).

There are no genes that are universally shared by all viruses (in contrast to cells, all of which share genes for ribosomal RNA). But there are

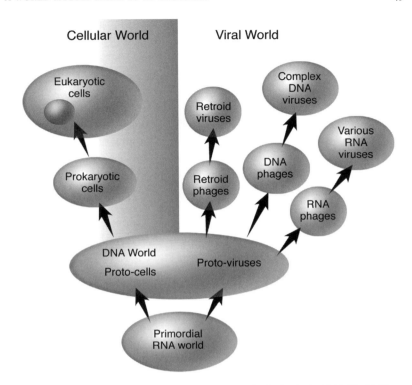

FIGURE 3.3. Coevolution of the cellular and viral spheres, beginning in the RNA World. Heavy arrows designate the predominant direction of gene flow (not vertical descent). Time's arrow runs from the bottom toward the top. (Adapted from Koonin and colleagues.)

genes that are widely distributed among the viral clusters (again, not all). Eugene Koonin and colleagues designate these as "viral hallmark genes" and see them as relics of a common stock that defined an ancient virus world.[19] Among the proteins these genes encode are several enzymes of DNA replication, and especially the "jelly-roll capsid protein" that makes up many outer shells. Koonin advocates an imaginative scheme in which viruses go clear back to precellular times. They may have preyed upon evolving protocells of the RNA World (fig. 3.3). Even then, it seems, the bigger bugs had smaller bugs upon their backs to bite 'em.

Such deep antiquity offers ample scope for creative interplay between viral and cellular streams of the biological universe. In this vein, Patrick Forterre argues provocatively that DNA may have been first invented by viruses, that the three cellular domains may represent three distinct episodes of replacement of RNA by DNA as the genetic material, and that the origin of the eukaryotic nucleus should be sought in infection of a protocell by something like the newly discovered giant DNA viruses.[20]

Giant viruses challenge not only the conventional perception of viruses as minuscule but also the boundary that has traditionally separated viruses from cells. The Mimivirus of the protist *Acanthamoeba* was discovered more than a decade ago but mistaken for an infectious bacterium; its viral nature was only recognized in 2003.[21] Mimivirus virions are 0.75 micrometers in diameter, larger than many bacteria and easily visible under the light microscope. They harbor a genome of bacterial size with nearly 1,000 genes. The genome encodes enzymes of DNA replication, some proteins involved in translation (but no ribosomes), a lipid membrane and even a peptidoglycan cell wall. Mimivirus "mimics" cells, hence the name. The virions are bona fide viruses insofar as they are enclosed in a shell of capsid proteins, require a cellular host and lose their structural identity during infection. But virus reproduction takes place in a specialized region called a "virus factory," produced jointly by virus and host, which has been described as a "transient microorganism." There is even a second virus called Sputnik that infects the factories, commandeers them for its own reproduction, and diminishes the yield of Mimivirus. The boundary between such viruses and prokaryotic parasites has begun to blur.

So should the Mimivirius be considered living? Jean-Michel Claverie thinks so, not the virion or propagule, but the virus factory; others disagree, as would I.[22] But even those who do not fall in with Claverie's argument will admit that Mimivirus stretches the bounds of the viral domain. So do some other sightings, notably the viruses of Archaea that come with altogether unexpected forms, appendages and reproductive strategies. Viruses may not be "living" in any conventional sense, but if it turns out to be true that they represent a biological universe distinct from but parallel to cellular life, we will have to re-examine many entrenched suppositions about life, the universe and all that.

CHAPTER FOUR

The Deep Roots of Cellular Life

The Common Ancestry of Living Things

The central question posed by the universal tree is the nature of the entity (or state) represented by its root, the fount of all extant life. Herein lies the door to the murky realm of cell evolution. — Carl Woese, "On the Evolution of Cells"

The Ancestor of Us All
A Portrait Blurred by Time
Alternate Histories

The physiology and phylogeny of organisms are provinces of empirical science, where data can be gathered and interpreted in a reasonably objective manner. The same cannot be said of the origin of cells themselves, which took place in the most remote past and has left few traces. Indeed, until recently the issue hardly arose, because the origin of cells was conflated with the origin of their molecular constituents and treated as a matter of chemistry. On the widespread (albeit unspoken) premise that once the chemical building blocks of life had come into existence they would spontaneously coalesce into primordial cells, there was no history to investigate. But that premise is surely incorrect: cells do not assemble themselves from preformed parts today, and they would not have done so in the past. The genesis of cellular organization is a problem in biological history, and must be understood as an evolutionary process.

What brought cell evolution to the forefront of inquiry was the advent of the universal tree of life with three major trunks. By directing attention to the origins of the three domains, this formative conception rephrased the mystery of how cells came to be and supplied a framework within which it could be investigated. The chapters that follow are grounded in the premise that cell evolution was a protracted historical process in which contingency played as large a role as did chemistry. The evolution of cells probably spanned much of the interval between the cessation of intense meteorite bombardment, some four billion years ago, and the appearance of full-fledged microbial communities about 3.5 billion years ago. Indeed, cell evolution continues still, but its most creative phase was that which brought forth the machinery of life and saw the birth of the major patterns of cell organization. Students of cell evolution seek the origins of the biological universals that all living things share in one form or another: proteins and nucleic acids, the apparatus for transcription and translation, metabolic pathways, energy-transducing membranes, cell walls, organs of motility, a cytoskeleton, and much else besides. Most of all we seek the origin of cells, those organized and purposeful molecular systems that are the indispensable vehicles of the living state.

How cells and their parts came to exist is, for all practical purposes, unknown. But it may not be altogether unknowable. The operations and architecture of contemporary cells hold numerous clues, enough to reassure one that ongoing efforts to portray the prehistory of cells in a plausible historical narrative are based on something more than unbridled speculation. For convenience and clarity of presentation, I have divided the continuous process of cell evolution into seven somewhat arbitrary segments. The present chapter considers the root of the universal tree, the putative ancestor of all life. Chapter 5 examines the generation and utilization of energy today and in the past. Chapter 6 seeks to reconstruct the origins of the molecular machinery and structural organization of contemporary cells, drawing chiefly on insights from genomics. Chapter 7 tackles the lesser of the great mysteries, how eukaryotic cells came to be. Chapter 8 returns to more solid ground, as we inquire into the transformation of endosymbionts into cell organelles. Chapter 9 explores what has been learned from the fossil record. Chapter 10 faces the ultimate riddle, seeking clues to the inception of entities that possessed the qualities of life.

The Ancestor of Us All

Ever since Darwin, biologists have cherished the belief that all living things share a common origin in "one or a few simple forms." That conviction was reinforced in the twentieth century by the recognition that all organisms share a common biochemical makeup[1] and became a certainty when all organisms were found to employ the same apparatus for the transcription and translation of genetic information. The discovery that the genetic code is uniformly the same (apart from some trivial exceptions) sealed the matter, but left the nature of that ancestral life form far beyond reach.

Our most remote ancestor became a proper subject for scientific inquiry when it acquired a habitation on the universal phylogenetic tree (see fig. 2.1). LUCA, the last universal common ancestor (also known as the cenancestor) is represented there as the first node, the bifurcation that divides the line of descent leading to Archaea and Eukarya from that of the Bacteria. Features shared by all organisms are likely to have been present in their common ancestor, and make up quite a long list: key molecules including ATP, the pyridine nucleotides and many others; K^+ as the chief cellular cation; the twenty standard amino acids and five central nucleotides; RNA, DNA, and basic procedures for gene replication, transcription, and translation; metabolic enzymes; plasma membranes with ion-translocating ATPases and redox chains; chemiosmotic energy coupling; and much else. LUCA could be imagined as a discrete and well-integrated entity endowed with the essential characteristics of modern cells from which all extant organisms derive by the familiar interplay of variation and natural selection, and this remains the conventional wisdom. If we could inspect LUCA under the microscope, we would probably designate her a prokaryote.

Reflection on this conception of LUCA as an organism much like the ones we know today raises vague but nagging misgivings. The image conflicts with the commonsense expectation that the ancestor of all life should be more primitive than modern cells, more makeshift and flexible. During the first half-billion years of life's history, more novelty arose than during the following three; is it credible to maintain that an advanced and fully fledged precursor brought forth three distinct patterns of cellular organization by a succession of small variations over a relatively brief span of geological time? To do so, LUCA would require precursors to all existing protein families, a repertoire far richer than that of any contemporary organism. Indeed, does common ancestry of molecular features necessarily imply a unique common ancestor? Ruminations along this line go back at least thirty years,[2] but it was left to Carl Woese to rethink the nature of LUCA from the ground up and to question just what the deepest branches of the universal tree really represent. His ensuing series of publications constitute the first penetrating and sustained inquiry into the nature of cell evolution; they supply a basis for continuing reflection on the deepest roots of biological organization and a framework by which we can assess new data and arguments.[3]

Woese traced the earliest stages of cell evolution far back to a time when cells in the contemporary sense did not yet exist. The story begins in the RNA World, a hypothetical stage in life's genesis when both heredity and catalysis were rudimentary and carried out by RNA rather than proteins (chapter 10). Its plot gathered speed with the invention of translation, which underpinned the transition to modern life forms that rely on proteins whose structure is specified by genes. Cell evolution proper began when life attained the "progenote" stage: mechanisms for the replication and expression of genetic instructions had come into being but were still roughcast. Transcription, translation, and replication were far less precise than they are today, and in consequence both genomes and proteins were much smaller; mutation rates were high, genotype and phenotype loosely coupled. By the same token, the sophisticated molecular machinery of contemporary cells still lay very largely in the future. Even the chemical nature of the genome, DNA or RNA, remained indeterminate. As to the physical nature and organization of progenotes, Woese was carefully and purposely agnostic, but he entertained the possibility that progenotes were precellular entities lacking even a bounding membrane.

Even if the entities in which the machinery of life evolved were recognizably cells, they would not have been organisms in the modern sense. The organisms we know pass genes vertically from parent to offspring; acquisition of a foreign gene is a rare event. By contrast, says Woese, in the early stages of cell evolution progenotes freely exchanged genes and incorporated these wild oats into their own genomes. Promiscuity was encouraged and facilitated by the loose organization of both genomes and cells, which made them imprecise but also tolerant of inaccuracy. Thanks to the inordinately high rate of lateral gene transfer, evolution was not channeled into discrete lineages as it is today. Indeed, there were no lineages: "Progenotes are cell lines without pedigrees, without long-term genetic histories."[4] Instead, evolution was a communal affair, as innovations that arose in one locale were quickly shared and widely distributed. Any single line of progenotes had a limited repertoire, but the community as a whole had every potential for open-ended advancement; high variability and the collective mode of evolution explain how so much novelty and adaptation were achieved in the relatively narrow window, only five hundred million years, between the cessation of intense meteorite bombardment and the first appearance of microbial communities in the fossil record.

Out of this era of rampant genetic mixing and matching, the three fundamental patterns of cellular organization emerged by a process somewhat akin to crystallization. Translation would have been the first to consolidate, followed by transcription. Other features, such as metabolism, which relies on free-standing ("modular") enzymes, would have remained fluid for much longer, and to some degree remain so today. As progenotic cells acquired improved genes they made better proteins, which in turn supported more accurate processing of information; cells became more autonomous, better integrated, and more efficient. Some lines became sufficiently successful that further incorporation of foreign genes was more likely to disrupt existing functions than to enhance the systems' capacity to reproduce and garner resources. At this point, natural selection would begin to favor the erection of barriers to the import of genes from the outside; consolidation and refinement come to prevail over foreign novelties. This is the stage represented by the deepest nodes on the universal tree, those that mark the emergence of lines that gave rise, first to Bacteria and subsequently to Archaea and Eukarya (see fig. 2.2). LUCA and the ancestors of the individual domains were not specific organisms but represent the initial consolidation of patterns of organization, which, over time, were elaborated and refined. Only then could discrete lineages begin to emerge, and therefore Woese designates these steps as "Darwinian thresholds." From this perspective, there never was a proper LUCA, no unitary organismic common ancestor. The three existing domains (and perhaps others that failed to survive) would have emerged individually and successively from the communal genetic pool over a prolonged span of time by a deep-seated change of "biological phase."

In the course of the vast span covered by the universal tree of life—three billion years—and more, the character of the tree and the very nature of evolution underwent profound transformation. In contemporary cells and organisms, evolution is almost entirely a matter of refining and elaborating established patterns of organization, and it takes place by the interplay of variation and natural selection confined to discrete organismic lineages. By contrast, at the progenote level and up to the Darwinian threshold, evolution was communal and powered by pervasive lateral gene transfer.

Vertical and lateral gene transfer represent opposite poles of the evolutionary scene. The former is chiefly responsible for the mounting specificity, integration, and complexity of cellular life; the latter accounts for the acquisition of entirely novel traits. The tree of life was not organismic at first but became so with the emergence of the three main stems. The deepest nodes represent not speciation events but the consolidation of basic designs to the point where organismic lineages could endure. With the change in the pattern, or mode, of evolution came a change in tempo.

Intense and frenzied early on, beyond the Darwinian threshold the rate of evolution settled down as the options for radical alterations were increasingly constrained by the integrated patterns created by earlier choices of design.

The time when rampant gene transfer overwhelmed vertical heredity and served as the chief source of innovation lies far back in the remote past, but its relics can still be made out. Woese finds evidence of that history in sequences of very ancient genes, particularly those of amino acid transfer RNA synthetases.[5] The canonical pattern of three stems can still be detected in this universal class of molecules, but it is seriously eroded by the lateral gene transfer that raged while their genes were taking shape. More recently, a reassessment of the nature of the genetic code by computer modeling led Vetsigian and colleagues to the conclusion that it could only have attained its present state of perfection in an environment that allowed communal, as distinct from familial, evolution.[6] Torrential gene transfer that precluded the establishment of lineages may also explain a peculiar feature of the universal tree, the long, bare branches that connect each of the domains to the primary bifurcation (see fig. 2.1).

The image of progenote genomes as fluid and transient constellations raises a fundamental question: what force would have promoted the assimilation and integration of all that random novelty to generate increasingly effective cellular systems? Woese is inclined to minimize the role of natural selection, favoring the idea that the process of gene invasion itself promoted adaptation. This suggestion has been sharply criticized by others,[7] and I too find it hard to credit. Hurricanes, as the cosmologist Fred Hoyle pointed out long ago, do not create order. In the absence of natural selection, evolution lacks direction. It follows that progenotic lineages, however fuzzy by the standards of contemporary cells, must have been sufficiently long lasting for natural selection to pick out the winners from the losers. If, indeed, an alternative to Darwin's ratchet does exist (by neutral evolution, perhaps, or by the direct selection of genes), it needs to be spelled out in detail. The implications would reverberate far beyond the tiny community of scholars exercised about the origin of cellular life!

A Portrait Blurred by Time

Carl Woese, and concurrently Otto Kandler and Wolfram Zillig, penned their musings on the universal ancestor just before the tide of genomic data came in. Now that the tide has crested we are awash in pertinent information, including formal proof that all living things do share a common ancestry,[8] but the cloud of unknowing hangs as thick as ever over the nature of that entity, or state. The logical principle still holds that characteristics widely shared by members of all three domains were likely present in their common ancestor. Gene sequences are now the criterion of choice. So ion-translocating ATPases were probably part of LUCA's endowment; flagella of the Bacterial kind were not; and the glycolytic pathway of glucose degradation, although common, also does not derive from LUCA because the enzymes that catalyze these reactions in the two prokaryotic domains are not homologous. I shall here set aside the solipsist argument that, thanks to relentless lateral gene transfer over the past four billion years, all significant traces of life's earliest stages have been lost beyond recall. Even so, one must note that lateral gene transfer introduces a pervasive note of uncertainty, and the findings can usually be interpreted in more ways than one. Economists ruefully admit that for every economist there is an equal but opposite economist, and one is tempted to say the same of molecular evolutionists.

In principle, one should gain information about LUCA by locating the root of the tree of life, the most basal bifurcation. Ribosomal RNA sequences place that root between Archaea and Bacteria, consistent with Woese's vision of LUCA as a progenote and an evolutionary progression from rudimentary to complex and refined. Other authors, drawing on different approaches, place the root of the tree elsewhere—among the diderm Bacteria, among the monoderms, even on the branch that led to eukaryotes.[9] Lake and his associates confidently exclude all possibilities except for a rooting between actinobacteria plus diderms on the one hand, and Archaea plus monoderms on the other.[10] What this suggests is not that one rooting algorithm is true and all the others erroneous but that neither LUCA nor the tree of life are as unambiguous as we once thought. In fact, if the genes that ended up on the three familiar stems were recruited from several sources at different times, there never was a discrete LUCA and no single root.[11] LUCA designates a stage in a protracted, continuous process, a work in progress (fig. 4.1).

Was LUCA a progenote? The answer depends on just what we take that term to mean. A substantial body of evidence suggests that she was not some nebulous acellular slime but an unequivocal biological system, the last universal *cellular* ancestor.[12] She seems to have possessed a substantial number of genes, as many as six hundred or even a thousand. The

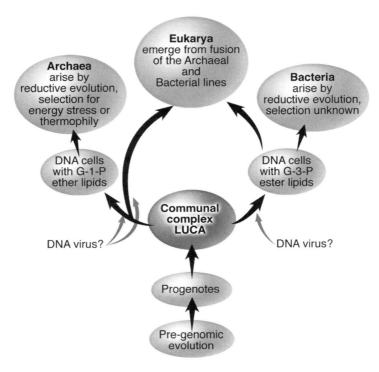

FIGURE 4.1. A legend of LUCA. The last universal ancestor is envisaged as a gene-swapping community of thermotolerant organisms that were genetically redundant, metabolically diverse, and structurally complex. The chemistry of both membranes and genomes remains undecided. The three domains emerged by assimilation of viral DNA polymerases. (Drawn from the writings of Forterre 2006b, Glansdorff et al. 2008, and Lake 2009, none of whom should be held responsible.)

basic mechanisms of transcription and translation were in place. The ubiquity of ion-translocating ATPases, some of which seem to have undergone duplication before Archaea diverged from Bacteria, points to the existence of a plasma membrane engaged in chemiosmotic energy transduction. The coupling of genotype and phenotype was not notably loose. On the other hand, several molecular features of translation and transcription differ between Bacteria and Archaea, and both their membranes and the mechanisms of DNA replication are strikingly different; LUCA was a progenote in the sense that fundamental mechanisms of information processing were still plastic.

DNA is the universal genetic material of cellular life; many viruses do likewise, but some rely on RNA. The ubiquity of DNA, the genetic code and the standard set of twenty amino acids all strongly suggest that LUCA's genome was made of DNA. Nevertheless, from the beginning, Woese and others envisaged LUCA as an offspring of the RNA World, with a genome made of RNA rather than DNA. This counterintuitive speculation is reinforced by the discovery that, although the molecular structure of DNA is universal, the molecular mechanisms that produce it are not.[13] DNA polymerases of Bacteria and Archaea are completely unrelated; those of Eukarya are different again, but resemble those of Archaea. This has led Patrick Forterre to propose that DNA evolved in an ancient virus world, a universe parallel to but distinct from the cellular one (chapter 3). DNA may initially have served viruses as a shield against the defense mechanisms of their protocellular prey but was subsequently adopted by the latter because it makes a chemically superior gene carrier. The facts that viruses harbor a diversity of DNA polymerases, and that in animal mitochondria the standard Bacterial DNA polymerase has demonstrably been replaced by a viral one lend credibility to this hypothesis. In Forterre's view, there is much more to this than another instance of the evolutionary arms race: he argues that the substitution of viral-type DNA polymerases for the primordial RNA polymerases happened three times on separate occasions and lies at the heart of the events that generated the three cellular domains (see fig. 4.1). Let me reserve judgment on the origin of the polymerases, but a DNA genome is in better accord with the evidence for a relatively sophisticated LUCA than one made of RNA.

Suppose we could examine a specimen of LUCA. Would we describe her as a "prokaryote" of sorts? That is what is implied by placing the root of the tree of life basal to the prokaryotes, or within them: life began simple and grew complex over time. But reflection on the nature of cellular genomes points to a rather different scenario.[14] LUCA, it now appears, will have had mechanisms for splicing introns out of RNA transcripts; strategies for regulating gene expression that resemble those of eukaryotes; and (if, indeed, the Planctomycetes recall something of LUCA's architecture, as suggested in chapter 3) a relatively elaborate system of cytoplasmic membranes. LUCA was neither simple nor rough-hewn; she was not a eukaryote in the modern sense but may have been a protoeukaryotic communal organism whose rich and dispersed genome threw offshoots both up the scale of cellular organization and down. Eukaryotes arose by the accretion of greater complexity, larger size, superior autonomy, and more efficient use of energy; these laid the foundations for their subsequent spectacular advances in physiology and morphology. Prokaryotes, by

contrast, are products of reductive evolution. Selection for rapid reproduction and the quick utilization of resources whenever they happened
to become available favored small size, specialization, and streamlined
genomes. Archaea in particular appear specialized for life at high temperatures and other stressful circumstances.

The proposal that LUCA was a protoeukaryotic entity takes a stand on
one of the perennial mysteries of early evolution: what lipids were their
membranes made of? As mentioned in chapter 3, the phospholipids of
Bacterial plasma membranes consist of fatty acids ester-linked to glycerol-
3-phosphate; Archaea make theirs of isoprenoids ether-linked to glycerol-
1-phosphate; and Eukarya notably group in this respect with the Bacteria.
The present interpretation sees acyl ester lipids as ancestral while isoprenoid ethers represent an adaptation to life at high temperature, and one can
make a reasonable case for this view.[15]

Few questions have generated so much debate as whether LUCA herself was a thermophile. The issue first arose because thermophiles and hyperthermophiles are commonly found on basal branches of the universal
tree of life.[16] Such phylogeny would be compatible with an origin of life
in a hot setting, including submarine hydrothermal vents; alternatively,
thermophiles may have been the only organisms to survive the devastating meteorite bombardment of about four billion years ago. In principle,
macromolecular structure holds clues to the temperature at which the organism flourished, and several attempts have been made to apply this approach to LUCA's habitat.[17] The evidence is not clear cut but now seems
to favor the claim that life originated in hot water and LUCA was a thermophile. This would be in accord with geochemical evidence that the Archaean ocean was much warmer than it is today.

So what kind of creature was LUCA? Specifically, how did she make
a living? We turn to this conundrum in the next chapter, but first we must
make room for dissenters who take an altogether different view of phylogeny, cell evolution and the nature of the last common ancestor.

Alternate Histories

History is incorrigibly untidy, evolutionary history doubly so. All our hypotheses are interpretations drawn from indirect and fragmentary clues,
which are open to alternative readings. The triple-stemmed universal tree
(fig. 4.2a) has become conventional wisdom in cell evolution, and enjoys

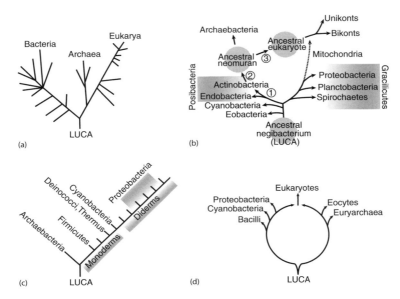

FIGURE 4.2. Four visions of life's history. (a) The universal triple-stemmed tree, after Woese et al. 1990. (b) The world according to Cavalier-Smith 2006a, 2010. Numbers mark major transitions: 1, loss of the outer membrane; 2, loss of the bacterial cell wall and its replacement by new walls; and 3, emergence of the eukaryotes by quantum evolution. (c) Prokaryotic evolution as envisaged by Gupta 2002. (d) The ring of life, after Rivera and Lake 2004.

broad allegiance (including mine) despite the problems already noted in chapter 2. But it remains a hypothesis open to challenge, and those who propose alternatives deserve a hearing, not merely out of scholarly courtesy but because advances in scientific understanding often come from the clash of conflicting hypotheses. The issues are fundamental: the status of the Archaea as a separate domain, the phylogeny of Bacteria, the nature of the last common ancestor, and whether eukaryotes turn the tree of life into a ring. The ongoing debates turn on technicalities that are best left to specialists, but they also let us glimpse alternative ways to imagine life's early history whose merits can be judges on their general plausibility.

Thomas Cavalier-Smith has been a relentless and unsparing critic of the triple-stemmed universal tree ever since it was first proposed. Some of his objections are directed at the statistical methods, which, he claims, have grossly distorted the ribosomal RNA tree and misplaced its root. But the thrust of his dissent is broader, faulting the narrow-minded reliance on macromolecule sequences to the neglect of pertinent findings from cell biology, comparative biochemistry, and even the geological record. For the purpose of reconstructing evolutionary history, he contends, cell structure and physiology are far more informative than the molecular mechanisms of transcription and translation. In a masterly display of both scholarship and polemics, Cavalier-Smith offers a comprehensive and iconoclastic re-examination of cell evolution,[18] complete with terminology all his own, which will be honored in this section.

According to Cavalier-Smith, archaebacteria do not constitute a separate domain but belong, together with all other prokaryotes, in the kingdom Bacteria (the only other kingdom is designated Eukaryota). He subdivides the Bacteria not along sequence lines as the rRNA tree does but by ultrastructure: Negibacteria, with two heritable ("genetic") membranes, are distinguished from Unibacteria, which have a single such membrane (these correspond respectively to the traditional gram-negative and gram-positive organisms and to diderms and monoderms). Archaebacteria possess but one membrane, the plasma membrane, and are therefore classified among the Unibacteria; they represent a highly derived but relatively recent offshoot from the bacterial tree.

Cavalier-Smith begins his account of cell evolution (fig. 4.2b) with Negi-bacteria, endowed with both an outer and an inner lipid bilayer membrane. That choice is not arbitrary but stems from his belief that life began with protocells that were topologically everted (chapter 10). He argues that the ancestral bacterium, and the last universal ancestor of all life, was probably an anaerobe that lived by photosynthesis; its closest living relatives would be the green filamentous bacteria, such as *Chloroflexus*. He derives the Unibacteria from Negibacteria by loss of the outer membrane, a portentous and unique episode in cell history that gave rise to the many kinds of gram-positive (monoderm) bacteria. The course and direction of that evolution can be inferred from a detailed scrutiny of cell physiology ("transition analysis").

Prokaryotic evolution took a momentous turn with the advent of a new bacterial clade, which replaced the peptidoglycan cell wall characteristic of true bacteria with one made of N-linked glycoproteins. The Neomura ("new walls") were to become the ancestors of both archaebacteria and eukaryotes (fig. 4.2b). The precursors of the Neomura must be sought among the gram-positive bacteria, perhaps the actinobacteria or one of the mycobacteria, because of two features that foreshadow later developments: proteasomes and the capacity for cholesterol biosynthesis. Glyco-protein walls may represent an adaptation to high temperature.

Archaebacteria and Eukaryota are sister taxa, sharing a large number of features that must have arisen quickly and almost simultaneously in their neomuran ancestors. Cavalier-Smith lists nineteen of them, including cotranslational secretion of N-linked glycoproteins, signal recognition particles with 7S RNA, core histones, and similar mechanisms for transcription, translation, and repair. How did these drastic differences from bacterial practices come about? In contrast to Woese, who categorically rejects the idea that archaebacteria could have arisen by the evolutionary transformation of eubacteria, Cavalier-Smith promotes exactly that. To him, the rise of the Neomura (followed by the eukaryotes) illustrates the phenomenon called quantum evolution: rapid and extensive evolutionary changes in response to intense selective pressure. Hundreds of genes and gene products were transformed, while others remained unchanged ("mosaic evolution") to create a new class of organisms.

The subsequent emergence of the archaebacteria, with their own set of unique characters in addition to those inherited from their neomuran progenitors, represents adaptation to settings hot and acidic. These adaptations include membranes made of isoprenoid ether lipids (less permeable to protons at high temperatures than the standard fatty acyl ester lipids), and flagella made of glycoprotein rather than flagellin (less prone to disassembly and conformational transition at acid pH). Lateral gene transfer was part of the genesis of archaebacteria and so were gene losses, particularly the loss of cholesterol synthesis. The ancestral archaebacterium may have been a heterotrophic, sulfur-dependent anaerobe and highly thermophilic. Adaptations to thermophily were retained when archaebacteria subsequently radiated into mesophilic habitats, such as the ocean, presumably because they were not too burdensome.

In Cavalier-Smith's opinion, molecular biologists have essentially ignored the fossil record and were thus led to believe, quite erroneously, that both archaebacteria and eukaryotes go back to deep antiquity. Bacteria clearly do, but Cavalier-Smith argues strenuously that the divergence of the Neomura occurred less than a billion years ago, and he dismisses the fossil evidence for far earlier archaebacteria and eukaryotes as unconvincing. To him, unambiguous eukaryotic fossils are not more than eight hundred million years old; more ancient large cells are likely to be giant prokaryotes. Archaebacteria have no fossil record of their own, but their obvious affinities with the eukaryotes strongly suggest that they arose only a little earlier. They are both the most eccentric and the most recent shoot of the ancient bacterial stock.

Cavalier-Smith's account of early cell evolution differs quite sharply from that inferred from rRNA sequences, and from genomics in general. Can it possibly be true? Alternatively, on what basis can it be rejected? The identification of archaebacteria and eukaryotes as sister taxa, that share many basic features because they descend from a common ancestor, is widely accepted and asserted in many versions of the universal phylogenetic tree (chapter 2). I find much merit in Cavalier-Smith's insistence on cell membranes as evolutionary stepping stones, especially the deep cleavage between bacteria with two membranes versus those with one (diderms and monoderms). Radhey Gupta makes the same point, but I am very skeptical of the propositions that double membranes came first and gave rise to singles by loss of the outer membrane, that Neomura then arose after loss of the cell wall, and that subsequent changes were driven by selection for thermophilic and acidophilic lifestyles. The stubborn insistence that archaebacteria are a recent branch of the tree of life conflicts with indirect but persuasive evidence that supports the antiquity of methanogenesis, a metabolic pathway found today only among archaebacteria.

The biggest stumbling block to Cavalier-Smith's thesis is the claim that archaebacteria derive from a eubacterium of the contemporary type, by descent with drastic modification under intense selection. No one can be sure what variation and selection can, or cannot, accomplish in a billion years. Nevertheless, eubacteria and archaebacteria are such very different creatures that transfiguration of one into the other simply does not seem credible; the alternative belief, that they shared a common ancestry in the remote past but diverged prior to the emergence of modern prokaryotic cells, is far more plausible. Perhaps this is why Cavalier-Smith's views, however forcefully argued, have garnered so little support from students of microbial evolution.[19]

Radhey Gupta mounts his challenge to the conventional order from a platform of conserved insertions and deletions ("indels") that occasionally occur in the amino acid sequence of cell proteins,[20] and his terminology is again different from the standard one. Many indels are sufficiently long to make it unlikely that they have arisen more than once; it is a reasonable presumption that two proteins that share an indel probably share a common ancestor as well. For example, the chaperone protein Hsp70 from all gram-negative eubacteria contains an insert of twenty-three amino acids at a particular position. No such insert occurs in Hsp70 from gram-positive eubacteria or from archaebacteria; the interpretation is that the latter two classes are related, and that the insertion occurred subsequently in the ancestor of the gram-negatives. This conclusion is reinforced by the distribution of indels in several more proteins, while a separate set of indels documents the distinctive nature of the archaebacteria. By the judicious use of indels, Gupta has identified major divisions within the bacterial world, most of which agree quite well with those inferred from rRNA sequences. Moreover, and unlike the rRNA data, the distribution of indels suggests a plausible sequence for the emergence of branches, with the gram-positive bacteria ("low GC," or Firmicutes) ancestral and the numerous proteobacteria most recent (fig. 4.2c). Gupta places the root of the prokaryotic tree deep within the gram-positive bacteria, separating archaebacteria from all the rest.

A family relationship between archaebacteria and gram-positive eubacteria is consistent with the ultrastructure of their cell envelopes. As noted in chapter 3, gram-negative bacteria are typically enclosed in a multilayered envelope consisting of the plasma membrane, an outer lipopolysaccharide membrane and a thin peptidoglycan mesh in between; they are "diderms," in Gupta's useful terminology. By contrast, both grampositive eubacteria and archaebacteria are "monoderms," endowed with but a single plasma membrane. Note, however, that the monodrem eubacteria also have a thick peptidoglycan wall, while archaebacteria usually have walls made of S-proteins.

In Gupta's view, archaebacteria descended from monoderm eubacteria, probably on more than one occasion. Their distinctive traits evolved under intense selection for resistance to antibiotics (produced by other bacteria, notably streptomycetes), most of which target proteins involved in the biosynthesis of cell wall or in the transcription and translation of genetic information. Lateral gene transfer played a role in this, but the candidates for transfer would be those that alter antibiotic susceptibility, such as ribosomal genes; just the ones that others include in the stable core of phylogenetics. Archaebacteria are products of the standard interplay of variation and selection; according to Gupta they do not deserve the status of a separate domain, and assuredly do not represent a third domain of life.

I am inclined to take Gupta's contribution more seriously than have most of his peers. The use of conserved indels to map the sequence of branches within the eubacterial stem makes good sense, both methodologically and in terms of its results. The fact that this procedure, unlike the rRNA tree, recovers the classical division of the eubacteria into monoderms and diderms give one confidence. I cannot swallow the claim that archaebacteria are nothing more than highly modified monoderm bacteria, nor that they are products of everyday evolutionary processes, but would entertain the suggestion that both these groups share a common ancestor in the remote past. With this revision, Gupta's finding would complement the conventional account, and remedy some of its deficiencies.

The "ring of life" is the captivating phrase fashioned by James Lake and his colleagues to highlight the unique genesis of eukaryotic cells, which puts eukaryotes quite outside the conventional tree of life. How eukaryotes came to be has long been a mystery, the more so because they combine aspects of both Archaea and Bacteria. In a nutshell, genes that encode informational processes commonly resemble those of Archaea, operational ones those of Bacteria. The most straightforward explanation of this and other observations is that eukaryotic cells are chimeric in principle, products of one or more fusions between Archaeal and Bacterial cells. In that case, LUCA becomes the last common ancestor of Bacteria and Archaea.

A novel statistical algorithm for phylogenomic analysis supplies evidence for a scenario of this kind. "Conditioned reconstruction" relies primarily on scoring the presence or absence of specific genes; the procedure is unaffected by lateral gene transfer and, according to its authors, "can rigorously identify the merger of genomes."[21] The critical conclusion is depicted in an eye-catching diagram (fig. 4.2d), which shows Eukarya as products of fusion of a proteobacterium with an "eocyte," Lake's term for what others call the Crenarchaeota. To be sure, it has been understood for decades that Bacterial endosymbionts were the progenitors of mitochondria and plastids, but Lake and colleagues have in mind something more dramatic, that the essential nature of the eukaryotic cell is rooted in a merger of genomes. We shall return to this subject in much more detail in chapter 7.

Eukaryotic cells often harbor prokaryotic endosymbionts, and the possible consequences of such partnerships have been part of the discourse about the evolution of the Eukarya ever since Lynn Margulis put that notion front and center forty years ago. Prokaryotes do not normally host other prokaryotes, but a few rather specialized instances have turned up. Besides, prokaryotes often form intimate associations that allow the consortium to exploit niches unavailable to the partners individually; such "syntrophic" companies between cells (or protocells) might serve as precursors to outright fusion that generates a novel kind of cell. James Lake, drawing upon genomic data analyzed by conditioned reconstruction, put forward the suggestion that symbiosis may be the origin of the double-layered envelope of diderm bacteria.[22] He envisages the merger of two monoderm bacteria, one related to the clostridia, the other to the actinobacteria. Details are sketchy, of course; cell physiologists will have strong reservations, and molecular geneticists have rejected the evidence outright.[23] And yet, if something of that sort did happen early in the evolution of Bacteria, many puzzling facts would fall into place.

Now glance once more at figure 4.2 with its dramatically different sketches of the course of cell history. Many serious scientists claim Archaea as a third kind of cellular life; others see them as derived from Bacteria. A case has been made for both monoderms and diderms as the root of the Bacterial tree, not to mention precellular progenotes. Eukarya may be one of the primary lines of descent, a late offshoot from the Bacterial tree, or products of a merger between two different prokaryotes. When doctors differ, who decides? I find the conventional doctrine of three domains persuasive and will continue to use it as the framework for this book, albeit in modified form. But in truth, all the issues remain open and the methods at hand seem incapable of resolving them in a compelling manner. Remember, it ain't necessarily so.

CHAPTER FIVE

The Perplexing Chronicles of Bioenergetics

Making a Living, Now and in the Past

What the scientist needs is a selective and nonautomatic memory and, even more, he needs plenty of empty spaces, as it were, between the reminiscences. Great scientific concepts have an entirely noninductive, dreamlike quality. So what a scientist requires more than anything is the ability to maintain empty spaces, both around and within himself. — Erwin Chargaff, *Heraclitean Fire*

Energy Coupling by Ion Currents
How Did LUCA Make a Living?
Towards a Prehistory of Energy Conservation
The Expansion of Bioenergetics

One can imagine life without the molecular and structural complexity of contemporary cells—even life without genes. Indeed, in the early stages of cell evolution all of life's machinery must have been primitive and simple. But at no stage can life have existed without access to a source of energy and the means to harness that energy to the purposes of life. The reason is that almost all biological activities, including biosynthesis, transport, and information processing entail the performance of work. They cannot take place without an input of energy. Energy is required to produce and reproduce life's fabric and to maintain its integrity in the face of incessant wear and tear; the very existence of living things depends on the flow of energy, and that must have been true from the beginning. To understand the evolution of cells, it is not sufficient to read their genomes; one must also work out how they make a living, now and in the past.

Energy Coupling by Ion Currents

The universe provides many potential sources of energy, but for practical purposes living organisms employ only three kinds. First and foremost is sunlight, most of which is captured and put to work with the aid of chlorophyll. Photosynthetic bacteria, eukaryotic algae, and plants are the primary producers of organic matter in today's world, and of its oxygen as well. The second is the breakdown of that organic matter, either by anaerobic reactions (fermentation) or by oxidation; most prokaryotes and many eukaryotes, including almost all animals, live off the organic matter produced by photosynthesis. The third category of energy sources consists of inorganic chemical reactions whose substrates are generated by geochemistry; examples include the use of hydrogen gas to reduce ferric iron to ferrous, or carbon dioxide to methane. Chemolithotrophic organisms, all of them prokaryotes, used to be considered a minor and esoteric class, but it turns out they are widespread and may be of particular relevance to the evolution of bioenergetics.

If we take energy sources as the supply side of the biological economy, the demand side is cell physiology. At the heart of what "living" means is the concept of "work," a term used by biologists to designate any process that runs counter to the spontaneous direction of events; the biosynthesis of complex molecules, the accumulation of ions and nutrients, and motility of cells and organisms all come under the heading of work. Most of biological work is chemical in nature and must be supported by energy presented in chemical form. In most instances, the energy comes from the breakdown of ATP, adenosine triphosphate, which participates chemically in the work function and undergoes hydrolysis to the diphosphate (ADP), or even the monophosphate (AMP). It is inaccurate but irresistible to think of this transaction as the expenditure of one or two "high-energy bonds" and of ATP as a biological energy currency (fig. 5.1). In actuality, matters are a little more complicated: many work functions are driven not by ATP itself but by some other donor of phosphoryl groups; and some, as will be seen shortly, fall outside the scheme altogether. But since all chemical energy donors are interconvertible, we can consider ATP to be the universal coin of biological energy; what the environmental energy sources do is to generate and replenish the ATP consumed during work.

Just how do cells capture the energy inherent in a pinch of sugar or a beam of light and harness it to the transfer of a phosphoryl group to ADP

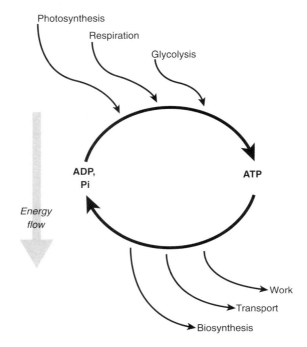

FIGURE 5.1. ATP serves as the central energy currency in all living organisms. Pi, inorganic phosphate.

so as to make ATP? The nature of "energy coupling" was the central issue in bioenergetics for much of the twentieth century, and its resolution by Peter Mitchell in the 1960s forged an intimate and unexpected link between energy transduction and cellular organization. In a nutshell, there are two ways to generate ATP, designated "substrate-level" and "chemi-osmotic." The former category is composed of reactions that produce ATP as an integral part of the reaction mechanism (more commonly, they produce a phosphorylated molecule that can donate a phosphoryl group to ADP, to make ATP). A case in point is the formation of ATP during glycolysis, a widely distributed pathway for the anaerobic degradation of glucose to lactic acid. The bulk of the world's ATP, however, is produced by a very different sort of chemistry, one that depends on a current of protons across a membrane to couple the energy source to a specialized enzyme, ATP synthase, which then generates ATP. Chemiosmotic energy transduction is the basis for ATP production by both respiration and photosynthesis, and it underlies a host of other cellular functions as well. It is one of biology's fundamental mechanisms and is critical to any understanding of cell evolution.

Like other facets of cell physiology, energy coupling by ion currents comes in many variations, some of which are illustrated in figure 5.2. Let *E. coli* serve as an example to introduce the theme (fig. 5.2a). The products of central metabolism (typically NADH or succinate) are oxidized by a cascade of redox proteins called the respiratory chain, which carries electrons to oxygen, reducing the latter to water. The redox cascade is deployed within and across the plasma membrane such that, as electrons travel to oxygen, protons are carried across the membrane from the cytoplasm to the exterior. Protons carry a positive charge, so proton extrusion results in the generation of an electrical potential across the plasma membrane (inside negative), and, to a lesser degree, a gradient of pH (inside alkaline). These two gradients add up to a combined "proton-motive force," or electrochemical potential gradient, that pulls protons back across the membrane into the cytoplasm. Now, the plasma membrane is impermeable to protons except where a specific pathway exists that allows protons to pass. One such is the ATP synthase, a complex piece of machinery that spans the plasma membrane, which couples the flow of protons to the chemical reaction between ATP, ADP, and inorganic phosphate (Pi). The coupling operates in both directions. Influx of protons down the electrochemical gradient elicits ATP synthesis; conversely, ATP hydrolysis causes proton extrusion and is an alternative way to generate a proton-motive force. It remains to add that the proton circulation drives, not only ATP synthesis, but also a long list of work functions including the transport of nutrients into the cell and of waste products out; and also the motor that turns the flagellum and thus propels the cell. Note the pivotal position of the ATP synthase in the cell's metabolic economy; it couples metabolic processes in the cytoplasm to the transport of ions and substrates across the plasma membrane and interconverts the two energy currencies, ATP, and proton-motive force.

It goes without saying that the foregoing is a sketchy summary of a complicated subject; readers are referred to the literature for a fuller account.[1] What matters here is that microorganisms have found many ways to apply the principles of energy coupling by ion currents to their particular circumstances. *Enterococcus hirae* (fig. 5.2b) lacks a respiratory chain and generates all its ATP by glycolysis; in this instance the ATP synthase operates in reverse, extruding protons to generate the proton-motive force

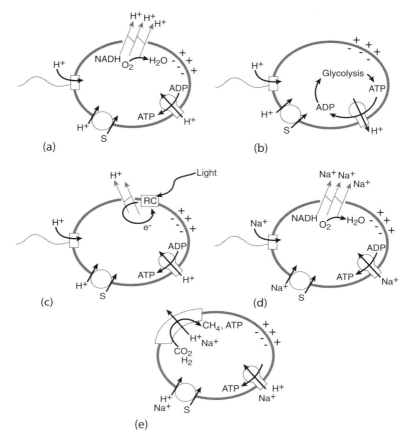

FIGURE 5.2. Variations on the theme of chemiosmotic energy transduction in prokaryotes. (a) An aerobic Bacterium. Note the respiratory chain, which translocates protons out of the cell. The proton circulation is completed by the ATP synthase, the flagellar rotor, and by an array of transport carriers that couple the movement of substrates (S) with that of protons. (b) Fermentative Bacteria lack respiratory chains and live by glycolysis. The ATPase extrudes protons, generating the proton circulation that supports uptake of nutrients and flagellar rotation. (c) A photosynthetic Bacterium. Light absorbed by the reaction center (RC) generates charge separation across the membrane. A redox cascade transduces that into a current of protons, which in turn supports ATP synthesis, transport, and motility. (d) A marine Bacterium in which sodium ions rather than protons carry the current. (e) Methanogenic Archaea live by a combination of substrate-level and chemiosmotic coupling.

that drives nutrient transport and motility. In photosynthetic Bacteria, light absorption generates the initial separation of electric charges across the membrane; a redox cascade converts the electrical potential into a flux of electrons and of protons, which ultimately produces ATP with the aid of a synthase (fig 5.2c). Most photosynthetic Bacteria require an external reductant (e.g., H_2S) to produce biomass. But cyanobacteria, and they alone, can utilize water as the reductant, generating oxygen as a byproduct of photosynthesis. Some Bacteria, particularly marine ones, use sodium as the coupling ion in place of protons (fig. 5.2d). Other organisms couple an inorganic redox process to the generation of an ion current; methanogenesis is a particularly intriguing case involving the translocation of both protons and sodium ions, with many details still uncertain (fig. 5.2e). But all prokaryotes, without exception, generate and utilize ion currents in one way or another, and all possess an ion-translocating ATP synthase (ATPase).

It is important to recognize that members of all three domains rely on ion currents to couple energy sources to physiological work, but sometimes deploy rather different mechanisms or molecules. Both Bacteria and Archaea feature a proton-translocating ATP synthase in the plasma membrane. However, the F_1F_0-ATP synthase of Bacteria differs considerably in structure from the A_1A_0 version found in Archaea (fig. 5.3; F_1 and F_0 refer to the catalytic headpiece and the membrane-spanning rotor, respectively; A_1 and A_0 designate the equivalent components in Archaea). The two ATP synthase families apparently arose as a consequence of an early gene duplication, prior to the divergence of the two prokaryotic domains. Duplication of the genes that encode subunits of the catalytic headpiece provided an opportunity to root the tree of life, with the root positioned so as to group Bacteria on one side, Archaea and Eukarya on the other (see fig. 2.1). Eukarya, whose anatomy is dramatically more elaborate than that of either prokaryotic type, drive multiple ion circulations and possess three kinds of ATPase, each with its own role (fig. 5.4). ATP generation is the task of mitochondria and plastids, operating in the Bacterial manner with an F_1F_0-ATP synthase. The endomembrane system utilizes ATPases of the V_1V_0-type, structurally related to the Archaeal A_1A_0 enzyme to pump protons but not to synthesize ATP. The transport of ions and nutrients across the plasma membrane is again linked to an ion circulation, but the P-ATPases that drive the current (carried by H^+ or Na^+) are quite unrelated in structure and mechanism to the F and A/V ATPases of prokaryotes.

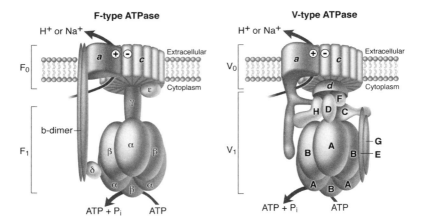

FIGURE 5.3. Schematic structures of the F_1F_0-ATP synthase (characteristic of Bacteria) and the V_1V_0-ATP synthase (characteristic of Archaea and eukaryotic endomembranes). The enzymes are shown operating in the hydrolytic direction. Note the catalytic headpieces (F_1 and V_1, respectively), the rotors F_0 and V_0, the shafts that couple those two, and the stators that keep the headpieces fixed. Subunits depicted in the same shape are thought to be homologous. (Adapted from Mulkidjanian et al. 2007, with permission of the author and Nature Publishing Group.)

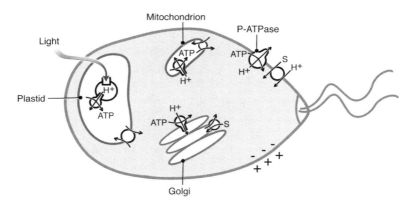

FIGURE 5.4. Multiple ion circulations in a generic eukaryotic cell. Mitochondria operate like aerobic Bacteria, producing ATP that is exported to the cytoplasm. Plastids operate in the manner of cyanobacteria and export products of photosynthesis. V-ATPase acidifies compartments of the endomembrane system (ER) and supports the accumulation of ions; and a P-ATPase generates a circulation of protons across the plasma membrane.

Chemiosmotic energy coupling is one of the universals of biology; what is the point of such elaborate arrangements, in preference to the direct production of ATP by a substrate-level reaction? The answer is that chemiosmotic coupling is more flexible in at least two ways. First, chemiosmotic economies allow organisms to "plug in" an unlimited variety of redox reactions: whether the substrate be glucose or iron, as long as a reaction can be structured so as to transport protons, it can be coupled to the generation of ATP or to the uptake of nutrients. Second, while substrate-level phosphorylations must pack enough free energy to make a molecule of ATP all at once, chemiosmotic ones need only translocate a proton, and protons can be accumulated incrementally. This feature is especially useful to Archaea and Bacteria that eke out a living from marginal sources of energy.

How Did LUCA Make a Living?

Whatever LUCA looked like, she was an organism of sorts: an organized, dynamic ensemble of myriad molecules sustained by a flux of energy. Of all the uncertainties that keep LUCA so elusive, none is more vexing than the mystery of where this energy came from.

Keep in mind that LUCA was, by definition, the last common ancestor of contemporary cells, not the first organism that one could recognize as living. By hypothesis, she was not a particular organism at all but an ecosystem, a community of quasi-cellular entities that exchanged genes and evolved collectively, and from which the progenitors of the three domains crystallized (chapter 4). Mobile genes probably rode on the ubiquitous viruses, themselves members of a parallel virus universe, as ancient as that of cells. Then, as now, the world would have been patchy; genes that ended up in one or another of the cellular domains may have been recruited at different times and gathered in separate locales. LUCA must herself have been the product of prolonged evolution that gave rise to the universal cellular devices (ribosomes, membranes, cytoskeleton, etc.), and the essentials of cellular organization, too. Presumably several modes of energy metabolism had appeared, dispersed among those protocells, and been available for sampling by rampant gene swapping. Can one make a stab at guessing what was on the menu?

It is easier to guess what was not on the menu. Methanogenesis, that intricate sequence of reactions by which carbon dioxide is reduced to methane with the production of ATP, is confined to one branch of the *Euryarchaeota*. Some have argued that this is an ancestral pathway of energy metabolism that played an important role in the origin of cellular life but survives today only among the Archaea. In principle, it is conceivable that today's restricted distribution is the result of selective gene losses or even of lateral gene transfer; but it is surely more parsimonious to consider methanogenesis as a later invention of the *Euryarchaeota* and to exclude it from LUCA's repertoire. By the same token, photosynthesis based on chlorophyll as the light-harvesting pigment occurs today only among the Bacteria and their descendants, and thus will not have been among LUCA's options. Oxygen-producing photosynthesis is more restricted still, confined to the cyanobacteria. Aside from minute amounts produced by photochemistry, oxygen was totally absent from the early atmosphere. Cellular life evolved in a world that was effectively anaerobic.

Historically, it has been assumed that the earliest cells were fermentative, relying on the anaerobic degradation of organic substances of abiotic origin (amino acids, perhaps, or simple carbohydrates), and that they produced ATP by substrate-level phosphorylation. The most familiar such pathway is glucose degradation by glycolysis, widely distributed among all forms of life. But there is good reason to doubt that glycolysis goes clear back to LUCA's day. First of all, organic matter—and sugars in particular—would only have become abundant after the rise of plant photosynthesis. Second, while glycolysis is found in both Bacteria and Archaea, the two versions are not the same and are catalyzed by enzymes that are not homologous.[2] The data suggest that glycolysis in these two domains arose separately and represents an instance of convergent evolution. But if not fermentation, what is the alternative? Support has been building for the counterintuitive notion that LUCA was a community of anaerobic chemolithotrophs that drew upon substrates of geochemical origin and resided in a warm or even hot setting. If so, by LUCA's day the basic mechanisms of energy coupling by ion currents must already have been in place. Among these will have been redox-based electron transport, ion-translocating ATP synthases and ion-coupled transport systems.

The open question is how LUCA put her endowment to work. On the premise that both reduced carbon compounds and oxygen were scarce, she could hardly live, as so many contemporary Bacteria and Archaea do, by oxidative phosphorylation with organic substrates. A more abundant energy source is light; it is at least conceivable that the role played in contemporary organisms by chlorophyll was in LUCA's day assigned to another pigment, possibly a rhodopsin (see page 84, below). But the most plausible sources of metabolic energy are those derived from geochemistry: the oxidation of inorganic substances such as H_2, H_2S, S^0, Fe^{2+}, and CH_4, with oxidants including CO_2, CO, S^0, NO, NO_3^- and SO_4^{2-}. There are many prokaryotes, both Bacteria and Archaea, that employ chemiosmotic mechanisms to extract useful energy from such inorganic reactions, and the molecular mechanisms are understood in some detail.[3] They include both full-fledged redox chains and simpler, one-step reactions that are coupled to proton translocation, such as hydrogenase and ferredoxin oxidoreductase.[4] It's a hardscrabble existence, for none supply energy in abundance, but they do sustain simple ecosystems both above ground and below. Many chemoautotrophs are also thermophiles,[5] which reinforces the perception that these metabolic patterns may have survived from the earliest stages of cell evolution.

One metabolic pattern that can serve as a model for primordial bioenergetics is the Wood-Ljungdahl pathway of carbon fixation, as practiced today by acetogenic Bacteria and (with variations) by certain Archaea. Strict anaerobes, these organisms employ hydrogen gas to reduce carbon dioxide, producing both metabolic energy and biomass. The pathway parallels that of methanogenesis; it has been extensively studied,[6] but it is still difficult to envisage the cells' energy economy as a whole. Very briefly, one molecule of carbon dioxide, after attachment to a tetrahydrofolate cofactor, is progressively reduced to a methyl group. That combines with a second molecule of carbon dioxide to yield the energy-rich thioester acetyl-CoA, which in turn generates acetyl phosphate by a substrate-level reaction. Acetyl phosphate is a key intermediate that donates its phosphoryl group to ADP, producing ATP. Carbon dioxide reduction is exergonic overall; it generates ATP but also consumes some, and so the overall yield of ATP suffices only to keep the reaction going but not to form biomass. The additional ATP needed for growth (and also the energy required to overcome a barrier to one step in the sequence) are supplied by a chemiosmotic ion circulation. This entails a sodium-translocating ferredoxin oxidoreductase plus a sodium-translocating ATP synthase. The sequence is anything but simple (even if one ignores the source of the complex cofactors) and is unlikely to represent the first biological economy, but could be descended from those metabolic pioneers. LUCA's scheme for energy production must have been quite sophisticated, and the outcome of protracted evolution.

The hypothesis that LUCA relied on geochemical substrates for her sources of energy bears directly on the question of her most likely habitat. Evolutionists have traditionally looked to Darwin's "warm little pond" or a quiet and sunlit marine bay, but such benign places would have been scarce on a young planet riven by geological tumult and battered by meteorites. Instead, attention has come to focus on sheltered sites where substrates to sustain a chemoautotrophic lifestyle emerge from the earth, and specifically on warm alkaline submarine vents. A prime example is the Lost City hydrothermal field, on the ocean floor just off the Mid-Atlantic Ridge.[7] In its current form, the proposal states that the last common ancestors of all life were truly precellular entities, molecular societies evolving in the crannies and crevices of mineral deposits laid down by such vents.[8] Life's devices, including metabolism, protein synthesis, and DNA replication would have taken shape within such more or less individual compartments, sustained by the bounty of the earth. The final step would have been the emergence of lipid bilayer membranes, which allowed the first true cells to break free from their inorganic hatchery. Bacteria and Archaea went their separate ways at last, enclosed in chemically different membranes. Proponents of this radically novel scheme are well aware of the objections to their arguments, its counterintuitive texture, and especially the conflict with solid evidence that points to a LUCA with functional membranes equipped with the machinery of energy transduction; the recent article by Lane and colleagues makes a valiant effort to put the doubts to rest.[9] We shall return to this subject in chapter 10, for all current hypotheses for the origin of cellular order face serious difficulties, and those porous mineral deposits do make the most attractive setting for the birth of biochemical systems. In the meantime, all of us would do well to recall the spoof of our pretensions in a poem by Hilaire Belloc, often and gleefully quoted by an earlier generation of microbiologists, which concludes: "So let us never, never doubt/what nobody is sure about."[10]

"The progenote is today the end of an evolutionary trail that starts with fact, progresses through inference, and fades into fancy."[11] In 1987, when Carl Woese wrote this, there was good reason to anticipate a flood of insight from the decipherment of gene sequences. That did happen, yet LUCA remains indistinct, seen through a glass darkly. As so often in the course of this writing, I am troubled by the sense that we have been searching under the light, and some essential truth has quite escaped our notice.

Toward a Prehistory of Energy Conservation

The devices modern cells use to capture and transform energy are inordinately sophisticated and linked to cellular organization. Reaction pathways are deployed so as to cross lipid membranes, energy is transferred both chemically and by means of ion currents, and ingenious ATP synthases connect the vectorial biochemistry of membranes to the scalar one of the cytoplasm. All this machinery, which appears to have been as much a part of LUCA's metabolic economy as of contemporary cells, must itself be the product of prolonged evolution. Can one shine any light into that murky past? In truth, not much, but the emerging consensus that the origins of bioenergetics, and of life, must be sought in its geochemical context encourages one to speculate on that prehistory from a new and unconventional perspective.

Both the capture and the utilization of energy revolve around transport systems for metabolites, and especially for ions; the roster of known ones runs into many hundreds. How and from what did transport systems evolve? Two decades ago, it was generally believed that transport systems arose from metabolic enzymes, but that seems to be a misapprehension. Milton Saier and his colleagues read innumerable gene sequences to construct a phylogeny of transport systems and concluded that they evolved from channel-forming peptides.[12] Today's profusion is the product of repeated mutations, duplications, trimming and rearrangement of membrane-spanning segments, together with the acquisition of regulatory binding sites. One can thus begin to imagine how plain pores, offering nothing more than a pathway of diffusion, came under metabolic control ("gating"). Many evolved into "carriers," proteins in which the substrate-binding site faces alternately one side of the membrane or the other. And some came under the control of an energy source, either the cell's ion circulation or ATP; the latter are thought of as "pumps" in which the energy released by the hydrolysis of ATP moves the substrate against a gradient of concentration and/or electrical potential. The greater mystery is what kind of membrane supplied the matrix for these transport carriers, and whence they came.

Redox processes, reactions in which electrons pass from a donor to an acceptor, supply much of biological energy and have long been thought to be among biochemistry's most ancient elements. While today's catalysts are, of course, proteins, the redox chemistry itself is characteristically mediated by clusters of sulfur and metal atoms—iron, nickel, copper, and molybdenum. It is conceivable that at the precellular stage of evolution, proteins were not involved at all and all catalysts were inorganic;[13] according to Bill Martin, bioenergetics has rocky roots. Some traces of the transition may survive, not so much in the amino acid sequences of redox proteins as in the persistence of their folding patterns. These have been taken as evidence for the antiquity of a "redox protein construction kit" composed of catalysts containing iron sulfide, nickel, and molybdenum; proteins such as cytochromes, ferredoxins, and Rieske factors may have been part of that kit. Once again, one wonders into what membranes ancestral redox carriers were inserted.

Of all the devices that participate in energy transduction, none so challenges the imagination as the rotary ion-translocating ATP synthase (fig. 5.3). This minute electrical machine is made up of some two dozen proteins organized into four functional elements: the headpiece, which catalyzes the chemical reactions among ATP, ADP and Pi; the rotor, which spans the energy-transducing membrane; the shaft that links those two; and the stator, which keeps the headpiece fixed in place. When the complex functions as an ATP synthase, protons (or in some cases, sodium ions) pass down the electrochemical gradient across the rotor, setting it spinning at some eight thousand revolutions per minute. The rotary motion is transmitted to the shaft and thence to the headpiece, which remains immobile. Instead, the alpha and beta subunits that make up the headpiece pass through a sequence of conformational changes; the catalytic sites located between the alpha and beta subunits successively bind ADP and Pi, fuse them and eventually release the ATP.[14] When the device operates as an ATPase, the sequence is reversed: ATP is hydrolyzed driving protons outward across the rotor and generating the electrochemical ion gradient. How on earth could this astonishing transformer have arisen? The answer is quite unknown, and only tenuous clues have emerged.[15] According to Mulkidjanian and colleagues, the ATP synthase may have evolved from a protein translocase, itself the offspring of a (hypothetical) RNA translocase that was part of the (hypothetical) RNA World. They also argue that, although in today's world the chief biological coupling ion is the proton, it was actually Na^+ in the world that preceded LUCA.

Reflection on individual transport processes highlights, but does not illuminate, the core mystery, which is the origin of a constellation of coupled vectorial systems that first generate and then exploit an electrochemical gradient of protons or Na^+. Here is again the place to take notice of geochemists' dream, which turns on the proposal that energy conservation (and indeed cellular life) began in alkaline hydrothermal vents on the bottom of the ocean.[16] Sites of that kind are built of rocks rich in catalytic metals. They discharge effluent laden with abiotic hydrogen gas, carbon monoxide, and carbon dioxide, even some simple organic compounds, that may have supplied both precursors and chemical potential for the evolution of chemo-autotrophic life. Unlike the well-known black smokers whose effluents are both acidic and extremely hot, those of alkaline vents reach only a moderate 70°C or so. Most intriguing is the fact that alkaline effluent, pH about 10, interfaces with relatively acidic seawater (ocean pH in the Hadean epoch is thought to have been as low as 6), within the nooks and crannies of mineral deposits laid down at the vent's edge: a natural chemiosmotic gradient of the correct polarity and magnitude. Could this be where energy transduction began?

For more than two decades, Michael Russell has been urging the proposition that the gradient of 3 to 4 units of pH, established across inorganic membranes consisting largely of iron sulfide, constituted the primordial electrochemical gradient that biological systems evolved to exploit.[17] Many of his followers are skeptical about that inorganic membrane but do envisage the emergence within that mineral honeycomb of all cellular biochemistry, including proper lipid membranes to bound and define protocellular compartments. If so, the primordial ATP synthase will have evolved in the context of energy transduction from the beginning, serving to couple the geochemical ion gradient to protometabolism within the cavity. Could this outrageous proposition possibly be true? Caveats abound, as they do for all explicit proposals about the locus and manner of genesis (see chapter 10), but it is noteworthy that no more plausible alternative has been advanced. As so often in the field of cell evolution, sensible people will cultivate an attitude of open-minded skepticism. But the origin of energy transduction is no ordinary problem, and is not likely to yield to conventional thinking.

The Expansion of Bioenergetics

On the premise that the last common ancestors of all life were already equipped with the basic mechanisms of energy transduction, one can sketch an outline of the subsequent evolution of bioenergetics as the three domains emerged[18] (fig. 5.5). Among the Archaea, the major innovation

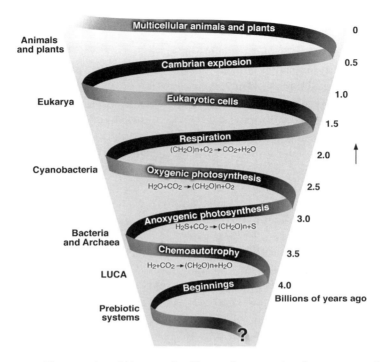

FIGURE 5.5. The expansion of bioenergetics. The growing repertoire of energy-conserving mechanisms is correlated with increasing metabolic and genomic complexity of organisms. (Adapted from Falkowski 2006, with permission of the author.)

would be methanogenesis, the reduction of carbon dioxide to methane with dihydrogen or formate as the electron donor. The pathway consists of about a dozen steps which support both the generation of ATP and the reduction of carbon dioxide to biomass; it calls for the participation of several cofactors seldom found outside the Archaea, and requires an ion current at one or two steps to generate ATP.[19] The sequence of reactions largely parallels those of acetogenesis, but the enzymes that catalyze methanogenesis are not homologous to those of acetogenesis. William Martin and others have made a case for methanogenesis as a primordial pathway of energy generation, but its restricted distribution suggests that methanogenesis was a uniquely Archaean invention that remains confined to the lineage in which it arose.

The chief driver of biological energetics today is photosynthesis, and we owe that momentous invention to the Bacteria. Not only did photosynthesis open up an unlimited supply of energy, it also generated oxidized forms of nitrogen, iron and sulfur, and eventually oxygen itself; these in turn supplied substrates for respiration. Photosynthesis characteristically relies on chlorophyll to absorb incident light. The chlorophyll is embedded in a reaction center complex that spans a membrane, and activation of the chlorophyll mediates the initial separation of charges across that membrane. That initiates a complicated sequence of redox transfers by which the energy of light is transduced, first into an electrochemical gradient of protons and subsequently into the production of ATP or the performance of work. Photosynthesis comes in several versions. Bacteria commonly possess either photosystem I or photosystem II; they carry out ATP synthesis by a cyclic process in which electrons, kicked out of chlorophyll by light, are passed down a chain of redox carriers, returning eventually to their starting point (fig. 5.2c); and they require an external source of reducing power, typically H_2S, to reduce carbon dioxide to cellular building blocks. The cyanobacteria, and they alone, possess both photosystems arranged in tandem; and only the cyanobacteria possess the molecular machinery to extract electrons from water and generate molecular oxygen. This gave cyanobacteria access to an unlimited supply of reducing power and incidentally changed the planet: thanks to oxygen, the anaerobic world in which LUCA arose was transformed into our own familiar aerobic one.

Where did all this novelty come from? The evolution of photosynthesis and its timing continue to be much debated without a clear resolution.[20] Bacterial photosynthesis is unevenly distributed, with representatives in only six broad groups: proteobacteria, green sulfur bacteria, filamentous phototrophs, the monoderm heliobacteria, acidobacteria (a recent discovery), and the cyanobacteria. There is good reason to believe that, despite the complexity of the photosynthetic apparatus, it occasionally dispersed from one phylum to another by lateral gene transfer, but only within the domain Bacteria. It is formally possible that genes for chlorophyll-based photosynthesis were already present in LUCA and were lost in all but a few lineages, but the burden of proof must be on the proponents. The two photosystems, photosystem I and photosystem II, probably arose by early gene duplication. But it is quite uncertain what organism invented photosynthesis, and whether the ancestral state consisted of a single photosystem or of two. Recent genomic studies by Mulkidjanian and colleagues suggest that, contrary to common belief, the original phototroph may have been an ancestral cyanobacterium. Though the specifics remain uncertain, the evolution of photosynthesis evidently depended on elements of the redox kit that may even antedate LUCA: iron-sulfur clusters, tetrapyr-roles, flavins, quinones, cytochromes, and much else. These supply possible precursors to the photosynthetic apparatus itself, including chlorophylls and ancillary pigments.

Most of the world's photosynthesis revolves around chlorophyll, but not all. Next time you fly into San Francisco across the Lower Bay, spare a thought for the brine ponds below your flight path. Their bright orange-pink color is caused by a dense population of halobacteria, salt-loving Ar-chaea that flaunt their possession of the protein bacteriorhodopsin with its carotenoid pigment, retinal. Some forty years ago, Walter Stoeckenius and Dieter Oesterhelt showed that this protein spans the plasma mem-brane and serves as a proton pump, translocating one proton for each pho-ton absorbed. The electrochemical potential so generated can be coupled to the performance of work, including the synthesis of ATP. Of all the variants of photosynthesis, this is by far the simplest: a single protein, un-connected to redox carriers, that carries out proton translocation driven by light. Related rhodopsins of Haloarchaea serve as chloride pumps and also to sense light.

Bacteriorhodopsins were thought to be restricted to extremely halo-philic Archaea, so it came as a considerable surprise to learn that related proteins are both abundant and widely distributed among Archaea and Bacteria that inhabit the sunlit waters of the open ocean.[21] Proteorhodop-sin genes appear to be cosmopolitan wanderers, readily transferred across all taxonomic lines, to be picked up by whatever organism finds them use-ful. The physiological functions served by proteorhodopsins are not fully understood, but it is likely that many do play an ancillary role in energy transduction. Might they also be an ancient class, relics of an earlier era in the evolution of bioenergetics? Possibly yes, but if so, proteorhodopsins turned out to be a dead end, destined to be relegated to a secondary role by the advent of chlorophyll-based true photosynthesis.

The beneficiaries of oxygenic photosynthesis include the innumerable organisms that live by respiration, with organic substrates as electron do-nors and oxygen as the final acceptor. Respiratory redox chains come in many variations but the general pattern is clear: elements of the redox kit are assembled into discrete complexes that translocate protons and generate a proton-motive force; the electrochemical potential in turn is captured by an ATP synthase, which transduces its energy content into a chemical form. The origins of these systems is no less obscure than that of photosynthesis, the key unresolved issue being whether respiration followed or preceded photosynthesis.[22] The contemporary version of respiration presumably came into existence only with the advent of oxygen, but earlier kinds may have drawn on traces of oxygen produced by photochemistry, or on inorganic electron acceptors such as Fe^{3+}, sulfate, and nitrogen oxides. The hallmark enzyme, dioxygen reductase, may have begun as a scavenger for traces of toxic oxygen. Genomic studies suggest a complex history, confused by multiple origins and some lateral gene transfer.[23] Dioxygen reductases arose on at least four occasions, one of which seems to antedate the separation of Bacteria from Archaea.

Where, in the widening gyre of biological energy conservation and transformation, do the eukaryotes find their place? It is certain that prokaryotes discovered all the energy sources and coupling mechanisms available to contemporary life. Eukaryotes rely on only two: the degradation of preformed organic matter, and the direct capture of light. Some useful energy is produced by fermentative reactions in the cytosol but the bulk of cellular ATP is forged by either respiration or photosynthesis, both of which are carried out by specialized organelles: mitochondria and plastids, respectively. Both operate in the prokaryotic manner, and both are descendants of Bacterial endosymbionts. There are no known chemolithotrophic eukaryotes and no evidence that such existed in the past. Taken together, these undisputed facts add up to strong evidence that eukaryotic cells could only have evolved after prokaryotic cells had appeared and begun to diversify.

Unlike prokaryotic cells, eukaryotic ones do not house energy-conserving redox chains in the plasma membrane. But they do drive currents of H^+ or Na^+ across the plasma membrane at the expense of ATP hydrolysis, which support the transport of ions and metabolites into the cell and out. The plasma membrane ATPases are completely unrelated in structure and mechanism to the F- and A-ATPases that play so central a role in prokaryotic energy transduction (ATPases of the V type do function in the eukaryotic endomembrane system, but not in the plasma membrane). It is not obvious what one should make of this. One possibility, which I find attractive, is that the distinctive mode of energy metabolism seen in eukaryotic cells is a relic of their origin in a separate line of cellular descent, as Woese and his disciples have maintained for decades. This is a highly controversial issue, which will be the central theme of chapter 7.

CHAPTER SIX

Life's Devices

On the Evolution of Prokaryotic Cells and Their Parts

The stuff of life turned out to be not a quivering, glowing, wondrous gel, but a contraption of tiny gigs, springs, hinges, rods, sheets, magnets, zippers, and trapdoors, assembled by a tape whose information is copied, downloaded and scanned. — Steven Pinker, *How the Brain Works*

Molecular Machines
Cell Continuity
Slouching toward Bethlehem to Be Born

L iving things are autopoietic systems: they make themselves. Proposed more than thirty years ago by Francisco Varela and Humberto Maturana, this remains one of the most concise and comprehensible definitions of the living state, and it has the particular merit of underscoring what organisms actually do. Self-making implies a multitude of activities directed towards the acquisition of matter and energy, production of the fabric, maintenance and repair, and ultimately reproduction. Not surprisingly, even the simplest cells are dauntingly complex systems made up of many thousands of molecules arranged into functional units. Over the past half-century we have become thoroughly familiar with the standard parts found (with variations) in all cells: enzymes and genes, transport systems and scaffolding, ribosomes and membranes, and organs of mobility. We know in general what they do and how they work, and how they contribute to the operations and architecture of the whole cell. By contrast, we know very little about how these devices, or the cell as a whole, came to be.

The origin of complex features has been one of the central topics in biological theory since the beginning. Charles Darwin devoted a section to "organs of extreme perfection and complication," and worried about the vertebrate eye: "To suppose that the eye, with all its inimitable contrivances . . . could have been formed by natural selection, seems, I freely confess, absurd in the highest possible degree."[1] But he argued forcefully that the eye, and all other complex organs, must be seen as products of "numerous, successive and slight modifications," each of which confers some small benefit, and biologists have grown ever more confident that this is correct. Sometimes this understanding is bolstered by fossils. More commonly, as in the case of eyes, we rely on the existence of intermediate forms that allow one to arrive at a plausible historical narrative. In recent years, this procedure has been greatly facilitated by computer simulation, as in the impressive reconstruction of the evolution of complex lens eyes in gastropod snails by Nilsson and Pelger.[2] In no instance can one recount the actual course of evolutionary history, but there is ample evidence that it proceeded by descent with modification channeled by natural selection; and that may be the best we can hope for.

Turning now to cells and their organelles, the problems are compounded by the fact that all these originated in the exceedingly remote past, leaving no visible trace. On the other hand, in the case of molecular devices the relationship between the product and the genes that specify its parts is much more immediate than for the eye. Thanks to molecular genetics, there is now extensive evidence that the molecular constituents of any one organelle commonly have homologs elsewhere, and that the corresponding genes have undergone variation, duplication, transfer, modification, and selection. Experimental evidence that heritable variations winnowed by natural selection can indeed generate complex functions comes again from computer simulations, leaving little doubt that Darwin's insights apply as much to cells as to eyes.

Before we venture into the morass of detail, several important constraints must be made explicit. First, life's devices are organelles; they have functions that confer benefits upon the cell or organism as a whole, and natural selection acts at the level of the collective, not that of the device or its molecular constituents. Second, cells and their organelles are composed of proteins, nucleic acids, and the products of their chemical activities. What we know of cell evolution has been drawn almost entirely from genomics and pertains to a time after contemporary mechanisms for the expression of genetic information had come into existence. Of what came before, genomics can tell us little or nothing. Third, an evolutionary narrative based wholly on the variation and selection of information encoded in genes is likely to be incomplete, perhaps grossly deficient. The reason is that the form and organization of cells is not explicitly spelled out in the genetic instructions, but arises epigenetically from the activities of the multitude of molecules that make up the cellular society (see chapter 1). The role of uncoded gene modifications, structural heredity, and self-organization is only now being explored. Our understanding of cell evolution is likely to be radically revised in the coming decades.

Molecular Machines

When Michael Behe made the Bacterial flagellum into an icon of "irreducible complexity," he surely never meant to do evolutionary biology a service; but the law of unintended consequences prevailed. He argued in 1996 that this exquisitely complex piece of machinery (fig. 6.1a), made up of some fifty interlocking parts, each of which is essential to the function of the whole, could never have arisen by the Darwinian process of small random steps, guided by natural selection.[3] It must have been designed as a unit, for the purpose of motility, by a superior intelligence. Behe's broadside, squarely aimed at the grand unifying theory of modern biology, elicited a burst of interest in Bacterial flagella and a spate of publications. We are still far from being able to supply a coherent account of how flagella evolved, but there is now ample evidence that they did so in the conventional manner, by descent with modification.

The diversity of prokaryotic flagella is already a strong indicator.[4] The flagella of Spirochaetes do not protrude into the medium, but are confined to the periplasmic space between the outer sheath and the plasma membrane. *Vibrio* species make two kinds of flagella, polar ones driven by a current of sodium ions and lateral ones driven by protons. Cells of *Salmonella* tumble when rotation stops, *Rhodobacter* halts in its tracks. Flagellin, the protein that makes up the flagellar filament, comes in numerous versions; *E. coli* alone boasts forty distinct antigenic variants. Proteins essential to flagellar function in one species may not even be present in another. There is clearly no single invariant design but a family of designs, and they do not employ a universal set of components.

The adjective "irreducible" is misleading also in another sense: flagellar components are not unique to that organelle but share homologies with other cell constituents. Bacterial flagella grow by the addition of subunits to the tip; the apparatus responsible for extruding subunits into the central channel that runs the length of the filament is clearly related to the needle complex employed by certain pathogenic Bacteria to inject toxin into their victims (the type III secretion system). These two devices evidently share a common ancestry. Most of the proteins that make up Bacterial flagella have homologs elsewhere: for example, subunits of the ATPase that drives flagellin extrusion are homologous to the alpha and beta proteins of the F_1F_0-ATP synthase. Links have also been discerned between the flagellar filament and adhesins, and between the motor and certain transport systems. Both gene duplication and the recruitment of genes from unrelated functions appear to have contributed to the making of flagella. What has been learned does not yet add up to a historical narrative, but it strongly supports the conclusion that flagella are products of evolution, just as wings and legs are. They were not designed by an engineer according to a plan, but cobbled together by a tinkerer.

Few investigators, if any, anticipated the most spectacular discovery in the field: the flagella of Archaea are utterly different from those of Bacteria in both structure and function, and they do not share a common ancestry.[5] Denizens of both domains possess rotating organelles of motility, but there the resemblance ends (fig 6.1). Bacteria power their flagella with the proton current; Archaea use ATP. The flagellar filament of Bacteria is made up of numerous subunits of the protein flagellin; Archaea employ an unrelated glycosylated protein. Bacterial flagella are assembled at the tip with the aid of a secretory complex; Archaeal filaments grow from the base by mechanisms clearly related to those involved in the extension of pili. So glaring are the differences that they justify designating the Archaeal organelle by a name of its own, the archaellum.[6] Curiously, the chemotactic machinery appears to be quite similar in the two domains.

One can imagine natural selection perfecting cell motility by honing and elaborating the flagellum, step by tiny step; but how could the rudiments of such an organ arise in the first place, for selection to work on? The word to conjure here is "exaptation."[7] At the incipient stage, the structure that evolved into flagella need not have had a role in motility but served an altogether different function (or none at all). If its presence had an adventitious effect on motility, natural selection could have recruited or exapted it as a foundation to build on. The diversity of Bacterial flagella affords a glimpse of what such a rudiment might look like. Aphids harbor an endosymbiont, *Buchnera aphidicola*, distantly related to *E. coli*. This intracellular bacterium is nonmotile but possesses the genes needed to synthesize the flagellar basal body and hook; the cells' surfaces are

A. Flagellum   B. Archaellum

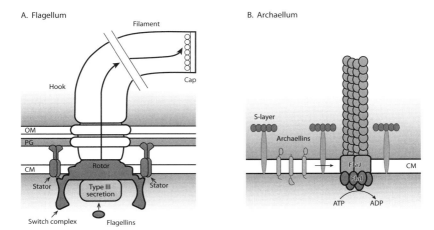

FIGURE 6.1. The organs of prokaryotic motility: complex and sophisticated but not irreducible. (a) The Bacterial flagellum. Rotation is driven by an ion current. Flagellins are produced in the cytoplasm, secreted into a channel that traverses the length of the flagellum, and assembled at the tip. (b) The Archaeal archaellum is both structurally and functionally quite different. Rotation requires ATP hydrolysis. Archaellins travel through the cytoplasmic membrane to be assembled at the base. CM, cytoplasmic membrane; OM, outer membrane; PG, peptidoglycan. (Adapted from Jarrell and Albers 2012, with permission of the author and Elsevier.)

studded with hundreds of them.[8] Though the function of these flagellar rudiments is unknown, it is reasonable to assign them a role in the export of proteins that are involved in the symbiotic relationship between the bacteria and their insect host. The *Buchnera* mini-flagella are derived rather than primitive, but highly suggestive; it seems likely that flagella evolved by the exaptation and elaboration of an ancient pathway for protein secretion.

Living cells have been described as computational entities ("wetware"), and that is often a useful and enlightening point of view;[9] but it is only one of several. The complementary perspective highlights the material, mechanical aspects of the living state. It takes a multitude of devices for cells to harvest energy and put it to use, to maintain and reproduce their structure, to receive information and respond to it. Not all devices are as strikingly machinelike as the flagellum with its rotary electric motor and propulsive filament, but they are identifiable, modular elements that can often be isolated. Familiar organelles of prokaryotic cells include the cell wall and plasma membrane, gas vesicles, half a dozen systems of cell motility, ribosomes and carboxysomes and magnetosomes, reaction centers,

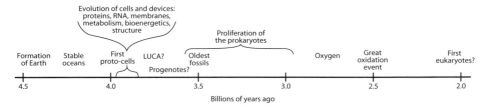

FIGURE 6.2. Early events in the evolution of life

The story begins with the origin of life, which I take to be synonymous with the emergence of entities commonly referred to as "protocells." Of their nature we know little, and nothing at all for certain, but I take protocells to have had a rudimentary capacity to assimilate matter and energy, to persist, and to multiply. The geochemical circumstances under which life began are debatable, and are presently the focus of particular interest. By definition, protocells lacked both genes and proteins specified by coded instructions, but they provided a platform for the evolution of those hallmarks of biology. This prologue to life's mystery play records the initial appearance of molecular systems capable of evolution; here is biology's ultimate riddle, to which we shall come in chapter 10.

The advent of translation, the process whereby information encoded in nucleic acids specifies the structure of proteins, marks the beginning of cell evolution proper. I remain agnostic as to whether this era began with a unique First Cell, a singular historical entity that first brought together in one membrane-bound vesicle some capacity for replication and energy coupling (as Koch and Silver propose) or grew out of a communal orgy of chemistry driven by energy flow and abetted by information swapping (as other students of the subject believe). In either case, we cross here a watershed between entities that are merely chemical, and their descendants, who display qualities that we customarily associate with life. We walk swiftly but warily, looking neither to the left nor to the right, for the ground quakes underfoot and there are no firm steppingstones.

What shall we call entities endowed with primitive versions of today's genes, proteins, and the machinery that allows the former to specify the latter? Woese referred to them as "progenotes," entities in which gene expression and replication were imprecise, proteins were short and variable, and genotype and phenotype were loosely coupled.[24] This term aptly describes the state of information processing but leaves out of account metabolic and structural aspects. I prefer to think of them as "stem-group" cells, just as biologists speak of stem lineages, extinct precursors of contemporary crown groups such as reptiles or conifers. By hypothesis, stem-group cells acquired not only primordial versions of information processing but also the requisite precursors of metabolism, membranes, and energy transduction; they supplied vehicles for an era of intense biochemical experimentation and invention during which the operation and organization of cellular life took shape. Rough beasts lumbering toward greater autonomy, individuality, and complexity, stem-group cells evolved and refined transcription, translation, replication, and also metabolism and bioenergetics. They also gave rise to the true cells that populate the three domains of life. This astonishingly creative phase spans the period from the origin of protocells, perhaps 4 billion years ago, to the first appearance of prokaryotic cells 3.5 billion years ago; whether that leaves ample time or feels a little tight remains a matter of personal judgment. The course of events is largely obscure, fitfully illuminated only by molecular genomics; that technology has generated a huge literature of data, ideas, and disputation, but not much certainty.

How can one explain this ancient burst of innovation, greater by far than in all the subsequent history of life? The familiar mode of Darwinian evolution, creeping at a petty pace by gene mutations and rearrangement, duplication and differential loss, seems too stately to account for that onrush of novelty. Woese argues that the characteristic feature of that time was pervasive lateral gene transfer, so that any useful invention would spread quickly and widely. Genomes and the structures and processes they encode were forged at a "high evolutionary temperature," in an era of incessant mutation and exchange and also, one must presume, of equally intense selection.[25] The community of transitional cells that made up this frenzied common market in genes will have evolved, not only rapidly but also collectively. Unlike the organisms we know, stem-group cells were not locked into vertical transmission of genes from parents to offspring. Time and again, vertical heredity was overpowered by lateral gene transfer, diverting the emergence of the organismic lineages that we now take for granted. The agents of genetic exchange are unknown, but viruses make plausible candidates. Let it be clearly understood that the foregoing is a hypothesis, not established fact or even a universal consensus, but it has to my ears the ring of truth.

The most conspicuous landmark in cell evolution is the establishment of the three domains, represented on the universal tree of life by LUCA, the last universal ancestor of all extant organisms. The nature of that hypothetical stage has been much debated; one should probably keep an open mind as to whether the "C" stands for common, communal, or cellular. My own preference favors the view that LUCA was a community of fairly complex entities—not some primitive slime but also not a singular organism—and that cells ancestral to the three domains emerged from the melee of stem-group cells by a process somewhat analogous to crystallization. As evolving stem cells acquired useful innovations and adaptations, they also became increasingly well integrated; eventually the point was reached when selection came to favor improvement of an existing pattern over continued acquisition of foreign genes. The emergence of true cells, cells that transmitted genes vertically from parent to offspring rather than by lateral transfer, will have been a process of some duration. The evidence suggests that intricate assemblages such as ribosomes and ion-translocating ATPases consolidated early; more modular ones, especially metabolic pathways, continued to undergo lateral exchange for longer, and retain significant propensity for transfer even today. The nature of that "crystallization" that gradually put the brakes on genetic promiscuity and replaced it with family values is entirely unknown. The term implies a dramatic change of phase, but in fact the episode may have entailed little more than a progressive shift in the balance between lateral and vertical transmission, in favor of the latter.

How and why did three distinct domains emerge from the melting pot? As usual, it is easier to pose the question than to answer it. There is some reason to suspect that the Bacteria were first to split out of the stem-group cell community, Archaea second. Eukarya, being chimeric by nature, must have come third. But their origin remains a mystery second only to the origin of life itself; we shall tackle that next, in chapter 7. It has become traditional to think of each domain as the elaboration of one particular cellular pattern or design. In the case of the Eukarya, all of which belong to a single clade with many molecular and structural features in common, this is surely an appropriate viewpoint. Bacteria and Archaea also represent distinct patterns of organization, but these are conspicuous only at the molecular level. In fact, aside from specific molecular features such as ribosomal RNA sequences and lipid chemistry, one would be hard put to define just what distinguishes the Archaeal design from the Bacterial. The answer may come from ongoing efforts to work out what environmental challenge each was required to meet at the time of their divergence.

True cells inherited the fundamental features of the living state from their stem-group ancestors from LUCA. But cell evolution continued apace, for one can draw up a lengthy list of features characteristic of each particular domain. Bacteria go with peptidoglycan cell walls, acyl ester lipids linked to glycerol-3-phosphate, F-ATPase, flagella built of flagellin, and photosynthesis based on chlorophyll. Archaea go with proteinaceous cell walls, isoprenoid ether lipids linked to glycerol-1-phosphate, A-ATPase, Archaeal flagella, and methanogenesis. These presumably originated at the time of crystallization or soon afterwards and remained confined to their domain of origin. Exceptions are best understood as the handiwork of lateral gene transfer and gene loss; they erode the trace of evolutionary history, but do not obliterate it. What each of the emerging domains was selected for is unknown (thermophily in the case of Archaea, perhaps, or energy stress?), and so are the environmental circumstances that favored the emergence. It is also anyone's guess why Archaea split early on into two or three major branches, while Bacteria diversified into a bush.

The emerging storyline seems plausible, but also leaves me vaguely dissatisfied. What we are learning centers on the descent and migration of genes; it is good as far as it goes but provides only a narrow and oblique view of the subject of the present inquiry, which is the evolution of cells. Tracing genes is what we know how to do, but cell evolution is concerned with the origin and transformation of spatial patterns on a larger scale, orders of magnitude above the molecular. The trouble is that genes specify only the structure of proteins and nucleic acids. All the higher levels of order, even molecular structures like Bacterial flagella or cell walls arise by self-organization of one kind or another. The molecular components are specified by genes but the spatial patterns are not, and the latter are often transmitted from one generation to the next independently of the genes. It follows that cellular form and architecture are properties of the system, the society of molecules as a whole, and cannot be inferred from genomic instructions alone. Significant progress in our understanding of the history of cell forms and functions will demand novel methods of inquiry, whose nature I cannot foresee.

This line of reflection raises an awkward question: how do we know whether cell evolution fits into the Darwinian framework, committed as it is to small heritable variations winnowed by natural selection? In truth, we don't. Over the years, serious scientists have proposed a wide range of mechanisms that could bring about large jumps in the organization and behavior of a molecular system. Orthodox ones call for mutations in regulatory genes, whose consequences reverberate through the system with large-scale results, and for the acquisition of new genes and membranes wholesale by endosymbiosis. More adventurous ideas invoke the emergence of order driven by geochemical energy in settings such as submarine hydrothermal vents, constructive neutral evolution without selection, or the self-organization of patterns that persist even in the face of natural selection. Such proposals are especially pertinent to the inception of cellular organization, and we shall revisit them in later chapters. For myself, I have been impressed by studies that use computer modeling to evolve complex functions, even the vertebrate eye, by small orthodox steps, just as Darwin envisaged it. Darwinian evolution can clearly work marvels, but that does not put to rest nagging doubts as to whether it fully accounts for the emergence of the very cellular systems on which evolution now depends. Besides, evidence that variation and selection suffice in theory does not exclude the possibility that life has taken advantage of other options as well. The realm of cell evolution is murky, and as yet thinly charted; if you go there, anticipate things never before seen, and pack a headlight that casts a wide beam.

CHAPTER SEVEN

Emergence of the Eukaryotes

The Second Mystery in Cell Evolution

I am concerned in this [chapter] to present the problems and not the solutions, because the solutions come and go but the problems remain. — Harald Hoffding

Roots and Shoots
What Any Pedigree Must Accommodate
Inventing the Eukaryotic Cell
Seeking the Roots of Complexity
A Durable Puzzle

"An abominable mystery." Darwin had in mind the genesis of flowering plants, but he might as well have been speaking of the origin of the eukaryotes. Eukaryotic cells, which make up the domain Eukarya, differ glaringly from those of Archaea and Bacteria, not so much at the molecular level as in their physiology and architecture (see fig. 1.1). Eukaryotic cells have true nuclei, bounded by a nuclear membrane and containing chromosomes that divide by mitosis. They display an elaborate cytoskeleton, discrete organelles and intracellular membranes, split genes, and many other special features, and they have evolved various kinds of motility not found in bacteria. As a rule eukaryotic cells are much larger than prokaryotic ones (though picoeukaryotes are now known to be abundant). Compared to prokaryotes, eukaryotic cells represent not only an alternative mode of cellular organization but one endowed with far richer evolutionary potential: only among Eukarya do we find integrated multicellular creatures that bear embryos, respond to music, and reflect on their own nature. Whatever meaning one assigns to the history of life, the advent of eukaryotic cells marked a radical and fateful transition.

How did eukaryotic cells originate? The problem is neatly defined, for all the Eukarya from the smallest amoeba to the tallest redwood make up a monophyletic group, a line of descent from a singular ancestral species. But consensus ends here. For more than a century, controversy has swirled around the best way to classify the unicellular eukaryotes and to map out their phylogeny. In recent years, the debate has come to center on the nature of the common ancestor of all eukaryotes, the origins of the eukaryotic gene complement and cell organization, and the driving force for the emergence of complexity. These unresolved issues are the subject of the present chapter.

Roots and Shoots

The eukaryotic cell is an abstraction. Reality is a multitude of organisms large and small, whose cells share all or most of the features depicted in cartoons of the "typical" eukaryotic cell (see figs 1.1 and 3.1). The great majority are unicellular microbes, historically classified in the kingdom Protista (or Protoctista). Some are familiar, at least to biologists—ciliates, amoebas, diatoms, foraminifers, and dinoflagellates—but most of the protists are known only to specialists. That is a pity, for it is among the protists that we must look for clues to some of biology's deepest riddles: the origin of the eukaryotic order, the beginnings of sex and multicellularity, and the reasons for the astounding evolutionary achievements of this cellular design.

Inquiry into all these matters is founded on the premise that the eukaryotes make up a single clade related by descent from a common ancestor. The reason for believing this is that eukaryotic cells share many complex features that are unlikely to have arisen more than once (for instance, undulipodia, the elaborate kind of flagella built of microtubules; nuclear membranes studded with specialized pores; a cytoskeleton based on actin, tubulin, myosins and kinesins; and the endomembrane system with its associated organelles). With the advent of molecular genetics, genes for the ribosomal RNA of eukaryotes as well as many characteristic proteins were mapped upon a single trunk of the tree of life. Recall that by this criterion the prokaryotes split into two distinct domains, Bacteria and Archaea. The essential unity of all the eukaryotes is indisputable; the devil is in their diversity.

The classification and evolution of protists, the most ancient and most varied members of the eukaryotic tribe, has been a bone of contention for a century, and so the advent of ribosomal RNA as a phylogenetic tool was greeted with high hopes. Indeed, in the 1980s, Mitchell Sogin and others used SSU rRNA sequences to bring a satisfying coherence to the evolutionary tree of the eukaryotes.[1] Figures 2.1 and 2.3 depict the emergence of the eukaryotes as a series of branches diverging from a central trunk giving rise successively to anaerobic flagellates, amoebas, dinoflagellates, ciliates, and so on. The phylogeny implies that the complex architecture of eukaryotic cells arose progressively in the course of their evolution. The most distal and therefore most recent lineages were designated the crown of the eukaryotic tree, which includes the more elaborate protists and all multicellular organisms: fungi, plants, and animals.

Of particular interest here is a group of lineages that diverge close to the putative root of the eukaryotic tree, all of them anaerobes or microaerophiles; some are known parasites of humans and animals, but they also have free-living relatives. The most familiar is the diplomonad *Giardia intestinalis*, notorious among travelers as the causative agent of a nasty and persistent diarrhea. Others are the parabasalia (for instance, *Trichomonas vaginalis*), and the microsporidia. All these organisms lack mitochondria and other standard organelles, and in 1983 Thomas Cavalier-Smith proposed a kingdom, Archezoa, to house them and other protists that were thought to have diverged at an early stage in the evolution of the eukaryotic cell. By then, it had been generally accepted that mitochondria are the descendants of bacterial endosymbionts, α-proteobacteria that had taken up residence in the eukaryotic cytoplasm; by the same token, chloroplasts are the descendants of cyanobacteria that joined in later (see the following section and chapter 8). The clear implication was that Archezoa derive from early eukaryotic cells that split away from the main stream prior to the acquisition of the endosymbionts. They lacked mitochondria but were already endowed with the essential attributes of the eukaryotic order, including a true nucleus and the capacity for phagocytosis.

This neat and plausible narrative unfortunately has not fared well. Organisms classified as Archezoa lack mitochondria but turned out to harbor genes of bacterial origin, including that for chaperonin Hsp60, which are typically associated with mitochondria. The inference, that the progenitors of the Archezoa had possessed mitochondria in the past but lost them over time, was strongly reinforced by the discovery that they still harbor membrane-bound bodies that appear to be degenerate mitochondria (hydrogenosomes and mitosomes; see chapter 8). There is now general—but not unanimous—agreement that all living eukaryotes either possess mitochondria or descended from progenitors that did, and the archezoan hypothesis has been formally abandoned by its author. There is no proof that primitive protists ancestrally devoid of mitochondria did not exist in the past, but despite an assiduous search no living descendants have been found.[2]

The RNA-based tree of the eukaryotes has encountered further problems that cut even closer to the bone. Long branches, such as those that lead to the diplomonads and other putative Archezoa, are subject to a statistical artifact called "long branch attraction." Its effect is to make such branches diverge closer to the root than they should, and thus it confers upon them a spurious appearance of antiquity. Indeed, it is now established that the microsporidia, far from being an ancient lineage, are in fact related to the fungi and are of relatively recent origin.[3] The status of the other deep branches is still under discussion. In general, it appears that ribosomal RNA sequences by themselves sometimes do not contain enough information to resolve the evolutionary relationships among groups, let alone the order of their emergence.

The deluge of data unleashed by the genomics revolution swept away that first edifice and cleared the ground for a new and quite different one. Over the past decade a consensus has begun to emerge that depicts the rise of the Eukarya as more of a burgeoning bush than a tall tree, and which places the root between unikonts and bikonts, the progenitors of animals and those of plants.[4] One current version divides all of eukaryotic diversity into just six "supergroups," roughly equivalent to the kingdoms of a less egalitarian era (some authorities prefer five or eight). These groups are Ophistokonta (animals, fungi, and their unicellular precursors, ancestrally endowed with a single posterior flagellum); Amoebozoa (diverse amoeboid protists); Archaeplastida (formerly Plantae, lineages whose plastids derive from the primary cyanobacterial endosymbiont); Chromalveolata (unicellular photosynthetic and nonphotosynthetic organisms whose plastids were acquired by secondary endosymbiosis); Rhizaria (a collection of protists, most of which possess reticulate or threadlike pseudopods); and the Excavata (unicellular organisms that share a distinctive ventral feeding groove, and their relatives) (fig. 7.1, listed clockwise from the left). The former Archezoa have found a new home among the Excavata, but it is not clear whether they can claim exceptional antiquity. Support for all the groupings varies, and if history is any guide the consensus may evaporate.

Biologists peering at the world through microscopes have described some 250,000 species of protists, with many still to come; they must have been chagrined to learn that they had altogether missed a whole wing of

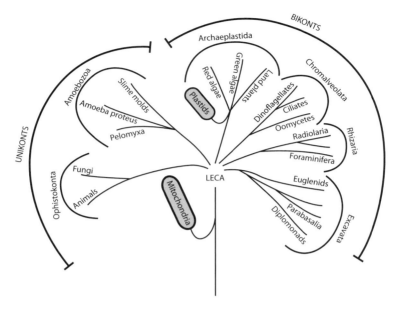

FIGURE 7.1. A current phylogeny of the eukaryotes. Note the six supergroups, the last common ancestor of contemporary eukaryotes (LECA), and the proposed entry of mitochondria and plastids. (After Baldauf 2003; Keeling et al. 2005.)

the eukaryotic universe. Picoeukaryotes, whose minute cells are comparable in size to bacteria (1–5 μm in diameter) turn out to be abundant, diverse, and scattered all across the tree.[5] Most have not been grown in culture and are only known from the amplification of environmental DNA, but that picture is changing. *Ostreococcus tauri*, a unicellular green alga 2–3 μm in diameter, currently claims the record as the smallest free-living eukaryote, yet it holds a mitochondrion, a chloroplast, and a nucleus with 20 chromosomes and over 8,000 genes.[6] Another tiny eukaryote is *Cyanidioschzyon merolea*, a red alga 2 μm long that inhabits certain hot springs. It also holds a chloroplast, mitochondrion, Golgi and a nucleus with 20 chromosomes, but only 5,331 genes. Both genomes have been sequenced.[7] Picoeukaryotes make a major contribution to biomass and photosynthetic production in the sea. Could they also change our understanding of cell history? So far, at least, this has not happened. Picoeukaryotes occur in all the supergroups, and there is no reason to suspect that they share a common ancestry; they appear to be products of miniaturization, not relics of an earlier stage in cell evolution.

The tree of the Eukarya as depicted in figure 7.1 lacks the narrative character of its predecessor, but it does feature a root, a primary bifurcation, which divides all eukaryotic organisms into two broad classes based on the arrangement and evolution of their flagellar apparatus.[8] Unikonts (animals, fungi and amoebozoa) possess a single undulipodium; bikonts (all the others) possess two, one posterior and the other anterior, and the root lies between them. I hasten to add that the real anatomy is much more complicated—some unikonts have lost their undulipodia, others have gained one, so the terms apply at best to an ancestral state. The partition of the eukaryotes into two subdomains is not apparent from phylogenetics. It was discovered, instead, by using shared features of molecular structure that are complex enough to make it unlikely that they originated more than once. In this instance, bikonts are united by a fusion that links the genes for dihydrofolate reductase and thymidylate synthase; the genes are separated in ophistokonts and also in prokaryotes. In addition, unikonts are united by possession of certain genes of the myosin family. It goes without saying that this is a hypothesis, with virtues, faults, and a legion of skeptics,[9] but so far, it seems to be holding up.

Beyond that initial bifurcation, eukaryotic supergroups and phyla emerge as a bunch, with few indications of a chronological sequence. It will be recalled that the same is true of the Bacteria (see fig. 2.3 and chapter 3), and once again it is not clear whether the apparent radiation is merely the result of insufficient resolution or a real biological phenomenon, a "big bang" caused by the emergence of a new body plan and the opening of new ecological niches. Taken as a whole, the process of proliferation in Eukarya seems to differ significantly from its prokaryotic counterpart. Lateral gene transfer plays a major role in the evolution of prokaryotes, both as a major source of innovation and as the chief force that erodes the phylogenetic trace. On the millennial timescale, all prokaryotes are linked into a single vast genetic web. This does not appear to be true for the eukaryotes. Lateral gene transfer has been documented, but is not as prominent as it is among prokaryotes.[10] Instead, it is symbiosis—both intracellular and extracellular—that plays the leading role.[11] First place goes to the acquisition of the α-proteobacterial endosymbionts that evolved into the mitochondria. Figure 7.1 puts that early, prior to the diversification of the protists, and makes it a unique event that played a major role in shaping the eukaryotic cell. Not everyone accepts this thesis, but it meshes with the prevailing wisdom that all extant eukaryotes either possess mitochondria or descend from progenitors that did. Plastids came

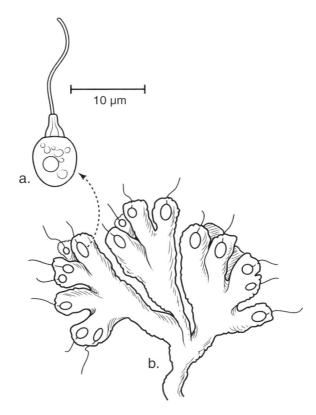

FIGURE 7.2. Is this what LECA looked like? *Phalansterium* is a genus of protists featuring
pear-shaped cells with a single undulipodium that emerges through a stiffened collar at the cell
apex; individual cells are about ten micrometers in length. Colonies hold dozens of individuals
in a gelatinous matrix. (Adapted from Patterson 1992, with permission.)

much later. There is broad agreement that this was another unique event,
following the acquisition of cyanobacterial endosymbionts by a member of
the line that then gave rise to the red and green algae, and later to plants.
The innumerable photosynthetic protists classified in other supergroups
obtained their plastids secondarily (chapter 8). Symbiosis in protists takes
many other forms, supplying houseguests that serve their host as a source
of useful energy, specific nutrients, and organs of motility;[12] but this subject takes us beyond the bounds of the present inquiry.

The base of the eukaryotic tree furnishes a habitation and a name for
the last eukaryotic common ancestor (LECA). Cavalier-Smith sketched

her portrait in 2002,[13] giving her the lineaments of a basic but full-fledged sexual eukaryote. She had a single posterior undulipodium that sprouted from a basal body or centriole, and was connected via a cone of mictrotubules to a proper nucleus with chromosomes and mitosis. LECA possessed the cellular machinery to form pseudopodia and made her living by phagocytosis, the uptake of particulate foodstuffs such as bacterial prey. This is important, for phagocytosis could have been the route by which the evolving proto-eukaryotic cell was invaded by the endosymbionts that eventually made mitochondria. LECA possessed mitochondria, and will have lived at a time when the atmosphere contained a significant level of oxygen. Organisms of this kind still exist, and Cavalier-Smith points to the unikont *Phalansterium*, a flagellated protozoan that consumes bacteria, as a plausible model for what LECA looked like (fig. 7.2). We don't know whether it was the new body plan that triggered proliferation or a change in the environment; whatever the cause, the rise of the eukaryotes made the world we see around us.

What Any Pedigree Must Accommodate

The last common ancestor of all living eukaryotes, as presently envisaged, was a fairly standard specimen, complete with all the essential attributes; it was in no sense primitive and must have been the product of prolonged evolution. Over the past three decades, LECA's progenitors have been avidly pursued, with forks and hope and occasionally with long knives. Expectations that evolutionary intermediates would be found lurking in dark corners have come to naught, leaving contemporary organisms and enigmatic fossils as the only sources of information and insight. The quest has not produced consensus, but it has generated a substantial body of observations that any serious proposal must incorporate.

(i) Eukaryotic cells share molecular features with both Archaea and Bacteria. The genetic core, comprising the transcription and translation of genetic information, is strikingly similar to that of Archaea (chapter 3). This affinity was first recognized from comparison of ribosomal RNA sequences but extends much further. For example, the eukaryotic RNA polymerase is made up of multiple subunits related to those of Archaea, while the Bacterial RNA polymerase consists of a single, unrelated subunit. DNA polymerase, histones, transcription factors, and proton pumps reinforce the kinship of Eukarya and Archaea. But the pattern is more complicated: eukaryotic plasma membranes are constructed of fatty acid esters linked to glycerol-3-phosphate, like those of Bacteria, and do not employ the isoprenoid ethers and glycerol-1-phosphate favored by Archaea. Furthermore, the metabolic enzymes of eukaryotic cells commonly resemble those of Bacteria.

(ii) Despite these affinities with both Archaea and Bacteria, Eukarya represent a distinct mode of cellular organization that cannot be directly derived from either of the prokaryotic domains. Cell biologists have long recognized characteristic eukaryotic features, including nuclei, chromosomes, mitosis, undulipodia (9 + 2 flagella), endomembranes, the secretory pathway, and split genes. The spatial organization of eukaryotic cytoplasm is conspicuously complex—far more so than that of prokaryotic cells.

(iii) Many eukaryotic genes have homologs among either Bacteria or Archaea but others do not, and these commonly encode proteins that mediate functions unique to the Eukarya.[14] In other instances a distant relationship exists, but is hard to recognize. A case in point is the cytoskeleton, with its characteristic proteins tubulin and actin, long thought to be a hallmark of the Eukarya. We now understand that both Bacteria and Archaea possess a substantial cytoskeleton that organizes cell division and morphogenesis and is based on distant homologs of tubulin and actin; that homology was not apparent from sequence analysis but could be inferred by comparing the three-dimensional structures of cognate proteins (chapter 6). The meaning of proteins and structures that exhibit a eukaryotic "signature" has occasioned much debate. Some authors take them as evidence for a very early divergence of the Eukarya from the common precellular stock, others attribute them to accelerated evolution under intense selection pressure; there are presently no objective criteria by which these competing interpretations can be assessed.

(iv) In contrast to their structural refinement, the metabolism of eukaryotes appears to be downright primitive. Its central feature is glycolysis, an anaerobic pathway of glucose degradation mediated by soluble enzymes. Respiration and photosynthesis, vastly more efficient pathways of energy generation that rely on ion currents and vectorial chemistry, are restricted to organelles derived from bacterial endosymbionts. Eukaryotes seem to hold no trace of the rich diversity of chemoautotrophic pathways that enabled Bacteria and Archaea to colonize inhospitable environments. The accumulation of ions and metabolites by eukaryotic cells relies on coupling to a circulation of protons or sodium ions across the plasma membrane, which is generated by a special class of ion-translocating ATPases. These

P-type ATPases are quite unrelated to the F- and A- ATPases which perform this function in prokaryotic cells, in mitochondria and plastids, and also in the endomembrane system of eukaryotes (chapter 5).

(v) It used to be one of the verities of cell biology that there are no organisms whose structure is intermediate between eukaryotes and prokaryotes; but such things do exist after all. As outlined in chapter 3, members of the Bacterial phyla Planctomycetes and Verrucomicrobia possess cytoplasmic membranes that define a "nuclear" compartment containing DNA and a separate cytoplasmic compartment (ribosomes are present in both). The membrane that divides them is studded with pores resembling those of eukaryotic nuclei, and some of these organisms contain both a true tubulin and calmodulin. It seems unlikely that these organisms are phylogenetic intermediates between eukaryotes and prokaryotes, but they do proclaim that the organization of prokaryotic cytoplasm is far more malleable than we had realized, and can include membrane-bound compartments.

(vi) The radiation of eukaryotic organisms as displayed in the fossil record began about a billion years ago; the age of the eukaryotic lineage is uncertain but may be far greater (see chapter 9). Fossils generally accepted as eukaryotic go back to about 1.7 billion years ago; molecular biomarkers may be older, and phylogenomics puts the origin of the nuclear lineage at least as far back as 2 billion years ago (chapters 2 and 8). Since mitochondria are derived from α-proteobacteria, the acquisition of mitochondria must have occurred after the differentiation of the Bacteria had begun, but the date of that fateful partnership is in doubt; LECA must be younger still. The plant lineage subsequently gained plastids, an event variously dated between 1.2 billion and 700 million years ago.

So much for the facts of the matter; we now turn to the continuing effort to extract from them the outlines of a historical narrative.

Inventing the Eukaryotic Cell

No one has yet put forward a comprehensive and satisfying account of the origin of the eukaryotic pattern, but there are plenty of data to tickle the imagination and to spawn a small zoo of competing hypotheses. The subject has a forty-year history, reaching back to the recognition that mitochondria and chloroplasts derive from bacterial endosymbionts; which in turn raised the questions of what sort of cell hosted those endosymbionts and where it came from.[15] Much of that literature has an edgy tone, which calls to mind Rowland Davis' quip that "scientists don't discover—they make discoveries, and then maintain them."[16]

Current discussions revolve around three themes, with substantial overlap among them.[17] One school of thought—probably the most popular at present—holds that some kind of a merger between cells of Archaea and Bacteria put evolution onto a novel track, which culminated in the emergence of the eukaryotic pattern of order. The idea comes in multiple versions, which differ with respect to the identity and number of the prokaryotic partners and the importance of mitochondria in generating complex organization. Their opponents defend the traditional view that eukaryotic cells arose autogenously from a prokaryotic line, so that mitochondria came late into a host cell that already possessed the basic eukaryotic capacities. The third proposition grows out of the premise that the nuclear lineage of eukaryotes represents one of the three basic domains of life, whose origin traces to an early stage in cell evolution, possibly as far back as do the prokaryotes. I find myself drawn to this romantic idea, for its physiological and ecological implications open onto fresh horizons.

TWO OR THREE INTO ONE.   The notion that the protoeukaryotic host that first acquired the precursor of mitochondria was itself the product of a merger between prokaryotic cells has considerable historical depth. Lynn Margulis proposed long ago that the nucleus began as a free-living archaeon related to *Thermoplasma*, while undulipodia descend from Spirochaetes; she maintained that position even though Spirochaetes have no genes for tubulins, but garnered few adherents.[18] Contemporary formulations of the chimera commonly begin from the well-established observation that eukaryotic cells share many "informational" genes with the Archaea, while "operational" ones group with the Bacteria. This has been taken as evidence for a fusion or merger between representatives of the two domains, which comes in quite a number of variations.

In one version, espoused by Lòpez-Garcia and Moreira, evolution of the eukaryotic cell began with a syntrophic association of a methanogenic archaeon with a metabolically versatile myxobacterium; the latter supplied its partner with molecular hydrogen.[19] The two cells became increasingly interwoven; they engaged in massive transfer of genes and membrane proliferation, gradually losing their former identities. The bacterium was the progenitor of eukaryotic cytoplasm and plasma membrane, the archaeon that of the nucleus. The consortium subsequently took in a third partner, the endosymbiont that gave rise to mitochondria and opened the door to respiratory energetics. Phagocytosis did not play a major role in the origin of eukaryotic organization but was presumably responsible for the subsequent acquisition of plastids.

William Martin and Miklòs Müller developed a more streamlined alternative, marketed as the "hydrogen hypothesis," which has attracted much attention.[20] Here again we have syntrophy in a predominantly anaerobic setting, this time of a methanogenic archaeon with a metabolically versatile α-proteobacterium. Initially, the proteobacterium supplied its partner with hydrogen, but its capacity for respiration came in useful later on; the merits of this transformation are worked out in detail. Increasing intimacy, gene transfer, and innovation produced a protoeukaryotic cell, but the two partners never quite merged. The archaeon became the precursor of both cytoplasm and nucleus, while the proteobacterium retained its identity and became the precursor of the mitochondrion. In this view, then, the origin of the eukaryotic cell is indissolubly linked to the genesis of the mitochondrion, the eukaryotic cell's powerhouse, which underpinned the rise of cellular complexity. Note that while the acquisition of the mitochondrion is envisaged as a unique event, the subsequent incorporation of the chloroplast is attributed to phagocytosis.

Hypotheses that place the merger of an archaeon with a bacterium at the very base of the eukaryotic stem, cleverly branded as the "ring of life," received a boost from the work of Rivera and Lake.[21] These authors devised a phylogenomic procedure that relies not on sequence similarities between individual genes (such as those for ribosomal RNA) but rather on the number of homologous genes shared by any pair of whole genomes. The method recovers the traditional separation of Bacteria from Archaea, but eukaryotic genomes group with both halves of the prokaryotic tree and are said to "seal the ring." The interpretation is that eukaryotic cells emerged from the fusion of an Archaeal genome, the "informational ancestor" with a Bacterial one (the "operational ancestor"). Such a scheme would mesh with either of the two versions sketched above, but absent some direct evidence for merger or fusion of the two cell lines, it does not carry us much beyond what we knew before: that eukaryotes share genes with both Archaea and Bacteria.

The quest for a genomic solution to the riddle of the eukaryotic order has spurred the development of more sophisticated algorithms, and the interpretation of that ancient partnership has altered in consequence.[22] Methanogens have fallen out of favor. Recent data suggest that the

Archaeal partner may have come from the ranks of the Crenarchaeota (a hypothesis first mooted by James Lake two decades ago) or from the Thaumarchaeota or from an unknown and extinct lineage that branched off prior to the divergence of the major Archaeal stems, and that it possessed considerable structural complexity. Eukaryotic genomes are chimeric, that is for sure, but as to how and when that came about, assurance remains elusive.

Scientists understandably play up their hypotheses' strengths while minimizing shortcomings; their colleagues are apt to be less charitable. The fusion of two very different prokaryotes does comfortably account for all those shared genes, and metabolism supplies a selective force to promote their association. It is not so obvious why eukaryotic cells would have preserved the informational core of the ancestral archaeon while utterly discarding its metabolic capacities, and conversely why they favored Bacterial metabolism and membrane chemistry while rejecting that genetic core. The fact that endosymbiosis is almost unheard of among prokaryotes (the few known cases are very specialized ones) argues against a key element of the merger hypothesis, leading Davidov and Jurkevitch[23] to revive the old idea that the partnership may have begun with predation rather than metabolic syntrophy. Other reasons for skepticism include the failure of eukaryotic informational genes to map with any known Archaeal lineage and the very fact that the merger hypothesis keeps shifting and comes in so many variations. As a matter of principle, I find it hard to believe that two mature cellular patterns fused and merged to generate a third, radically different kind of cell. But having said all that, it remains true that a merger of genomes is the most straightforward interpretation of the presence in eukaryotes of both Bacterial and Archaeal genes; and if a hundred million years sufficed for birds to arise from dinosaurs and whales from quadrupeds, can anyone say for certain what evolution can and cannot accomplish in a billion years?

AN AUTOGENOUS ORIGIN.   Fifty years ago, the microbial world seemed cleanly divided into prokaryotes and eukaryotes, with the latter having evolved from the former. This traditional stance has recently been fully articulated and strenuously defended by Thomas Cavalier-Smith.[24] In his view, protoeukaryotic cells evolved directly from bacterial ones, specifically from the actinobacteria, adopting a phagocytic way of life made possible by loss of the cell wall. Eukaryotes originated relatively recently, less than a billion years ago, together with their sister group, the Archaea (or rather, archaebacteria, since Cavalier-Smith would deny them the status of a separate domain; see chapter 4). The many genes that eukaryotes share with prokaryotes bespeak their common ancestry, not genome fusion or mergers. The sharp differences he attributes to intense selection pressure, first for a thermophilic lifestyle and then for phagocytosis; the emergence of the eukaryotes is the most dramatic instance of quantum evolution. The endosymbiont that gave rise to mitochondria will have colonized cells that had already achieved basic eukaryotic organization; the colonist's contribution was significant but not vital.

Of all the hypotheses presently on the books, this is the one most fully spelled out. Cavalier-Smith explains in detail why cell history must have begun with eubacteria—the gram-negative ones in particular—and how some lost first the outer membrane and subsequently the wall. He takes account of many biochemical details, such as the presence of sterols in the actinobacteria, and of the many inventions required to turn eubacteria into archaebacteria and eukaryotes (his nomenclature). I find it plausible, as do others,[25] that phagocytosis lies at the core of the eukaryotic mode of cell organization and supplied a mechanism for the acquisition of symbionts that eventually turned into mitochondria. Yet the argument for direct filiation from a bacterial ancestor begs too many questions. It ignores or minimizes the radical differences between Bacteria and Archaea, brushes aside the mounting evidence for the antiquity of eukaryotes, and demands a degree of structural transformation that is hard to credit.

THE THIRD DOMAIN.   The third viewpoint grows directly out of the universal tree of life with three discrete stems, all of which reach clear back to the earliest stages of cell evolution (see chapter 2). According to Carl Woese and a handful of supporters, the eukaryotic stem emerged, as did both prokaryotic ones, from the genetic free-for-all that preceded modern organisms (see chapter 4). The eukaryotic order did not evolve directly from either of the prokaryotic ones, singly or by fusion, except for the later acquisition of the precursors of mitochondria and plastids. The reason that the genetic core of the Eukarya shares many features with that of the Archaea is that both "crystallized" from parallel levels of the genetic melting pot. However, Eukarya also harbor numerous proteins that are specific to their lineage and have no homologs among either Archaea or Bacteria. Such "eukaryotic signature proteins" testify to the independent origin of the eukaryotic line.[26]A rudimentary kind of phagocytosis probably evolved early and opened up a novel and rewarding way of making a living; eukaryotes may have begun as "social cheaters" in microbial bio-films and later turned professional predators.[27] The role of mitochondria in generating eukaryotic organization will be considered below.

The hypothesis that eukaryotes represent a unique primary lineage has been attacked on both technical and ideological grounds and fervently defended. Like the other hypotheses, it does not immediately explain why Eukarya retain informational genes akin to those of Archaea, while their operational ones resemble those of Bacteria. There is no positive evidence that the eukaryotic line reaches clear back to the dawn of cellular life. Eukaryotic genes that lack prokaryotic homologs are not necessarily ancient; they could have lost all traces of their origin under the intense mutation and selection that forged the new cellular structure. A skeptical attitude is surely in order, as it is towards all hypotheses that purport to account for the origin of cells; yet the central idea has much merit, and will find a place in a comprehensive synthesis.

Seeking the Roots of Complexity

Eukaryotic cells are far larger than prokaryotic ones and are more complex by any standard. They have more genes, and their DNA content is larger by orders of magnitude. Genes are split into regions that code for portions of the protein product and others that do not (introns), necessitating machinery to splice out the introns from the transcript ("spliceosomes"). Besides, the great bulk of the DNA does not encode anything that is obviously useful; coding DNA makes up less than a tenth of the total, and the functions of the remainder are unclear. The hallmark of the eukaryotic cell is a true nucleus, enclosed within a nuclear membrane. In consequence, transcription and translation are separated in space and time; newly formed and edited messages are translocated into the cytoplasm via specialized pores and meet the ribosomes there. The nuclear membrane is continuous with a system of endomembranes that extends out to the plasma membrane and mediates both secretion of gene products and the uptake of particulate matter. Many protists, and also some cells of higher animals, live by phagocytosis, which depends on the endomembrane system and organelles linked to it, such as lysosomes. Cytoplasmic space is integrated by an intricate cytoskeleton, a meshwork of several kinds of fibers built around either actin or tubulin, which provides mechanical support and tracks for the transport of cargo through the cytoplasm. Cell division is much more elaborate than it is in the prokaryotes: genes are arrayed on chromosomes that are duplicated, rearranged, and distributed between the daughter cells by the courtly ballet of mitosis. Eukaryotic cells display several modes of motility not seen in prokaryotes, including crawling and swimming powered by undulipodia; these eukaryotic organs of motility are utterly unlike prokaryotic flagella both in structure and in function. Last but by no means least, eukaryotic cells possess prominent intracellular organelles including mitochondria, peroxisomes, and (in the plant lineage) chloroplasts.

Where did all this structural and functional complexity come from, and what selective advantage drove its evolution? Whatever may have been the provenance of the eukaryotic genome, the architecture of the eukaryotic cell remains to be accounted for, and that has proven to be equally opaque. It is useful at this point to note that eukaryotic cell organelles fall into two classes, those that are derived from endosymbionts and those that are not. Mitochondria, plastids, and perhaps peroxisomes fall into the first category and deserve a chapter unto themselves; our subject here is the remainder. If the current formulation of the eukaryotic tree is correct, all the basic elements of eukaryotic cells were already present in LECA, the last common ancestor of all extant eukaryotes (fig. 7.1). Their evolution seems to have left no traces in the ultrastructure of contemporary eukaryotes, whose immense variety reports the diversification of the basic design rather than its origin. Instances of simpler organization, such as the rudimentary Golgi found in *Giardia*, or, indeed, the many known picoeukaryotes, should be understood as products of reductive evolution rather than as stages in the rise of complexity. For clues to the latter there is no place to turn but the genome.

The eukaryotic genome appears to be a mosaic of genes drawn from multiple sources. Those related to information processing often (but not universally) have affinities with the Archaea; yet they cannot be traced to any known Archaeal lineage. Genes that specify cell operations are likely to map with Bacterial ones, though not with any single lineage. And then there are many hundreds of genes that have little sequence homology to any prokaryotic gene, or none at all, but seem to be characteristic of the domain Eukarya and help to specify just what is unique about eukaryotic cells. Eukaryotic "signature proteins" are instrumental in making the cytoskeleton, the endomembrane system, the signaling network, nuclear membrane pores, nucleoli, spliceosomes, and others; though here again, genes of prokaryotic genres contribute as well.[28]

One of the most instructive examples is the cytoskeleton, which was traditionally held to be a defining feature of eukaryotic cells and absent from prokaryotes. We now know that a cytoskeleton is present in prokaryotes after all, made of homologs to eukaryotic tubulin, actin, and other proteins. Remarkably, the functions served by the prokaryotic proteins are almost opposite to those of their distant eukaryotic cousins. FtsZ, unquestionably a homolog of tubulin, does not polymerize into fibers but into a contractile ring that brings about division of the cell at its midpoint. By the same token, MreB, clearly homologous to actin, has nothing to do with cytokinesis or motility, but guides the positioning of enzymes that produce the wall of rod-shaped bacteria and thus serves a crucial role in the morphogenesis (chapter 6). Now then, are the eukaryotic proteins to be regarded as descendants of the bacterial ones? It's conceivable, of course, but not plausible. Far more likely, the genes that specify those proteins share a remote ancestor, more ancient than either prokaryotes or eukaryotes as we know them, and they diverged so greatly in response to selective forces impinging on increasingly different cells.

Among the features that distinguish eukaryotes from prokaryotes, none is more conspicuous than the endomembrane system. The point is not the possession of internal membranes, per se. Many bacteria, including cyanobacteria, contain extensive systems of cytoplasmic membranes that serve to increase the surface area available for energy transduction. The central function of eukaryotic endomembranes is not bioenergetics but transport: they mediate both secretion and the uptake of particulate matter, and communicate (usually) by means of cargo-carrying vesicles that bud off one compartment and fuse with another. Genomics has shed some light on the origin of the system. According to Dacks and Field, there are few if any prokaryotic antecedents for the machinery of endomembrane transportation, suggesting that it evolved autogenously early in the history of eukaryotic cells.[29] Elaboration of the system entailed proliferation of the cognate genes by duplication and divergence, and the production of multiple proteins with similar functions but different locations.

Undulipodia, another of the eukaryotic signature structures, occur in representatives of all the phyla and were clearly among the endowments of LECA.[30] More than that, LECA was most probably a flagellated protist, endowed with the characteristic complex made up of a nucleus, undulipodia, and a connector. The intimate linkage between the nucleus and the motility device is underscored by the centrioles (almost ubiquitous but not quite), which serve alternately as the origin of the mitotic spindle and of the undulipodium. Undulipodia are not related in any discernible way to Bacterial flagella (Lynn Margulis's claim to have found tubulins in Bacteria notwithstanding), and their origin is unknown. Jèkely and Arendt make a reasonable case for descent from a microtubule-based secretory pathway; the early stages may have served a sensory function, with motility coming later.

What, then, of the nucleus, that *sine qua non* of the eukaryotic order? What selective benefit is conferred by sequestering the genome in an enclosure? The question remains open, but some reasonable proposals have been made.[31] For Cavalier-Smith, evolution of the nucleus was a response to the mechanical problems that arose from the increasing size of the primitive phagocyte. A larger cell requires a larger nucleus, which favors the accumulation of DNA even if it codes for nothing; and an active cytoskeleton generates stresses on the genome that a nuclear membrane would alleviate. Lane has suggested that it may help protect the chromosomal DNA from free radicals generated during respiration. Martin and Koonin, as well as Lòpez-Garcia and Moreira, take yet another tack.[32] The essential consequence of a membrane-bound nucleus is that transcription and translation are now separated in space and in time. They argue that this was useful in controlling a pernicious byproduct of the acquisition of a Bacterial endosymbiont, the mobile introns that the latter carried in its own genome. The nuclear membrane would allow the spliceosomal machinery to excise these harmful interlopers from the transcripts before they reach the ribosomes. All this depends, of course, on the existence of nuclear membrane pores, whose origins are equally shrouded in fog.

When all is said and done (and there has been plenty of both), the genesis of eukaryotic cell organization remains as enigmatic as the provenance of its genes. It seems clear that the common ancestor of all eukaryotic cells alive today already had all the standard parts, at least in essence. These were not inherited from any prokaryotic ancestor but were the fruit of innovation. No clear sequence in the development of eukaryotic complexity can be discerned, and it is likely that its elements all evolved in concert over a relative brief span of time. But the when, how, and why remain as mysterious as ever.

A Durable Puzzle

Concerning the origin of eukaryotic cells, it sometimes seems that the more we learn the less we understand. Two decades ago, when the tide of phylogenetic data first began to roll, there was good reason to hope that long-standing conundrums would soon be resolved. That did not happen, as shown by the proliferation of hypotheses (more than twenty at last count), all now rooted in ever-more sophisticated statistical analysis of gene sequences and gene distributions. So we fall back on a method more traditional in evolutionary biology, construction of a tentative historical narrative whose explanatory merits are then assessed by confrontation with whatever data can be gathered. The narrative sketched below is one that presently seems to me most plausible; none of its elements is original, but it has the virtue of combining ideas from different camps. I take it that cell organization comes in three modes, but these are not of equal status. The presence in eukaryotes of genes of both Archaeal and Bacterial affinities smacks of a merger of genomes, but not necessarily the genomes of contemporary organisms. The merger of genomes implies the fusion of cells, possibly by mechanisms ancestral to contemporary phagocytosis; once established, that lifestyle became central to the economy of the protoeukaryotes, and the foundation of their subsequent proliferation. As to mitochondria, it is certainly possible that they appeared relatively late, after the outlines of the eukaryotic order had been established; but I shall here adopt the idea developed by William Martin and Nick Lane that mitochondria appear early and were a prerequisite for the emergence of the eukaryotic mode of organization.

Scenarios of cell evolution are necessarily fanciful accounts of what may have happened and should be held every bit as lightly as Kipling's *Just So Stories*; the following, O Best Beloved, is mine. In concert with Carl Woese and his followers, I regard the eukaryotic lineage as an ancient one that reaches deep into the past, nearly as far back as the prokaryotic ones. What set this lineage apart was its reliance on organic foodstuffs, supplied by the other denizens of its ecosystem. Emergence of scavengers and predators in the biofilms, stromatolites, and undifferentiated slimes inhabited by the primary producers (prokaryotes and/or their precursors) seems likely, even inevitable.[33] Phagocytosis as we know it must have been the product of prolonged experimentation. However, long before the savage but noble "unicellular raptor" celebrated by Kurland and his colleagues,[34] there would have been humble scavengers capable of secreting digestive enzymes and absorbing the products. The key postulate is that among these were "urkaryotes," distinctive from the beginning in both organization and metabolism: wall-less, with membranes and cytoskeleton capable of engulfing fluid and particles, but devoid of the energy-transducing redox chains that sustained early prokaryotes. Instead of competing with the primary producers, they ate them. Relics of that metabolic order can perhaps still be discerned in the fermentative economy of eukaryotic cytoplasm, and in the characteristic suite of membrane ATPases that drive a circulation of protons or sodium ions across the plasma membrane and support the accumulation of soluble nutrients.

Should those urkaryotes be regarded as a primary domain, on par with Bacteria and Archaea? Probably not, though they will not have been standard-issue prokaryotes, either. The well-established kinship between the informational genes of eukaryotes and those of Archaea is best interpreted as evidence that those domains share a common ancestry; recent genomic studies point specifically to Crenarchaeota, Thaumarchaeota, or to an unknown organism that flourished prior to the diversification of modern Archaea.[35] This assignment is strongly supported by the discovery of bona fide actins and tubulins in members of those phyla (as mentioned in chapter 6). Curiously, Archaeal proteins often have homologs in the endomembrane system of eukaryotes rather than in the latter's plasma membrane. The evolving protoeukaryotes also acquired numerous Bacterial genes, apparently from several different Bacteria and over a prolonged period of time. Likely sources include genetic material derived from their fodder and assimilated intact. Perhaps the largest infusion of genes came with the acquisition of endosymbionts.[36] If this is correct, then the Eukarya, unlike the two primary domains, make up a secondary or derived domain. Moreover, the eukaryotic genome is not a product of either vertical descent or straightforward cell fusion, but a collage of disparate elements picked up over time; in Ford Doolittle's memorable quip, they became what they ate.

How, when, and why would the (hypothetical) urkaryotes come by the endosymbionts that eventually gave rise to mitochondria? As scavengers, urkaryotes may have been preadapted to evolve towards true phagocytosis, the capacity to ingest particulate food and digest it intracellularly. But it seems quite unnecessary to insist that they traveled the whole way before acquiring endosymbionts. Early steps along this road may have been sufficient to admit colonists by way of uptake, or as predatory invaders like *Bdellovibrio*. It is not clear just what features allow eukaryotic cells to host endosymbionts, whereas prokaryotes almost never do so; nor can one say what benefits accrued to either of them from their partnership. But one can approximately locate its establishment in geological time. Since mitochondria descend from α-proteobacteria, the acquisition of endosymbionts must have come after the diversification of the Bacteria had begun. By 2.5 billion years ago, oxygen produced by cyanobacteria was beginning to accumulate, at least locally. The striking resemblance of mitochondrial respiratory chains to those of some versatile aerobes, notably *Klebsiella* and *Rickettsia,* makes it plausible that the endosymbionts had respiratory capacities, and that the key benefit to the urkaryotic host was augmentation of its energy supply.

It is quite remarkable that all eukaryotes examined in detail either possess mitochondria or retain genetic traces of having possessed them in the past. To be sure, the observations may be misleading or incomplete, but if we take them at face value they suggest that mitochondria came early and played a crucial role in the genesis of the eukaryotic order. Why should that be so? Lane and Martin have recently made a convincing case for the proposition that the fundamental difference between eukaryotes and prokaryotes is a matter of energetics.[37] A prokaryotic cell requires all the energy it generates just to sustain the activities called for by its own genome; by contrast, in eukaryotic cells dozens of prokaryotic "power packs" are harnessed to the service of a single, masterful central genome. If that is correct, it makes sense to look to the origin of that partnership for the roots of the eukaryotic design. I shall not try to cobble together yet another speculative scenario. Suffice it to postulate that a population of urkaryotes, having established an intimate relationship with a prokaryotic endosymbiont (possibly for reasons quite unrelated to energetics) found itself at a metabolic advantage as its world turned oxygenic. With surplus energy at its disposal, that urkaryote's genome was free to expand and take on new functions. And then all things became possible: larger size, active motility, intracellular transport, phagocytosis, and all the sophisticated energy-intensive physiological machinery of eukaryotic cells. Indeed, as argued persuasively by Lane and Martin, endosymbiosis is the only way in which the constraints on prokaryotic energetics could have been overcome.[38]

Like all the other tales of the imagination, mine has loose ends; and plausibility, like beauty, is mainly in the eye of the beholder. Why, for example, do the membrane lipids of eukaryotes (and also the enzymes of anaerobic metabolism) group with those of Bacteria rather than Archaea? Just what is the provenance of eukaryotic signature proteins? Did remnants of those primordial scavengers survive into our own day, and if not, why not? I am not entirely comfortable with the implication that a long period of stasis intervened between the primordial scavengers and the birth of LECA a billion years later; nor, for that matter, with the popular notion that eukaryogenesis was a unique event—effectively a miracle. In

2006, in the introduction to a thoughtful review on the origin of the eukaryotes, Embley and Martin observed somewhat bleakly that "the evolutionary gap between prokaryotes and eukaryotes is now deeper, and the nature of the host that acquired the mitochondrion more obscure, than ever before."[39] Today, it seems fair to say that some consensus has been reached on a host broadly related to the lineage that also gave rise to the Archaea, and perhaps even on a major role for mitochondria in generating eukaryotic organization. But for all the sophisticated informatics and subtle dialectics, the emergence of the eukaryotic cell still passes understanding.

CHAPTER EIGHT

Symbionts into Organelles

Mitochondria, Plastids, and Their Kin

Nature does nothing in vain, nature is pleased with simplicity, and affects not the pomp of superfluous causes. — Sir Isaac Newton, physicist

Nature seems to ignore our intellectual need for convenience and utility, and is very often pleased with complexity and diversity. — Santiago Ramón y Cajal, biologist

How the Endosymbiont Hypothesis Was Proven
Forging the Partnerships
Metamorphosis
The Curious Case of *Paulinella chromatophora*
Evolution by Reduction

There is still something outrageous about the proposition that mitochondria and chloroplasts, those stalwart and stolid workhorses of metabolism, are in fact descendants of bacteria that had taken up residence in the cytoplasm of an ill-defined host cell and suffered permanent enslavement. The idea had been under discussion since the end of the nineteenth century, but few took it seriously until the late Lynn Margulis marshaled the evidence forty years ago.[1] Resistance was intense: Margulis's original 1967 paper was rejected by fifteen journals before it found a home in the *Journal of Theoretical Biology*, and it took another decade for the idea to enter the mainstream. Lewis Thomas, in his charming meditation "The Lives of a Cell," sounds resigned, but still a little queasy: "I was raised in the belief that those were obscure little engines inside my cells, owned and operated by me or my cellular delegates, private, submicroscopic bits of my intelligent flesh. Now, it appears, some of them and the most important ones at that are total strangers. . . . I only hope I can retain title to my nuclei."[2]

No one doubts any longer that this is what happened. In fact, in all the annals of cell evolution, the endosymbiotic origin of mitochondria and plastids may be the one generalization that no one disputes ("plastid" is the general term for photosynthetic organelles; "chloroplast" refers to the plastids of higher plants). It is also an episode of surpassing importance: the acquisition of mitochondria and plastids supplied the evolving eukaryotic cells with abundant energy and enabled the emergence of complex organization and all the wonderful world of higher organisms. All the oxygen in the atmosphere is a byproduct of photosynthesis carried out by cyanobacteria and plastids. But the bacterial ancestry of these most emblematic of eukaryotic organelles is the beginning of a long tale, not its end. The transfiguration entailed drastic changes in both symbiont and host and continues to raise endless questions that revolve around how, when, and why. These have generated a large and lively literature that is the foundation of the present chapter.

How the Endosymbiont Hypothesis Was Proven

The resemblance of size and form between bacteria and cell organelles that first led microscopists a century ago to speculate on their common origin might have turned out to be spurious, but the accumulation of molecular correlates soon became compelling. In principle, mitochondria and plastids could either have arisen autogenously, from within the evolving eukaryotic cell, or else be derived from exogenous sources such as bacteria. Gray and Doolittle, surveying the field in 1982, came down firmly in favor of the latter.[3]

The photosynthetic apparatus of plastids from algae and higher plants matches that of cyanobacteria. Both consist of two photosystems arrayed in tandem, with water as the reductant and oxygen gas as byproduct. Photosynthetic reaction centers, redox chains, and the proton-translocating ATP synthase are visibly similar. They are embedded in membranes that contain cyanobacterial galactolipids and beta-barrel proteins. Plastids contain DNA that encodes a subset of plastid proteins, notably those central to energy transduction. They also harbor ribosomes of the Bacterial kind, as judged by their size, protein content, and antibiotic sensitivities, all quite unlike those of Archaea and eukaryotic cytoplasm. Among the molecules specified by plastid DNAs are the ribosomal RNAs, once again Bacterial in size and structure and quite distinct from those of eukaryotic ribosomes. The same holds for RNA polymerase and mRNA; chloroplast mRNAs, like those of Bacteria, generally lack the polyA tails characteristic of eukaryotes. As a rule, plastids lack cell walls of the prokaryotic kind, but the "cyanelles" of glaucophyte algae retain peptidoglycan, the very hallmark of Bacterial cell walls. It would take a very tortuous argument to make these parallels consistent with an autogenous origin, whereas they are all smoothly accommodated by the endosymbiotic hypothesis.

An analogous case was made for mitochondria. The respiratory chain of animal mitochondria is very similar to that of many aerobic α-proteobacteria, and almost identical to that of *Paracoccus denitrificans*. Oxidative phosphorylation proceeds in the prokaryotic manner, with protons translocated outward by the respiratory chain. They complete the circuit via a proton-translocating ATP synthase that closely resembles its Bacterial equivalent in molecular structure and mechanism. Mitochondria harbor DNA that specifies a subset of mitochondrial proteins, especially the core of the redox chain and ATP synthase. Their membranes contain cardiolipin and porins, as Bacterial membranes do, but lack the cholesterol of eukaryotic membranes. They also hold an apparatus for protein synthesis, including ribosomes of Bacterial (not eukaryotic) size and architecture. The parallels between mitochondria and α-proteobacteria are not quite as close and straightforward as those that link chloroplasts to cyanobacteria. Mitochondrial DNA varies alarmingly in size, base composition, and gene organization. The mechanism of transcription is also variable, and even the genetic code deviates a little from the universal one. Such peculiarities do not challenge the derivation of mitochondria from Bacterial symbionts, but they suggest that this process may have had unexpected twists and turns. Part of the explanation may be that mitochondria were acquired much earlier than were plastids, and share with their hosts a lengthier history of mutual accommodation.

Just a few years later molecular sequences became available, and whatever doubts still lingered were laid to rest. Ribosomal RNA sequences firmly map chloroplasts among the cyanobacteria, and mitochondria among the α-proteobacteria.[4] With that accomplishment, interest shifted from whether the organelles evolved from bacteria to just how that transfiguration took place.

Forging the Partnerships

All eukaryotic cells are chimeras, products of mergers among distinct kinds of cells that mingled genomes, metabolic patterns and membranes; indeed, many scholars believe that a primordial fusion of Archaea and Bacteria lies at the very root of eukaryotic organization (chapter 7). Be that as it may, it is indisputable that three episodes of this kind figure prominently in the history of the Eukarya (fig. 8.1). The first was an association between a host cell, whose nature is still in doubt, and a Bacterium related to present-day α-proteobacteria. The partnership is conventionally described as an endosymbiosis, but given all the uncertainties it may be well to keep an open mind; what we know is that the proteobacterium was the progenitor of mitochondria. Mitochondria were among the endowments of LECA, the last eukaryotic common ancestor, and probably played a major role in shaping eukaryotic organization. Some time later an ancestral heterotrophic protist took up a cyanobacterium, made it captive, and eventually turned it into a plastid. Three photosynthetic lineages—green algae, red algae, and the somewhat obscure glaucophytes—descend directly from this primary endosymbiosis. Subsequently, oxygen-generating photosynthesis spread widely across the eukaryotic world by a process termed secondary endosymbiosis. In this kind of partnership, a heterotrophic protist engulfed not a prokaryotic cell but a whole eukaryotic alga possessed of a primary plastid and reduced that ensemble to the status of a secondary plastid. Note that such an organism, a meta-alga, comprises contributions from five genomes (the nuclear and mitochondrial genomes of the heterotrophic master of the house, plus three from the secondary symbiont). There are still more elaborate combinations, tertiary endosymbioses that merge additional genomes into yet more complex cells. The inescapable conclusion is that symbiotic relationships, which become progressively more intimate as symbionts turn into organelles, lie at the heart of the entire eukaryotic order.[5]

What is surprising is that there appear to have been so few of them. A single primary endosymbiosis is thought to account for the origin of all the world's mitochondria. Another unique endosymbiotic episode is believed to be the source of all the world's plastids, with just one or two minor exceptions. Even the number of secondary endosymbioses, which broadcast the descendants of the first plastid among innumerable species of protists, seems to have been small; depending on the authority, it ranges from

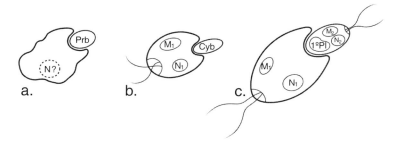

FIGURE 8.1. Three modes of endosymbiosis. (a) Association of an α-proteobacterium (Prb), progenitor of mitochondria, with an ill-defined urkaryote. (b) The primary plastid: uptake of a cyanobacterium (Cyb), progenitor of all plastids, by a heterotrophic eukaryotic protist. (c) Secondary plastids: a heterotrophic eukaryotic protist endowed with nucleus, N_1, and mitochondria, M_1, takes up a whole photosynthetic eukaryotic alga (complete with a nucleus, N_2, mitochondria, M_2, and a primary plastid 1^0Pl).

an absolute minimum of two to six or seven. The reason for the extreme rarity of organelle genesis is not at all obvious. Contemporary protists quite commonly harbor prokaryotic endosymbionts, including proteobacteria, cyanobacteria, and Archaea.[6] In most cases the terms of trade are unknown, but some endosymbionts are known to contribute products of photosynthesis, or fixed nitrogen; others participate in the defense of the consortium, or power its mobility. All this suggests that the first steps in forging a partnership are not difficult and have been taken many times by diverse eukaryotic lineages; the formidable obstacles must come at the level of integration of endosymbionts into the fabric of its host, or master. These high hurdles have been surmounted on very few occasions, which mark revolutionary punctuations in the even tenor of cell history.

FIRST PARTNERS.   The road that leads to mitochondria begins deep in the murky past of the eukaryotic lineage. Even though mitochondria of contemporary organisms differ greatly with respect to their morphology, ultrastructure and genome organization, the consensus is that all descend from a single endosymbiotic event.[7] The evidence comes from the sequences of genes retained by mitochondria, including those for ribosomal RNAs and other essential components, but the identities of the partners are ever in dispute.

The nature of the host is especially uncertain, as we saw in the preceding chapter. Was it a primitive phagocyte already endowed with all the key features of eukaryotic organization? A prokaryote, a bona fide member of present-day Archaea? Or a urkaryote, a distinct primordial lineage of scavengers and opportunistic predators? The identity of the symbiont is also unknown but better defined, since SSU rRNA sequences and other phylogenetic criteria firmly map mitochondrial genomes to the α-proteobacteria. The question is which of them is likely to be the closest extant relative of mitochondria? *Paracoccus denitrificans* first attracted attention forty years ago because its respiratory chain is almost identical to that of liver mitochondria.[8] Other candidates include the *Rickettsiales* and *Bdellovibrio*, a tiny predatory bacterium that burrows into bacterial cells and flourishes in the periplasm, suggesting a possible mechanism of invasion.[9] Cavalier-Smith has argued the case for a photosynthetic bacterium, such as *Rhodospirillum rubrum*, a versatile organism that can grow anaerobically in the light but oxidizes organic substrates when oxygen is present; its convoluted intracellular membranes make plausible precursors for mitochondrial cristae.[10] There is a good precedent for photosynthetic bacteria as useful endosymbionts in anaerobic protists,[11] but if indeed the precursors of mitochondria had photosynthesis, all traces of this capacity appear to have vanished. The difficulty with all these proposals is that the origin of mitochondria lies far back in the past, and there is really no reason to believe that their bacterial progenitor is closely affiliated with any contemporary bacterium.

How far back in the past? Did the endosymbionts come in relatively recently, say a billion years ago, just prior to the radiation of eukaryotic life, as Cavalier-Smith has strenuously maintained for many years, or earlier, 2.5 to 2 billion years ago, when oxygen was beginning to accumulate in the atmosphere? Or do they go back further still, to the anaerobic phase of cell history, and played a part in shaping eukaryotic cells? Each of these positions has advocates as well as detractors (see chapter 7), and one's choice is more a matter of temperament than of evidence.

Another unresolved conundrum is what could have been the metabolic basis of the endosymbiosis? Historically it was thought that the host's metabolism was anaerobic and fermentative, while the symbiont was an aerobe that oxidized end products of fermentation, such as lactic acid. The trouble is that unless the invaders share their bounty with the host, they will quickly outgrow him; they would be pathogens, not symbionts. Contemporary mitochondria are specialized for respiration and the generation of ATP, which they export to the cytoplasm by means of a specific transport system, the ATP/ADP antiporter. Such a carrier would not have been present in the free-living symbiont but must have been acquired in the course of its enslavement; it cannot be called upon to explain the initial benefits of the association. An alternative idea comes from

Andersson and colleagues, who argued that the endosymbionts' capacity for respiration would have been of benefit to their anaerobic host by lowering the oxygen tension in the cytoplasm;[12] this, too, is dubious, because respiration generates free radicals that are known to be a major source of damage to cellular membranes and genes. All these difficulties are evaded by the proposal, expounded by Martin and Müller and steadfastly defended, that the symbiont was a metabolically versatile proteobacterium that generated hydrogen gas, thus sustaining its host, a methanogenic member of the Archaea.[13] Subsequent evolution in a world increasingly dominated by oxygen led to loss of the anaerobic metabolism and its replacement by respiration; but this hypothesis also encounters difficulties, notably the lack of any evidence for a methanogenic host.

There is, then, plenty to be uncertain of; and under that heading we can include the surprising variability of mitochondrial genomes, whose significance is not understood.[14] None of these doubts undermine the belief that mitchondria descend somehow from bacterial symbionts, but they are a reminder of how hard it is to make out evolutionary events that took place in the far distant past when the world and its inhabitants were quite unlike those that we are familiar with.

PLASTIDS GALORE.   The origin of plastids is better defined than that of mitochondria, because the events are more recent and the metamorphosis of bacteria into organelles has not progressed as far. The historical narrative is clear and relatively simple (fig. 8.2). It begins with a population of heterotrophic protists, early eukaryotes complete with mitochondria and probably biflagellate, that preyed upon cyanobacteria, which they gobbled up by phagocytosis and then digested. From time to time the prey would have escaped from the phagocytic vacuole to take up residence in the host's cytoplasm. In some instances, what began as an infection prospered and matured into a relationship that was mutually beneficial. This scenario must have played out on many occasions, but the evidence suggests that only once did it progress to the point where the endosymbiont lost its identity and turned into a photosynthetic organelle. This was an earth-shaking event, second only to the domestication of mitochondria hundreds of millions of years earlier.

Subsequent diversification of that first consortium gave rise to three primary photosynthetic lineages: unicellular glaucophytes; red algae (*Rhodophyta*), which run the gamut from single cells to large and common seaweeds; and numerous green algae (*Chlorophyta*), which in turn

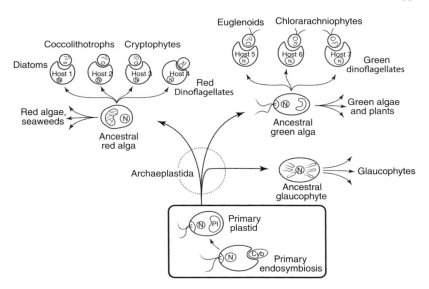

FIGURE 8.2. The dispersal of plastids, a simplified scheme. In the beginning, capture of a cyanobacterium by a heterotrophic protist generated the first photosynthetic eukaryote. That
lineage soon split, giving rise to the ancestors of glaucophytes, red algae, and green algae, each
of which then diversified. These three clades contain primary plastids and compose the supergroup *Archaeplastida*. Further plastid dispersal resulted from the capture of both red and
green algae by various nonphotosynthetic hosts, initiating new protistan lineages endowed
with secondary plastids. Additional lineages formed by tertiary endosymbioses have been
omitted.

gave rise to land plants. This triple ensemble makes up the *Archaeplastida* (figs. 7.2 and 8.2). But that first radiation did not exhaust the creative
powers of symbiosis. Secondary endosymbiosis, the engulfment of a whole
photosynthetic algal cell by a heterotrophic protist (fig 8.1c) gave rise to
many tribes of photosynthetic protists, some based on a green alga and
others on a red (fig 8.2). And tertiary endosymbiosis, the replacement
of one established secondary plastid by another, added further layers of
complexity.

Plastids differ enormously in their complement of pigments, their ultrastructure, and their place in the host cell's anatomy. A quarter of a century ago, most biologists considered it likely that cyanobacteria had been
turned into plastids more than once, perhaps many times. The counterintuitive notion, that at the base of all that diversity lurks a single unique
episode of primary endosymbiosis and a mere handful of secondary ones,

developed gradually from several lines of inquiry. A crucial advance
was the recognition that some plastids derive from the assimilation of
a prokaryotic endosymbiont while others descend from the uptake of a
eukaryotic one, and that these two classes can be distinguished by the
membranes in which they are enclosed.[15] In a nutshell, primary plastids,
which descended directly from a prokaryotic symbiont, lie free in the cytoplasm and are surrounded by only two lipid-bilayer membranes. Both derive from the endosymbiont—the inner one from the plasma membrane,
the outer from its outer membrane. Secondary plastids, whose progenitors
were full-fledged eukaryotes, are usually located within the host's endoplasmic reticulum and feature four (sometimes three) concentric membranes (the two inner ones from the original prokaryotic endosymbiont
one from the plasma membrane of the eukaryotic symbiont, and the outermost one from the host's phagocytic vacuole; fig. 8.3). This argument
groups together the glaucophytes, red algae, and green algae as the direct
offspring of primary endosymbiosis.

So do all primary plastids share a unique origin in a single endosymbiotic event? The argument, which has now persuaded almost all students of
plastid evolution, comes from molecular phylogeny.[16] The genes that encode plastid ribosomal RNAs and their metabolic enzymes, solute carriers,
and the machinery of protein import all resemble each other so closely as
to make multiple origins highly unlikely. Nuclear genomes are consistent
with the claim that all three primary lineages share a single origin, though
the identities of the partners remain under discussion. The eukaryotic host
has no known living relatives. The cyanobacterial endosymbiont, long elusive, has recently been linked by Deschamps and colleagues to subgroup
V (e.g., *Chroococcales*), nitrogen-fixing organisms that share with plastids
the capacity to produce true starch.[17] This kinship also suggests a possible
metabolic basis for that fertile union: the exchange of reduced nitrogen
compounds for a precursor to starch. Phylogenomic analysis, however, favors a more ancient cyanobacterium, one that diverged prior to the diversification of most of the contemporary lineages.[18]

Shortly after the establishment of the primary plastid, some 1.5 billion years ago,[19] that lineage began to differentiate (fig. 8.2). The primary
offshoots were probably the glaucophytes, a relatively unfamiliar class
of unicellular algae that harbor large and prominent plastids historically
know as cyanelles. Electron micrographs document the cyanelles' ultra-
structural affinity with cyanobacteria, but their DNA content is only a
fraction of that of free-living bacteria, and the propagation of cyanelles is

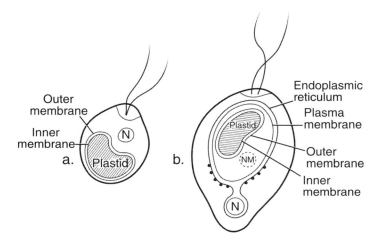

FIGURE 8.3. Plastids and their membranes. (a) An algal cell containing a primary plastid surrounded by two bilayer membranes, inner and outer, both derived from the endosymbiont. (b) An algal cell containing a secondary plastid, surrounded by four membranes. From the inside outward, these derive from the endosymbiont's inner and outer membranes, the plasma membrane of the captive eukaryotic alga (plasma membrane), and the endoplasmic membrane of the host cell, studded with ribosomes (endoplasmic reticulum). Note the nucleomorph (NM), nestled between the outer plastid membrane and the captive cell's plasma membrane; and the secondary host's nucleus (N).

controlled by the eukaryotic host. The discovery forty years ago that cyanelles contain peptidoglycan, an obvious remnant of the endosymbiont's cell wall, lent powerful support to the endosymbiotic hypothesis of plastid origins, which was still controversial at that time. Subsequent speciation events gave rise to the red algae and the green, and all entailed conspicuous changes in the structure and operations of the plastids.[20] For example, the original endosymbiont would have possessed peptidoglycan walls and characteristic bodies called phycobilisomes that house accessory pigments involved in light gathering. Glaucophytes alone retain peptidoglycan; the others lost it. Phycobilisomes still occur in red algae but were replaced in green algae by chlorophyll b embedded in stacked thylakoid membranes. And all the lineages underwent extensive changes at the genetic level.

The dispersal of secondary plastids may have been under way by 1.3 billion years ago, and it too entailed drastic changes. In most cases nothing remains of the eukaryotic algal cell that had been engulfed, enslaved, and assimilated, aside from its plastid and some genetic traces in the host nucleus. But in a few cases a remnant of its nucleus survives, now termed a nucleomorph (fig 8.3b) but still containing chromosomes and densely packed with genes.

Secondary plastids clearly arose more than once, since some have "green" affinities and others "red," but just how many times? That remains a contentious issue. Three unrelated clades of algae harbor "green" plastids: the euglenoids, chlorarachniophytes, and certain dinoflagellates, each apparently representing an independent instance of secondary endosymbiosis. "Red" plastids are widespread, being found in cryptophytes, haptophytes, many diatoms, dinoflagellates, and others. Some years ago Cavalier-Smith made a case for uniting all of those as descendants of a single event, and some predictions of that proposal have been born out. Nevertheless, the current consensus favors multiple separate events, perhaps as many as four[21] (fig. 8.2). A further layer of complexity is represented by "tertiary" plastids, cases in which one secondary plastid was replaced by a fresh one from a different eukaryotic endosymbiont. I shall venture no deeper into this minefield, which is already sufficiently littered with the remains of unwary travelers, and leave it to the experts to sort out the particulars.

Metamorphosis

The transformation of free-living bacteria into organelles entailed changes all around, but especially to the junior partners; as early as a century ago, the Russian biologist Konstantin Mereschowski understood it as a kind of enslavement. The bacteria were fitted with transport carriers that made products of their metabolism available to their host's cytoplasm. They lost more than 90 percent of their genome, partly by the transfer of genes to the host's nucleus. In consequence the great majority of organellar proteins—a thousand or more—are now encoded by nuclear genes, produced on host ribosomes, and transferred to the organelle by a dedicated import apparatus. Organelles also suffered far-reaching changes in physiology and ultrastructure, in keeping with their subservient role. Examples include the proliferation of energy-transducing membranes and host-controlled division. What the organelles retained suffices to document their microbial ancestry but not to support independent existence; it's much too late to "free the organelles"! In this section, we shall survey the molecular basis of the transformation.[22]

It is surely significant that mitochondria and plastids suffered parallel modifications, particularly with regard to protein import. One possible interpretation holds that the plastid machinery evolved from an earlier mitochondrial one. A more intriguing argument begins from the premise that the cells that acquired cyanobacteria were full-fledged eukaryotes. The parallels between mitochondrial and plastid transformations could then be a hint that, whatever may have been the origin of the eukaryotic lineage, they were well on their way by the time they domesticated mitochondria.

MAKING MITOCHONDRIA.   The most conspicuous role of mitochondria in the economy of eukaryotic cells is to generate ATP, most of which is exported to the cytoplasm and consumed there. It is likely that the first step in the integration of the endosymbiont was the insertion into its plasma membrane of a set of transport carriers, typically antiporters, that exchange a mitochondrial product for its "spent" counterpart in the cytoplasm. A case in point is the ATP/ADP antiporter. Such a device would be useless or worse to a free-living microbe, and so it is not surprising that sequence homologies clearly show the ATP/ADP antiporter to be of host provenance, a member of a class of eukaryotic carriers that normally operate in the endoplasmic reticulum.[23]

The origin of the machinery for protein import is more complicated and is the subject of much debate.[24] Protein import requires at least two molecular devices, one to pass nascent proteins (translated on cytosolic ribosomes) across the outer mitochondrial membrane and another to traverse the inner membrane (commonly referred to as TOM and TIM, respectively, for translocon outer/inner mitochondria). Each of them consists of about a dozen subunits, some of which appear to be of prokaryotic ancestry; others have no obvious homologs among prokaryotes and probably come from the eukaryotic partner. By contrast, a third device called SAM (for sorting and assembly machinery), which inserts and assembles the proteins targeted to the outer mitochondrial membrane, derives from the symbiont.

Integration or enslavement of the proteobacterium by its host was accompanied by a drastic reduction of the former's genome.[25] The free-living Bacteria would have possessed some four thousand genes, as their contemporary relatives do today. Mitochondrial genomes are tiny, ranging from sixty-seven genes (in the amoeba *Reclinomonas*) to as few as three. Many of the vanished prokaryotic genes wound up in the host nucleus and can still be recognized there; but the majority was simply lost. Most of the transferred genes continue to support mitochondrial functions, having somehow acquired the targeting sequences that allow their protein products to be recognized by TOM and TIM and imported into the organelle. Others may now serve directly in other facets of the host's metabolic economy.

The catchy phrase "enslavement" implies that the transfer of genes from symbiont to host nucleus was in some sense "imposed" on the former, but this is not necessarily the case. From the inception of the symbiosis, the host nucleus would have been bombarded by prokaryotic genes derived from lysed, or digested, guests. Some of these genes will have found a place in the nuclear genome, and come to support a duplicate of their original function. Natural selection would have favored the nuclear version, as the organelles were increasingly removed from the pressures of competition; the result was that the mitochondrial version was eroded by mutation, and eventually lost. Incidentally, the storyline sketched above implies that all the genes that service mitochondria descend either from the α-proteobacterial symbiont or from the eukaryotic host. Curiously, that is not the case: among the nuclear genes whose products travel to the mitochondrion, there are many that look to be of prokaryotic origin but have no homologs among the α-proteobacteria.[26] The reason is not clear; one possibility is that the genomes of α-proteobacteria have also undergone evolutionary changes and may have come a long way since their remote ancestors colonized their host two billion years ago.

Let us now turn from what was lost in making mitochondria to what was preserved. Mitochondria retain both structural features and physiological functions of the bacterial cells from which they descend. They contain DNA, ribosomes, and a capacity for protein synthesis. Mitochondria are the locus of the tricarboxylic acid cycle and some of its feeder routes, of cellular lipid synthesis and others. Their two membranes, the outer and the inner, are distinctly Bacterial in composition, not eukaryotic: they are made of Bacterial-type phospholipids, lack cholesterol, and are studded with an array of prokaryotic proteins including porins and the machinery of energy transduction. The proteins that make up respiratory chains and the F_1F_0-ATP synthase, not to mention chemiosmotic energy transduction by means of a proton current, are all essentially Bacterial in nature. The genes that encode these proteins have, with just a few exceptions, wound up in the nucleus. But the function, the purpose of the enterprise, remains linked to its Bacterial progenitor.

Mitochondria invariably harbor a small genome whose contents, albeit variable, are far from being a random sample of the genes that the symbiont brought into the partnership.[27] Typically, mitochondrial genes encode subunits of the respiratory chain and the F_1F_0-ATP synthase, core elements of the machinery of oxidative phosphorylation. Those genes are transcribed and translated in situ, on mitochondrial ribosomes, and some of the genes required to specify those operations are again housed in the mitochondria. Genes for the RNAs of both subunits are typically present, and so are some for tRNAs. Making proteins is costly; why would mitochondria retain all that machinery just to produce a dozen or so proteins, when hundreds of others are routinely imported? Several explanations have been offered, none completely satisfying. I favor the hypothesis advocated by J. F. Allen, who argues that mitochondrial genes encode a set of proteins whose synthesis must remain under local rather than centralized control.[28] The gist of his argument runs as follows. Respiration proceeds most rapidly and with minimal production of free radicals if all the components of the chain are poised near the midpoint, half oxidized and half reduced. Should one of them fail, it must be replaced. This is quite feasible if production of the key components is controlled by a local genome but would be a serious problem if all components had to be produced under the control of a central nucleus and then targeted to a particular mitochondrion.

But this hypothesis is but the tip of a much larger iceberg, for mitochondrial genomes may turn out to be an indispensable prerequisite for the evolution of eukaryotic complexity. In chapter 7, I argued, following Lane and Martin, that organelles—and only organelles—can generate essentially unlimited supplies of energy harnessed to the service of a single nucleus.[29] This abundance of energy allowed eukaryotic cells to expand their genome, and thus widen the repertoire of physiological functions a single cell can perform; the rise of complex cells depends on possession of organelles. For this to work, organellar genomes must be minute: just large enough to maintain redox control but not so large as to compete with the authority of the central nucleus.

Research on mitochondrial evolution is presently dominated by what can be learned from genomics and the merits of that approach are beyond question. It has shed a flood of light on what was involved in turning bacteria into organelles, and delineated early and essential stages in that metamorphosis. But it must also be said that it produces a partial and narrow view, like peering at a landscape through an arrow slit in the castle walls. Our burgeoning knowledge of gene shuffling and protein trafficking does not tell us how it came about that the ultrastructure of mitochondria looks nothing like that of a prokaryotic cell, dominated as it is by dense assemblies of membranous cristae. It contains no hint that, in many organisms, mitochondria lose their individuality and coalesce into a network of tubules that ramify throughout the cytoplasm. There is no immediate explanation for the fact that most (not all) mitochondria have jettisoned the prokaryotic mode of division in favor of an entirely different one based on the eukaryotic proteins of the dynamin family.[30] We have learned a fantastic amount in the past decade, and it is good; but the accumulation of genomic data also highlights the wide gap between what the genes encode and the products of cell evolution.

CYANOBACTERIA INTO PLASTIDS.   The course of events that turned free-living photosynthetic bacteria into symbionts and then reduced them to organelles runs generally parallel to the genesis of mitochondria. However, perhaps because plastids were acquired some five hundred million years later than mitochondria, the transformation has been less thoroughgoing; plastids are still recognizably similar to their prokaryotic ancestors and never surrender their cellular individuality. The subject has been studied in great detail,[31] but a brief summary must suffice here.

The conversion probably began with the insertion, into the bacterial plasma membrane, of a transport carrier that delivers products of photosynthesis to the host's cytoplasm. The triose phosphate/Pi antiporter was established in the common ancestor of the *Archaeplastida*. Like the ATP/ADP antiporter of mitochondria, it belongs to a family of eukaryotic transporters commonly associated with the endoplasmic reticulum, reinforcing the belief that the host cell was a full-fledged eukaryote. The apparatus for protein import came later, evolving in parallel with the transfer of prokaryotic genes to the host's nucleus. Initially, protein import may have relied on a mechanism derived from the host's secretory pathway, but most plastid proteins today travel by a dedicated import apparatus strikingly similar to that used by mitochondria. TOC and TIC (translocon outer/inner chloroplasts) are complexes made of multiple subunits that carry nascent polypeptides across the plastid's outer and then inner membranes. Like their mitochondrial cousins they are of mixed ancestry, with the core elements probably of eukaryotic provenance. Whether the plastid pathway evolved from the mitochondrial one or arose independently remains to be ascertained.

As in the case of mitochondria, a major feature of early plastid evolution is the reduction of the bacterial genome. Plastid genomes vary in size but retain only a fraction of the genes presumed to have been present in their free-living ancestor. The missing majority includes many that were simply lost but also hundreds that were transferred to the host nucleus. Their protein products, now made by cytoplasmic ribosomes, subsequently acquired leader sequences ("transit peptides") that are recognized by TOC and TIC; they provide handles to drag the protein into the plastid and are subsequently cleaved off. However, many of the genes that now service the plastid originally came from the host, not the cyanobacterium; and conversely, many of the symbiont's genes were transferred to the host nucleus and now serve functions unrelated to the plastid.

All plastids retain some of the symbiont's genes, and while these do vary they are not a random assortment. In every case, the plastid holds genes that encode key subunits of the photosynthetic apparatus, the electron-transport chain and F_1F_0-ATP synthase. The machinery of protein synthesis is also represented. The reason, as in the case of mitochondria, remains uncertain, but I am impressed by J. F. Allen's proposal that the genes retained by plastids are required to ensure local control of redox poise.[32]

What else did plastids keep? Even though the majority of the genes that encode the capture of light energy, its transduction into a proton current and the generation of ATP are now located in the nucleus, the functions themselves remain firmly bound to the plastids. The reason, of course, is that the mechanisms of chemiosmotic energy transduction, in photosynthesis as in respiration, require membrane-bound compartments. Those membranes themselves are another feature of the erstwhile symbiont that survived its transformation into an organelle: the inner and outer membranes of plastids share lipids and polysaccharides with the cyanobacteria, and the proteins that carry out energy transduction are unquestionably of bacterial provenance. The persistence of prokaryotic membranes in both plastids and mitochondria supplies the prime examples of membrane heredity; we shall return to this neglected principle in chapter 11.

The conversion of cyanobacteria into plastids matches that of α-proteobacteria into mitochondria, but beyond that point their historical trajectories diverge. Mitochondria have been inherited vertically since they were established, and even organisms that no longer possess conventional mitochondria still cling to relict membrane-bound compartments (see the final section of this chapter). By contrast, primary plastids spread laterally by a succession of secondary endosymbioses, "weaving a tangled web across a very large fraction of the eukaryotic tree."[33] Each of these entailed the evolution of transport carriers, apparatus for protein import and massive migration of genes to the new host cell. All have supplied challenging and significant issues for research by specialists, but fall outside the bounds I have set for this general treatment.

The Curious Case of *Paulinella chromatophora*

In biology, hardly any general assertion gets by without qualification, and the origin of plastids is a case in point. All plastids trace their origin to a unique, primary endosymbiotic partnership that formed roughly 1.5 billion years ago, except for those that don't. The chief exception is an unobtrusive unicellular alga called *Paulinella chromatophora,* which harbors plastids that result from an independent episode of endosymbiosis, and which are still in the making.[34]

Paulinella was first described more than a century ago and is classified in supergroup *Rhizaria* among the testate amoebas. These protists construct for themselves a delicate silica shell (or test), and feed by means of filamentous tentacles that emerge from an opening in the test. *Paulinella* has lost this apparatus, but acquired photosynthetic organelles that allow it to grow autotrophically in the light. The "chromatophores" are strikingly similar to cyanobacteria in ultrastructure and retain the peptidoglycan cell wall, but cannot be cultured in isolation. They reside in the host's cytosol, are not enclosed in a food vacuole, and divide in synchrony with the host cell. Ribosomal RNA sequences showed chromatophores to be unlike all other plastids, and specifically related to the *Synechococcus-Prochloron* clade of the cyanobacteria. Certain relatives of *Paulinella,* which lack chromatophores, prey on cyanobacteria of just this kind. The presumption is that on one occasion the bacteria escaped digestion, established themselves in the cytoplasm, and became partially integrated with the host cell.

The status of that relationship emerged more clearly from the full sequence of the chromatophore genome. This confirmed the kinship with *Synechococcus* and documented the changes that assimilation entailed. Free-living cyanobacteria of this group possess some 3,300 genes, whereas conventional plastids can muster at most 250. Chromatophores have 867, an intermediate number consistent with other indications that they derive from an independent and more recent association (still more than 60 million years ago!). Chromatophore genomes code for all the essentials of the photosynthetic machinery and ancillary metabolic pathways, but lack the genes required to make essential amino acids and cofactors. This may be a clue to the terms on which the partnership was concluded: products of bacterial photosynthesis in exchange for critical nutrients. Over time, symbiont and host merged, genes were transferred from the former to the latter, and now several proteins required for photosynthesis are produced by host ribosomes and targeted back to the chromatophore.[35] *Paulinella* is much more than just another biological curiosity. It demonstrates that conversion of symbionts into organelles, while rare, requires no unique events.

Evolution by Reduction

The idea that evolution is inherently progressive, tending over time to produce ever more complex and advanced forms, is deeply embedded in biological thought; but it is at best a partial truth. Everyone knows about the blind cave fish that have lost their eyes because they are useless in the dark. Parasites notoriously evolve simplified forms that depend on their victims for many metabolites that free-living organisms make for themselves. *Rickettsia*, the causative agent of typhus, even imports cellular ATP to fuel its metabolism. Mitochondria and plastids are prime examples of reductive evolution, having lost their autonomy and most cellular functions to become organelles controlled by their host. Remarkably, some organisms have carried this trajectory still further, generating diminished derivatives of both mitochondria and plastids.

Eukaryotic cells are typically aerobic, but as many as a thousand species of protists and also some fungi inhabit places that have little oxygen, or none at all. How then, do they produce useful energy? Some have found a substitute for oxygen as terminal electron acceptor, such as nitrate, but most have learned to make a living by fermentation. Some protists, especially ciliates, lack mitochondria but contain previously unknown organelles called hydrogenosomes.[36] These, too, are loci of energy production but lack both respiratory chains and the F_1F_0-ATP synthase. Instead, they decarboxylate pyruvate to generate acetate, carbon dioxide, and hydrogen gas (hence the name), and couple this reaction to the production of ATP by substrate-level phosphorylation. Hydrogenosomes are comparable to bacteria in size, are surrounded by two bilayer membranes, and were suspected to be descendants of bacterial colonists, perhaps clostridia, which carry out a similar fermentation.

Contemporary research instead has firmly placed hydrogenosomes in the evolutionary lineage of mitochondria.[37] They have lost the apparatus of oxidative phosphorylation but retain that for protein import. As a rule hydrogenosomes lack a genome but one at least, from a ciliate that flourishes in the hindgut of termites, features a residual genome that clearly resembles mitochondrial ones. Hydrogenosomes have apparently arisen independently on many occasions, as eukaryotic cells explored diverse anaerobic niches. Nor are hydrogenosomes the end of the line. A number of anaerobic protists, including such familiar parasites as *Giardia* and *Entamoeba,* lack both mitochondria and hydrogenosomes but contain certain genes normally associated with mitchondria. Their protein products were found to be localized in minute membrane-bound organelles, previously overlooked, that are now called mitosomes.[38] Mitosomes lack genomes and produce neither ATP nor hydrogen gas, but do import proteins by the canonical pathway. They fuel their activities with ATP imported from the cytoplasm. Just what functions mitosomes perform is still uncertain; the only known one is the synthesis of iron sulfide clusters, an important cellular process required for the maturation of diverse proteins, normally localized in mitochondria. Mitochondria, hydrogenosomes, and mitosomes are evidently members of an extended family that is sure to grow as more anaerobic protists are examined in detail. At the present time, the only features common to them all appear to be the biosynthesis of iron sulfide clusters and the possession of two membranes, all of bacterial ancestry. It will be interesting to see whether any eukaryotic cells have dispensed with their mitochondrial legacy altogether.

Reductive evolution is a feature of the plastid lineage as well.[39] Photosynthesis has, of course, been lost many times in representatives of all the green phyla, but most of these still feature plastids that supply their host cell with fatty acids, isoprenoids, and the tetrapyrrole compounds that are essential elements of electron carriers. In addition, some lineages that lack overt plastids but were suspected to be derived from plastid-bearing ancestors, have recently been found to contain cryptic plastids. A case in point is the apicoplast of apicomplexan parasites, which include the causative agent of malaria. Here again, the functions of this residual organelle include the provision of fatty acids, isoprenoids, and heme. Can plastids be lost altogether? The issue is important for understanding the phylogeny of several prominent taxa including ciliates, which show no trace of a plastid but may perhaps be descended from plastid-bearing ancestors.[40] If so, did any of their genes survive, and what became of their membranes?

CHAPTER NINE

Reading the Rocks

What We Can Infer from Geology

There is something about walking, and walking long distances, which can lift you out of the cares of the modern world and put you in touch with an older, quieter way of life. — Edwin Mullins, *The Pilgrimage to Santiago*

Snapshots of Two Billion Years Ago
Whispers from the Beginning
In Search of the First Eukaryotes
A Change in the Air
A Timeline for Cell Evolution

Lake O'Hara, a favorite haunt of all who love the Canadian Rockies, is nestled at the head of a long valley in the heart of Yoho National Park. A gated road gives access to the water and the lodge, but past those all ways lead up, steeply up into the alpine. My choice for today is the track that climbs to Lake Opabin, well beyond the timberline; on the far side, above the glacier, Mount Hungabee soars another 1,200 meters to touch the clouds at nearly 3,500 meters. It's a great place to sit, doze a little, and think about time.

The rocks around here are paleozoic marine sediments, laid down in shallow water and occasionally inlaid with fossils; the trilobite beds of Mount Stephen are just a few miles away, the Burgess Shale not much further. How long would it take to accumulate a slab the thickness of Mount Hungabee? Well, assuming a net deposition rate of 1 millimeter per century and neglecting losses due to erosion, 1,200 meters call for 120 million years, a fair span even as geologists measure time (but still only a quarter of the time elapsed since these rocks were laid down). To reach the era of cell evolution, say 1.2 billion years ago, we would have to pile 10 Hungabees, one on top of another, and that only gets us one-third of the way to the beginnings of life. All the metaphors and verbal gymnastics we contrive to make the remote past feel more homelike wither before the sheer immensity of geological time.

It does help some to divide eternity into smaller segments, and these have traditionally drawn on the fossil record to supply a timeline (fig. 9.1). By convention, the time of visible life, from the present to the beginning of the Cambrian era 543 million years ago, is designated the Phanerozoic eon. Earlier times are informally labeled "Precambrian" and divided into the Proterozoic eon (Greek for "earlier life"), 2.5 billion to 543 million years ago; the Archean eon, 4 to 2.5 billion years ago;[1] and the Hadean eon, the time of earth's accretion, which began about 4.5 billion years ago. The Proterozoic was the time when microbes flourished, diversified, and ruled alone; only at the very end of that eon did the first multicellular organisms make an appearance. Life must have originated during the Archean eon, which is therefore of particular interest to students of cell evolution; unfortunately, this is also the epoch whose life left the least and most ambiguous remains.

Sedimentary rocks of great antiquity still exist, but very few escaped metamorphosis by burial, heating, and compression. Not surprisingly, the fossil record of early life is sparse, spotty, hard to read, and full of gaps. But it is of the utmost importance, for geology is our only source of information that does not depend on contemporary organisms. And for all its deficiencies, the testimony of the rocks is unambiguous on the central point: the earth has hosted life, specifically microbial life, for almost as long as physical circumstances allowed, at least 3.4 billion years. All through its long-drawn history, life evolved in concert with its physical environment, shaped by geology and reshaping the earth in turn. The works of life include gigantic limestone reefs made up of the shells of innumerable marine protists; deposits of major commercial importance including oil, coal, sulfur, and iron; and most significantly, all the atmosphere's oxygen. That in turn altered the composition of the ocean, changed the pattern of weathering, and brought on global ice ages. We inhabit a living planet whose fortunes have been interwoven with the evolution of life.

This chapter does not pretend to offer a systematic introduction to early life or planetary history. That has been done most enjoyably by Andrew Knoll, whose fingerprints can be spotted all over this chapter;[2] more tech-

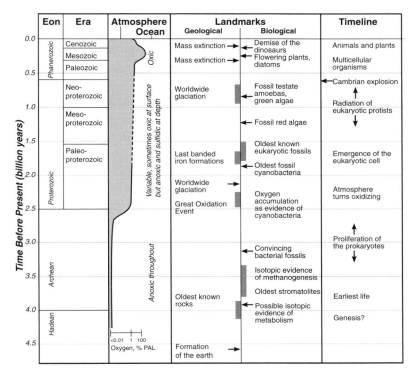

FIGURE 9.1. A timeline for cell evolution. (After Brocks and Banfield 2009.)

nical treatments are available in several symposia and a current textbook of astrobiology.[3] Nick Lane has written a lively yet scholarly introduction to oxygen and the way it shaped the world we now take for granted.[4] Suffice it here to consider what we have learned from paleontology about the earliest stages of biological evolution; readers may find figure 9.1 helpful in keeping track of the argument.

Snapshots of Two Billion Years Ago

Elso Barghoorn first made his name as a paleobotanist working on coal and was drawn into the search for fossil evidence of microbial life by collaboration with a field geologist, Stanley Tyler. Then, in 1954, the gods smiled on them: they reported the discovery of fossilized bacteria in the

Gunflint formation on the northern shore of Lake Superior and instantly quadrupled the length of life's known history. The minute fossils are embedded in a matrix of chert, commonly known as flint, a dense, hard, and fine-grained sedimentary rock composed almost entirely of interlocking silica crystals. Chert takes a sharp edge (hence its popularity with stone-age toolmakers) and can be cut into thin slices suitable for examination under a high-powered microscope. The Gunflint cherts, deposited some 1.9 billion years ago in a shallow tidal flat, host an abundant and varied assemblage of bacterial forms including cocci, chains, and filaments (fig. 9.2a). Among them are thin tubes coated with iron oxide that resemble contemporary iron-loving bacteria. The fossils are often trapped among the laminations of banded structures called stromatolites (fig. 9.2b), which are thought to represent fossilized microbial mats built up over hundreds of generations to macroscopic size.[5]

Cherts and shales dated between 2.1 billion and 1.8 billion years ago are relatively common, and many harbor well-preserved microfossils. Specific identification is seldom possible, but cherts from the Belcher Islands of the Canadian Arctic, Siberia, and elsewhere contain organisms that seem identical to modern cyanobacteria[6] (see also fig. 9.2c.) This identification is reinforced by the occasional presence of large specialized cells that look just like the resting bodies of cyanobacteria, such as *Anabaena*. These again are found tucked into the fabric of stromatolites, just like ones built by cyanobacteria to this day. Chemical biomarkers tell the same story: organic matter extracted from other cherts contains 2-methylhopanes, products of the degradation of hopanoids characteristic of cyanobacteria and seldom produced by other bacteria.[7] To be sure, neither cell morphology nor biomarkers document metabolic patterns, especially not the all-important capacity for oxygen-producing photosynthesis. Nevertheless, taking the findings at face value, it seems almost certain that by two billion years ago cyanobacteria had evolved their special mode of photosynthesis, and much of their present-day morphological diversity. The capacity for nitrogen fixation also seems to go back at least that far.[8] All these fossils are clearly prokaryotic; rocks of the early Proterozoic have not yielded fossils that can be convincingly interpreted as eukaryotic.

The fly in the ointment is, of course, convergent evolution. Is it not possible that early prokaryotes quite unrelated to today's cyanobacteria evolved deceptively similar morphology because they inhabited similar environments, or even acquired those structures by lateral gene transfer? Possible, but not likely. Resting bodies of this sort are found today only

FIGURE 9.2. Proterozoic and Archean fossils. (a) A slice of Gunflint chert, jam-packed with bacterial microfossils (Knoll 2003, courtesy of the author). (b) Stromatolites in Gunflint chert. Each column is about 1 inch wide (Knoll 2003, courtesy of the author). (c) Filamentous cyanobacterium preserved in Siberian chert, 1.5 million years old (Knoll 2003, courtesy of the author). (d). Archean bacteria, about 2.7 billion years old (Westall 2005). The letters identify rod-shaped bacteria (R), cocci (C), and filaments (F). (© 2006, The Geological Society of America, Inc. [GSA]. All rights reserved.) (e). Archean acritarch of uncertain affinities, 3.2 billion years old (Javaux 2010, courtesy of the author, with permission of Nature Publishing Group).

among cyanobacteria, and on their most recent branches at that.[9] Were they products of convergent evolution or subject to transfer, one would expect them to be scattered around the microbial tree. The implication is that cyanobacteria, ancestral to those of today and morphologically as well as physiologically similar, flourished more than two billion years ago. If we accept that argument, at least provisionally, two further implications follow. One is that lateral gene transfer, however common, is not so rampant as to erase lines of descent over two billion years; the other is that cyanobacteria exhibit a degree of evolutionary stasis that puts to shame those conventional paragons of conservatism, horseshoe crabs and brachiopods.

Did those microbial communities of the early Proterozoic eon include members of the Archaea? The question is critical, because the choice between two major versions of the tree of life hinges on whether Archaea came on stage early or late (chapter 4). Unfortunately, neither morphology nor biomarkers can specifically identify the fossil remains of Archaea. The assertion that Archaea were indeed present, and in fact enjoyed great prominence during the later Archean and early Proterozoic eons (say, 2.5 billion years ago) rests on a subtle geophysical clue: the C^{13}/C^{12} ratio of organic material formed at that time.[10] Very briefly, enzymatic reactions commonly discriminate between the carbon isotopes, generally preferring C^{12}. Fixation of atmospheric carbon dioxide by living organisms therefore produces organic matter that is relatively depleted in C^{13}, when compared with carbonate rock produced by inorganic reactions. The effect is especially pronounced for methanogenesis, whose products can therefore be distinguished from those of photosynthesis. The discovery that organic material extracted from certain Archean sedimentary rocks is strongly depleted in C^{13} is taken as evidence that they were produced by methanogenic Archaea.

In many respects, the world of the Gunflint and Belcher microfloras resembled our own. Geologists believe that the ocean was salty, much as it is today, and its temperature about the same or somewhat warmer. Tectonic processes were underway, and there is evidence for at least one worldwide glaciation about 2.2 billion years ago. The great nutrient cycles of carbon, sulfur, and nitrogen were in operation, including nitrogen fixation, though at a slower pace because biological productivity was far lower than at present.[11] The most conspicuous difference between that world and our own is that the Gunflint atmosphere was still very low in oxygen. The strongest evidence comes from a class of iron-rich rocks, the banded iron formations, which can only form in the absence of oxygen. Banded iron formations are plentiful in Archean rocks but then peter out; some of the youngest were laid down in Gunflint times. Incidentally, the prominence of iron-loving bacteria in the Gunflint chert implies that the water was rich in iron, a circumstance that again requires the absence of oxygen. The geological indicators suggest that, by two billion years ago the oxygen level had begun to rise, but was still no higher than 1 percent of the contemporary one. That should not be taken to mean that there could not have been organisms that respired oxygen. Within the lamina of stromatolites constructed by cyanobacteria, oxygen levels may well have been higher. Besides, it now appears that some bacteria, including the familiar *E. coli*, can grow (very slowly) on nanomolar levels of oxygen.[12] So there was something new in the air of Gunflint times, the transformation of the atmosphere had begun, but the oxygen-rich air that we take for granted today still lay far in the future.

Whispers from the Beginning

As we delve deeper into the remote past sedimentary rocks that have escaped destructive metamorphosis become ever harder to find, and truly ancient ones are very rare indeed. Only three localities take us beyond 3 billion years ago. The most productive have been cherts and shales from the Pilbara Hills in northwestern Australia and the Greenstone belt of Barberton Mountain Land in South Africa. Both fall well into the Archean, 3.5 to 3 billion years ago; they are geologically quite similar and may have once been parts of a single province that was severed by tectonic forces. Both have been subjected to heating and compression, sufficient to degrade fossil remains but not to destroy them altogether. The Isua and Akilia formations of southwestern Greenland are substantially older, 3.7 to 3.8 billion years, but have suffered drastic alteration. Some authorities claim that they nevertheless hold isotopic evidence for the existence of microbial life in the early Archean, others dismiss the findings and even question whether those were sedimentary rocks to begin with.[13] As far as we know, all earlier rocks have fallen victim to tectonic mayhem, leaving us without any geological traces of the origin of life.

Despite the limitations, there is general agreement that microbes flourished as far back as 3.4 billion years ago, and probably well before that. The evidence is of the same nature as that which testifies to the Proterozoic biosphere, but much less abundant; relics are less well preserved, and all conclusions are hedged with caveats.[14] The clearest evidence takes the form of stromatolites, and while similar objects can sometimes be produced by inorganic processes, the majority of Archean stromatolites are almost certainly genuine fossils.[15] Microfossils that look in every respect like bacterial cells, including cocci, rods, and filaments (fig. 9.2d) have been retrieved from several locations, including some that may have been part of a hydrothermal vent. Curiously, such fossils are seldom found embedded in stromatolites. Skeptics point out that inorganic processes can produce remarkably lifelike pseudofossils and rightly urge caution.[16] Organic biomarkers are not found in these rocks, all of which have been heated to more than 400°C, but the claims of Archean life are supported by isotopic evidence. The C^{13}/C^{12} ratios of fossilized organic matter ("kerogen," insoluble carbonaceous material) and of carbonate rock differ by about thirty parts per thousand, an observation most readily rationalized by regarding the kerogen as a degraded product of photosynthesis. Here again, certain inorganic processes can bring about isotope fractionation, but consistently high levels point to the operation of some biological process.

Of the nature of the organisms that lived in the early Archean, and of their metabolic capacities, little can be said with confidence. A decade ago, the mood was much more positive. William Schopf and his associates interpreted certain inclusions from Australia's Warrawoona cherts as cyanobacteria, 3.45 billion years old and the world's most ancient fossils.[17] There was always cause to be suspicious, both of the fossils themselves and of their extreme age that predates any indications of oxygen production by nearly a billion years, yet the claim was generally accepted and featured in books (including mine). No longer! Martin Brasier and colleagues mounted a fierce challenge, concluding that the alleged "cyanobacteria" were nothing more than abiogenic artifacts, products of crystal growth from amorphous organic matter, and probably not nearly as old as advertised.[18] Schopf has rebutted some of the criticism, but informed opinion has turned against him. So if not cyanobacteria, who were the mat builders of the deep Archean? Most probably bacteria carrying out anoxygenic photosynthesis, perhaps like today's *Chloroflexi* and *Chlorobi*, which are thought to represent very ancient lineages. Judging by minute bubbles of methane gas trapped in Australian quartz veins billions of years old, methanogenic Archaea must have been present as well.[19]

The latest entry into the Archean bestiary comes from Emmanuelle Javaux and colleagues, who reported the discovery of large spherical fossils up to about 300 micrometers in diameter in South African siltstones and shales some 3.2 billion years old (fig. 9.2e).[20] All the evidence is consistent with the claim that these are indeed biological fossils, albeit quite unprecedented in size and age. They have no obvious affinities with any extant microorganism, and the authors stoutly resisted the temptation to speculate; but Roger Buick, under no such constraints in his accompanying note,[21] floats the suggestion that they may possibly represent primordial eukaryotes. To be continued, for sure!

The more recent Archean is richer in fossil material, including unambiguous bacterial body fossils (both dividing cells and matting filaments) about 2.5 billion years old from South Africa's Transvaal. But the most spectacular microbial sightings came from the Fortescue and Hammersley shales of northwest Australia: biomarkers 2.7 billion years old, including both 2-methylhopanes and steranes. These are characteristic of cyanobacteria and eukaryotes, respectively, and despite some reasons for concern, the findings were taken as strong evidence that both these flagship groups were already present in the Archean eon.[22] That claim was called into question when Rasmussen and colleagues found that the critical biomarkers were not indigenous to the rock in which they were found, but more recent contaminants. At the same time, however, Waldbauer and colleagues applied the best available technology to drill cores from the Transvaal, and reaffirmed the presence of both steranes and hopanes at the end of the Archean eon. The same conclusion was reached by George and colleagues, who analyzed microscopic droplets of oil trapped inside crystals of quartz from somewhat younger Canadian sandstones, deposited about 2.45 billion years ago.[23]

So, where do we now stand in the quest for the earliest living things? The balance of the evidence clearly favors the claim that the planet hosted life early in the Archean, 3.5 billion years ago or more, though the organisms present at that time remain largely conjectural. Plausible candidates include Bacteria that lived by anoxygenic photosynthesis. Whether cyanobacteria also reach so deep into antiquity is still uncertain. Molecular biomarkers (hopanes) testify to their presence as early as 2.5 billion years ago, and it is very likely that they originated well before that. As to eukaryotic cells, there is presently a large gap between the oldest body fossils, dated to around 1.8 billion years ago, and the presence of steranes, products of the degradation of sterols, in sediments as old as 2.6 billion years.[24] It makes one wonder what creatures produced those steranes, and whether those fossil molecules can be taken as reliable evidence for eukaryotic cells.

In Search of the First Eukaryotes

Three billion years after their debut, prokaryotes remain what they had been in the Archean: small, relatively simple, biochemically venturesome, but morphologically almost unchanged. Even the cyanobacteria, larger than most others and endowed with internal membranes and specialized cell types, have remained resolutely prokaryotic to this day. However, in the course of the Proterozoic eon evolution changed gears. Organisms ancestral to modern animals appear "suddenly" in the geological record about 550 million years ago but, as Darwin suspected, a lengthy history of innovation and transformation preceded the Cambrian explosion of animal body plans. The Ediacaran fossils, some 50 million years earlier, apparently represent a first (and ultimately unsuccessful) attempt at multi-cellular life; and by then the flowering of the protists was well under way. Those ancient protists report, not only a change of evolutionary tempo but of mode, the emergence of a novel kind of cellular organization. The origin of eukaryotic cells is a conundrum second only to the origin of life itself but without the latter's aura of impenetrable mystery; and it falls squarely within the purview of cell evolution (see chapter 7). Paleontology offers glimpses of eukaryotic prehistory that complement—and also challenge—the view from genomics.

Protists that can be assigned to existing lineages such as dinoflagellates, diatoms, or coccolithotrophs appear quite late in the fossil record: in the Paleozoic era, or even the Mesozoic.[25] Molecular biomarkers confirm the conclusion that the protists which dominate today's oceans are not especially ancient. But the earliest fossils of any given lineage provide only a minimum estimate of that lineage's age, and both molecular phylogeny and palaeontology indicate that the evolution of protists began in the Proterozoic, more than a billion years ago. Indeed, among the fossil assemblages recovered from mid-Proterozoic rocks one finds organisms that are at least distantly related to contemporary protists. One such group consists of vase-shaped objects, about a hundred micrometers in length, commonly found in certain shales deep in the Grand Canyon dated to seven hundred to eight hundred million years ago (fig. 9.3a). They are identical in size and shape to the shells, or tests, constructed by several groups of amoebozoa today, and can therefore be taken as evidence that the amoeboid way of making a living goes back at least that far.

Equally persuasive are the multicellular filaments that Nick Butterfield found in the Canadian Arctic, in rocks dated to 1.2 billion years ago (but possibly younger). These closely resemble modern bangiophyte red algae and have therefore been designated *Bangiomorpha* (fig. 9.3b); if that assignment is correct, we can conclude that algal photosynthesis (and the endosymbioses that underlie eukaryotic photosynthesis) go back to mid-Proterozoic times. A very recent reevaluation of a fossil biota from the

2 billion years; biomarkers and genomics suggest still greater antiquity. For a billion years or more, eukaryotes of some kind flourished and diversified but without any obvious gain in structural complexity. Then, quite suddenly, came the burst of innovation associated with the radiation of modern eukaryotic forms in the Phanerozoic eon. What held them back? Were they waiting on some critical invention such as mitochondria, calcified skeletons or sex? Did eukaryotic radiation require a minimal level of oxygen or depend on some other environmental parameter such as the availability of nitrogen or trace metals? Alternatively, is it possible that all those early fossils have been badly misinterpreted and are actually prokaryotic, as Thomas Cavalier-Smith has claimed?[30] There is always room for more and better data, but an imaginative new synthesis would also be welcome. In the meantime, the hunt for the most ancient eukaryotes continues.

A Change in the Air

Given the premise that life originated here on earth, it must have done so in a world quite devoid of free oxygen, and in that setting brought forth all the basic mechanisms of cellular and molecular biology. The primordial source of energy is one of the unsolved mysteries of cell evolution (see chapters 5 and 10), but it is a reasonable presumption that life began with geochemical processes, and burgeoned after the invention of photosynthesis (fig. 5.4). The first substantial ecosystems, taking hold in shallow seas, were most probably based on organisms of prokaryotic grade that reduced carbon dioxide to organic matter with the aid of abiotic electron donors, such as hydrogen, hydrogen sulfide, or ferrous iron, and with light as the source of energy. The metabolic economy of organisms of this kind can be described by simple equations, such as $CO_2 + 2H_2S \rightarrow [CH_2O] + H_2O + 2S$, (where $[CH_2O]$ is shorthand for organic matter), and they continue to flourish today. Their operations rely on energy transduction by ion currents, a complex assemblage of membrane-bound devices (see chapter 5) that must have arisen very early in the history of life. Other organisms were presumably present as well, exploiting secondary niches such as the reduction of sulfur generated by the primary producers.

A drastic transition took place approximately 2.5 billion years ago, with the invention of a novel mode of photosynthesis that employed water as electron donor and produced oxygen as byproduct: $CO_2 + H_2O \rightarrow [CH_2O] + O_2$.

This requires 2 photosystems operating in tandem and 2 quanta of light rather than 1, but it made available an unlimited supply of reducing equivalents. Oxygenic photosynthesis seems to have been invented but once in all of earth's long history, by precursors of the bacterial clade that we call the cyanobacteria. Oxygen began to accumulate, locally at first and then globally, and the state of the atmosphere shifted from mildly reducing to mildly oxidizing. This is known in the trade as the great oxidation event, and it heralded the coming of a far richer and productive environment than its anoxygenic predecessor (fig. 5.4). Specialists disagree on precisely when the shift occurred; for present purposes we can peg it loosely between 2.3 and 2.5 billion years ago, with the advent of cyanobacteria and the onset of oxygen production somewhat earlier—perhaps much earlier.[31]

How can one know the composition of the atmosphere billions of years ago? There is abundant geological evidence that for the duration of the Archean eon, the atmosphere was entirely devoid of oxygen.[32] One item is the presence in Archean sediments of grains of pyrite and other minerals that are never found in later sediments because oxidative weathering destroys them. More spectacular testimony comes from the banded iron formations, massive deposits of iron ore formed by the periodic oxidation of dissolved ferrous iron (Fe^{2+}) probably with the aid of bacteria. The details are uncertain, but it is understood that banded iron formations no longer arise today because atmospheric oxygen oxidizes soluble Fe^{2+} to Fe^{3+}, which promptly precipitates as insoluble ferric oxide. The precise level of oxygen attained during the great oxidation event is uncertain, but was probably less than 1 percent of today's level. As to the source of that oxygen, only biology can supply sufficient oxygen to turn the atmosphere oxidizing. In consequence, atmospheric oxygen is grossly out of chemical equilibrium with the rest of the atmosphere: oxygen is perennially being consumed by respiration and weathering, turns over with a half-life of a few thousand years, and is maintained by continual photosynthesis. This is so remarkable that, should we ever detect oxygen in the atmosphere of some distant planet, it would instantly become a prime candidate to host life.

The rise in atmospheric oxygen was far from uniform. As shown in figure 9.1, there seem to have been at least two discrete steps, the first around 2.45 billion years ago and a second right at the beginning of the Phanerozoic eon, 540 million years ago. Atmospheric oxygen attained its present level just in time to support animal evolution. The trajectory in between is not well established but seems to have included at least one pronounced drop. There was also a time, some 300 million years ago, when the oxygen content of the atmosphere was about 30 percent, substantially higher than today's 21 percent. Oxygenation of sunlit shallow waters tracked that of the atmosphere, but the ocean depths lagged behind: for most of the Proterozoic, deep waters seem to have remained anoxic, dominated by hydrogen sulfide and hostile to all but anaerobic life. The causes of the fluctuations in oxygen level are, as usual, a matter of debate, but seem to be as much geological as biological.[33] The rates of oxygen production and consumption are nearly in balance. Oxygen accumulates when some of the organic matter generated by photosynthesis is withdrawn by burial in deep sediments, as the result of tectonic activity at the continental margins. The consequences of this small imbalance then reverberate across the evolution of life.

The two major increments in atmospheric oxygen levels correspond to conspicuous transitions in the organization of life: first the rise of the eukaryotes, which appears to have begun about 2 billion years ago, and then the appearance of multicellular animals. It seems unlikely that these correlations are coincidental. Two good reasons suggest that oxygen was a prerequisite for the evolution of eukaryotic cells. One is the well-known fact that eukaryotic cell membranes contain sterols, which modulate membrane fluidity and are probably indispensable for the membrane deformations that accompany endocytosis. Oxygen in trace quantities is absolutely required for sterol biosynthesis; the biosynthesis of hopanoids, analogous lipids found in prokaryotic membranes, does not require oxygen.[34] We can therefore take the presence of steranes as a proxy for oxygen as well as eukaryotic cells, with the implication that both go back at least 2.6 billion years. The other reason to suspect that oxygen was needed for eukaryotes to appear is the recognition that all extant eukaryotes either have mitochondria or had them in the past. Whatever may have been the metabolic basis of the symbiosis that gave rise to mitochondria, the end result was an oxidative organelle that enormously enhanced the provision of metabolic energy. I take these as evidence that the Eukarya had to await the availability of oxygen, at least in locales such as microbial mats. By the same token, the energy-consuming lifestyle of mobile animals may have had to await a sufficient level of atmospheric oxygen. This line of reflection is reinforced by the well-known fact that the carboniferous era, when oxygen levels were at their highest, also hosted giant insects with wingspans like that of seagulls. Oxygen did not drive evolution, but permitted it to take place.

Progressive oxygenation will also have affected the level of methane in the atmosphere, and with it, earth's temperature. Methane is today a trace constituent of the atmosphere, just a few parts per million, with a short residence time. Methane is readily oxidized when oxygen is present, and there are bacteria that derive energy from catalyzing this reaction. Gases emitted by volcanoes and submarine vents include abiotic methane, but the bulk of atmospheric methane is generated by methanogenic prokaryotes, all of them Archaea, which use hydrogen gas or organic electron donors to reduce carbon dioxide to methane and live by the energy liberated thereby. As in the case of oxygen, the level of methane in the atmosphere is determined by the balance of production and consumption; and there is evidence that methane was much more abundant during the Archean eon than it is at present. Indeed, biogenic methane was probably the greenhouse gas that kept earth from freezing over in its early stages, when the sun was less luminous than it is today. That era of abundant methane left traces in the isotope composition of Archean organic matter;[35] and very recently, Ueno and colleagues sampled methane gas sealed in rocks 3.5 billion years ago and demonstrated its biological origin.[36] With the beginnings of oxygen accumulation the balance shifted and methane vanished from the atmosphere; that may have been the trigger for the first global ice age 2.2 billion years ago. There is no direct evidence that Archaea were prominent 3.2 to 2.7 billion years ago, but that is the simplest interpretation.

Nitrogen, which makes up 80 percent of earth's atmosphere today and was probably always its chief constituent, does not track life's history as oxygen does but was indispensable to life from the beginning. Major molecular players, including proteins and nucleic acids, contain nitrogen; when life began, they must have been made from abiotic precursors, such as hydrogen cyanide or ammonia. But this is hardscrabble biochemistry. The proliferation of life required the means to "fix" nitrogen gas in usable form. This is done by the enzyme nitrogenase, which employs hydrogen gas to reduce nitrogen gas to ammonia. The reaction is exergonic but entails enormous activation energy, and therefore requires as many as sixteen molecules of ATP per molecule of nitrogen. Nitrogenase is scattered among both Bacteria and Archaea, but is never found in Eukarya; eukaryotes that fix nitrogen, such as legumes, do so with the aid of bacterial symbionts. And what nitrogenase grants, other enzymes take away: a network of reactions, both aerobic and anaerobic, converts ammonia to nitrogen oxide, nitrite and nitrate, even back to nitrogen gas, each with a particular prokaryote that relies on this reaction as a source of energy.

The nitrogen cycle, and others that turn over sulfur and phosphorus, must all have been established early on; isotopic evidence traces them to the Archean eon.

A Timeline for Cell Evolution

What the historian needs to know, first of all, is what happened and when; only after that does it make sense to inquire into the how and why. Until it is anchored in time, even the most sophisticated phylogeny lacks authority and meaning. That has proven hard to do, and so the literature is peppered with dates of all sorts, not all of them mutually compatible. Figure 9.1 presents a consensus timeline for cell evolution based on the most common interpretation of the geological record. There is room for caveats and doubts, and dissenters will have their due shortly.

Chronological information pertinent to cell evolution comes from two sources: fossils and sequences. Dating fossil-bearing strata used to be an art liberally laced with guesswork, but novel methods based on the decay of radioactive constituents now supply hard and precise data. The most familiar method is based on the decay of C^{14}, with a half-life of 5,730 years. That has been invaluable to archaeologists, but palaeontologists need a slower clock. They found it in zircon, a uranium-bearing mineral commonly present in granite and volcanic rocks in the form of minute crystals. Zircons contain two radioactive isotopes of uranium, which decay to isotopes of lead with half-lives of 4.5 billion and 700 million years, respectively; from the measured proportions one can calculate the age of the crystals (in practice, it is necessary to sample micrometer-thick layers of each crystal individually using a special ion microprobe). Note that the procedure dates, not fossils themselves but associated layers of volcanic material; that can be a cause for concern, since zircon crystals sometimes shift from one stratum to another. In general, paleontologists always worry whether the physical dating really refers to the biological material that is their focus; we had a cautionary example in the preceding section, in which biomarkers, extracted from rocks reliably dated to 2.7 billion years ago, had in fact been washed down from younger strata above.

Molecular data are even more riddled with problems. Sequences of nucleic acids and proteins change over time, but the rate of change is not constant and varies also from one locus to another. Algorithms have been devised to extract dates from sequences, with results that are often interesting but also controversial. I shall not describe these procedures, which fall quite outside my area of competence, but would note that they all require calibration with data points that come from the fossil record. At the end of the day, all chronology rests on palaeontology and geophysics.

The chronology outlined in figure 9.1 is consistent with the present state of knowledge, and probably enjoys widespread acceptance. It posits that prokaryotic cells originated early in the Archean, 3.5 billion years ago or perhaps even earlier, in a world that was strictly anoxic; all the major groups and metabolic pathways originated during the Archean eon. That specifically includes methanogenic Archaea, which seem to be among the most ancient of organisms. By contrast, cyanobacteria are more recent: they arose about 2.6 billion years ago or somewhat earlier, and brought about the first accumulation of free oxygen. That in turn touched off a worldwide ice age, perhaps thanks to the destruction of the atmospheric methane that had kept the earth warm until then. Though drawn primarily from the fossil record, this chronology is at least broadly consistent with inferences from molecular biology (chapter 2).

The place of the eukaryotes in cell history remains uncertain and controversial. Molecular phylogeny puts the divergence of the Eukarya from the Archaea early, at least 2.5 billion years ago (see chapter 2). That date may pertain to a (hypothetical) nuclear lineage, related to or rooted in the Archaea (see chapter 7). Fossils plausibly interpreted as eukaryotic appear 1,600 to 1,800 million years ago, but the presence of steranes in older rocks suggests that eukaryotes may have come on stage much earlier, 2.6 to 2.8 billion years ago. Recent attempts to attach a date to LECA with the help of sophisticated phylogenomic techniques calibrated with the fossil record give divergent results; means range from 1.1 to 1.8 billion years, within wide windows of uncertainty.[37] The molecular biology of organelles (see chapter 8) is broadly consistent with pegging the rise of the Eukarya at about 2 billion years ago. Mitochondria were probably in place by 1.8 billion years ago, plastids around 1.5 billion, and the secondary endosymbioses that produced the rich diversity of photosynthetic protists had got underway by 1.3 billion years ago.

By contrast, the profusion of protists that swarm in today's soil and water is quite recent, five hundred million years or less, and it is not clear how they link up with their Proterozoic precursors. Stepping back from the details, there is general agreement that the eukaryotic lineage is far older than today's representatives. Where does that leave the numerous

Proterozoic fossils that look plausibly eukaryotic but cannot be assigned to modern groups? Some may have been misdated; some may, in fact, be the remains of giant prokaryotes; but most of them probably belong to ancient stems of eukaryotic evolution that bear only very remote kinship to existing protists.

Consensus soothes the spirit but science thrives on controversy, particularly when it pits one explicit hypothesis against another. The counterpoint to the consensus chronology sketched in figure 9.1 is the radical alternative promoted and steadfastly defended by Thomas Cavalier-Smith (see also chapter 4).[38] From the beginning, Cavalier-Smith has rejected the status of archaebacteria as a separate domain of life; to him, these are highly derived bacteria, and relatively recent at that. Archaebacteria arose as little as a billion years ago, and so did their sister clade, the Eukaryota. Cavalier-Smith dismisses both the isotopic evidence for early archaebacteria and the putatively eukaryotic fossils of the mid-Proterozoic (prokaryotes, in his view), and has crafted a fully articulated narrative history to describe how today's microbial world came to be. His ideas have garnered no more than spotty support and look increasingly at odds with both molecular phylogeny and paleontology. In principle, both the conventional view and its opposite are falsifiable: unequivocal evidence for the antiquity of archaebacteria, for example, would do nicely to show the short chronology erroneous. In practice, such evidence remains elusive.

It would be agreeable at this point to proclaim progress, and to call for more research funding and better data. Those would surely not come amiss, but the hard reality is that erosion has worn away all but a handful of relics from that unimaginably distant time. In paleontology, the daughter of time is not truth, but uncertainty.

CHAPTER TEN

Ultimate Riddle

Origin of Cellular Life

A riddle wrapped in a mystery inside an enigma — Winston Churchill, speaking of Russia

Prebiotic Chemistry
The Beginnings of Order
Protocells
From Protocells to Progenotes
Enigma

The difference between a puzzle and a mystery is that the former can be solved within the framework of known principles, while the latter cannot. Over the past sixty years, dedicated and skillful scientists have devoted much effort and ink to the origin of life, with remarkably little to show for it. Judging by the volume of the literature, both experimental and theoretical, the inquiry has thrived prodigiously. But unlike more conventional fields of biological research, the study of life's origins has failed to generate a coherent and persuasive framework that gives meaning to the growing heap of data and speculation; and this suggests that we may still be missing some essential insight.

The search for answers is firmly constrained by the insistence that acceptable hypotheses be formulated in naturalistic terms, excluding *a priori* any intervention by supernatural forces. Scientists' refusal to grant some space to the mind and will of God may strike the majority of mankind as arbitrary and narrow-minded, but it is essential if the origin of life is to remain within the domain of science. A nudge from the divine would help us clear some very high hurdles; but once that possibility is admitted there will be no place to stop, and soon the settled principle of evolution by natural selection would be thrown into doubt. Scientific students of genesis are also inclined to set aside the possibility that life had an extra-terrestrial cradle or that life began with some exceedingly rare chance event or fluctuation—effectively, with a miracle. As a practical matter, we assume that life originated here on earth by a natural and probable out-growth of the chemical and physical circumstances prevailing some four billion years ago. The origin of life can therefore be construed as a problem in geochemistry; but it is also the black hole at the root of biological organization, and there's the rub.

What we have learned supports the conclusion that, once the organization and machinery of contemporary cells had come into existence (genes and their expression, catalytic proteins, phospholipid membranes, metabolic pathways, and so on), evolution was safely in Darwin's hands. Many puzzles remain to be solved, but there is no mystery: cellular life was shaped by the same evolutionary forces still at work today, including the transmission of genes and the creation of new ones, mutation, symbiosis, and cell heredity, all channeled by natural selection towards rising complexity and autonomy. But how could life ever get to that cellular stage? The answer must turn on the spontaneous emergence of molecular systems that first drew matter and energy unto themselves, maintained their integrity, reproduced their own kind, and evolved over time into cells as we know them. What little we have learned of these matters is the subject of the present chapter.

Life's origin has been most ardently pursued by chemists, apparently on the unspoken premise that once the molecular building blocks are on hand, cellular organization will take care of itself. That premise is surely incorrect. Modern cells do not assemble themselves from preformed constituents, and they would not have done so in the past. Instead, if we exclude explanations not grounded in natural law, we must envisage a pro-tracted phase of prebiotic chemical evolution that generated most of the molecular parts *pari passu* with their increasingly purposeful and elabo-rate organization. How this might have come about remains one of the deep mysteries of biology. It has engaged the attention of numerous capa-ble scientists and generated an enormous literature, including a number of recent book-length treatments.[1] Yet we seem to be little closer to enlight-enment than were A. I. Oparin and J. B. S. Haldane, who pioneered this line of inquiry seventy years ago.[2] The timeline sketched in figure 6.2, with its stately progression from a stable earth through prebiotic chemistry to protocells to the diversification of cellular life may look like a hypothesis, but it is not. Its purpose is merely to frame the questions and help readers keep a handhold on uncertainty.

Prebiotic Chemistry

A pioneering experiment published half a century ago laid the foundation and set the tone for almost all the subsequent research into the origin of life.[3] The late Stanley Miller, then a graduate student in the laboratory of Harold Urey at the University of Chicago, set out to test ideas formulated in the 1930s by J. B. S. Haldane and A. I. Oparin, who argued that abiotic chemical processes on the lifeless earth would generate a large stock of sundry organic molecules. In the absence of living things to consume them, these substances would accumulate in ponds and in the sea and supply all the essential raw materials for the assembly of primitive cells. These earliest protocells would be heterotrophs, reliant on the "primordial soup" for both constituents and energy. Miller constructed an ingenious apparatus in which mixtures of carbon dioxide, methane, ammonia, and hydrogen (thought at the time to resemble the composition of the early earth's atmosphere) were subjected to electric discharges (to simulate lightning), while the products were collected in a water trap representing the ocean. Over a period of days, organic matter accumulated in the trap. Most of it consisted of black tar, but there was also a significant yield of soluble small molecules, including several of biological relevance: glycine, alanine, glutamic, and aspartic acids and traces of other amino acids. There were also products that play no role in contemporary organisms, such as alpha-aminobutyric acid. The door to the origin of life had, it seemed, been cracked open.

From a contemporary perspective, the details of Miller's experiments are less important than their role in seeding a new branch of organic chemistry, sometimes called "prebiotic synthesis." Over the past fifty years, dozens of biological molecules have been produced under conditions that the authors, at least, consider to be consistent with circumstances that prevailed on earth four billion years ago.[4] These include a suite of the simpler amino acids, purines, and pyrimidines, an array of sugars, and simple carboxylic acids. Fatty acids and alcohols are harder to make, but complex mixtures of lipoidal substances can be extracted from stony meteorites; remarkably, these assemble spontaneously into membrane-bound vesicles.

Theoretical arguments suggest that members of the citric acid cycle were likely to have been available, especially under the conditions that prevail in hydrothermal vents. The roster of metabolites is far from complete, but growing. Recent discoveries include a route for the formation of pyrimidine nucleotides and the discovery of small organic molecules of biological interest in certain meteorites[5].

Macromolecules and molecular assemblages have also been generated by abiotic procedures. Mixtures of amino acids, when heated dry, polymerize into "proteinoids," chains linked together by peptide bonds, though not necessarily in the alpha position, as biological proteins are. Activated nucleotides adsorbed onto montmorillonite, a kind of clay, link up into polymers akin to RNA, with a length of up to forty residues. The operative principle here is that the catalyst, like heating and drying, promotes polymerization by driving the removal of water to form the bonds between monomers. Clays are also effective in promoting the self-assembly of alcohols and fatty acids into vesicles.[6]

The holy grail of prebiotic synthesis is the production of a linear macromolecule capable of replicating its own sequence, or that of an external template. In the eyes of many students of the origin of life, self-replicating RNA makes the most plausible starting point, and the quest has generated a devoted following.[7] Nevertheless, efforts to generate such a molecule abiotically, in the absence of some kind of biological catalyst, have not been successful. The most promising approach relies on the selection of a catalytically active RNA, a ribozyme, out of a large pool of random RNAs generated with the aid of enzymes. One such ribozyme proved capable of extending a primer, directed by an external RNA template, by as much as fourteen monomers, with a fidelity of 97 percent. More recently, Wochner and colleagues succeeded in engineering a ribozyme capable of synthesizing catalytically active RNA molecules ninety-five nucleotides in length, while Lincoln and Joyce produced a system in which a pair of ribozymes catalyze each other's synthesis.[8] However, it has thus far not been possible to isolate an efficient, self-replicating species of RNA; the exercise is clearly far more difficult than had been anticipated, and may not be feasible at all. This bleak assessment has sparked a search for other kinds of self-replicating structures that might have preceded RNA, again with partial success, but the relevance of those products to the origin of life remains to be demonstrated.

Even if the next issue of *Nature* heralds the successful isolation of a straightforward self-replicating species of RNA, the notion that the first protocells assembled themselves spontaneously from a generous menu of precursor molecules conveniently supplied by abiotic chemistry (or imported by way of comets and meteorites) is now widely recognized as simplistic and effectively has been abandoned.[9] Among its most cogent critics are experienced masters of the art of prebiotic synthesis, who are well aware of the shortcomings of many of the proposed routes and of the wide gap between the range of molecules that living things employ and those that can be made in the laboratory. The significance of the "prebiotic" molecules is not that they served as the raw materials for life's assembly but that they illustrate chemical structures that nature could readily discover and put to use. Almost certainly, life's chemical complexity did not spring directly from lifeless precursors, but arose in the context of an evolving cellular organization.

The Beginnings of Order

"The great mystery of life's origin lies in the huge gap between molecules and cells."[10] We do not presently understand how emerging life bridged this gulf. However, if we reject both the spontaneous coalescence of prefabricated components and supernatural intervention, we must envisage a protracted transition that began in geochemistry and led to molecular systems that exhibited at least some of the qualities of life. It's a tall order, and the more you reflect on it the taller it grows. In this section we shall consider not so much *how* this might have happened but *why*. Let us grant that a prebiotic broth of diverse organic molecules had come into existence. What natural causes might have ignited the rise of complex systems, and promoted the emergence of biological functions, autonomy and purposeful action?

THE POWER OF REPLICATION.   Of all the qualities that distinguish living things from inanimate objects, none is so spectacular as the capacity of the former to multiply and reproduce their own kind. Sprinkle a few bacterial cells into a flask containing a solution of salts and a pinch of sugar; next morning the sugar has been consumed and the medium teems with bacteria, each one a replica of the few that began it all. How bacteria grow and divide is a study in complicated cell physiology, but embedded deep in its core is a more elemental chemical process, the replication of those cells' genetic material. Like the multiplication of cells, replication of nucleic acids is an autocatalytic process: the product catalyzes its own formation, and so in principle the reaction proceeds at an ever-accelerating rate until it exhausts its substrates. The idea that life began its long journey with the appearance of a self-replicating molecule of some kind is both powerful and seductive; and it has come to exercise almost a stranglehold on the biological imagination.

The standard version of the hypothesis was formulated some forty years ago by Manfred Eigen, who proposed that life took its first baby step when a self-replicating molecule, probably RNA, arose by chance in the prebiotic soup of organic molecules.[11] Popular from the start, the hypothesis was heavily bolstered in the 1980s by the demonstration that RNA is not only a carrier of sequence information but a chemical catalyst as well. RNA thus embodies the essential qualities of both genes and enzymes, making it a prime candidate for the precursor to all the marvels of biology. That first clumsy replicator evolved thanks to imperfect replication ("mutation") and chemical selection for enhanced replication and stability (a function of the relationship between the rates of production and of degradation). Replicators clothed themselves in accessory molecules, including abiotic peptides, and eventually became protogenes. With the invention of translation and the coding mechanism, naked protogenes turned into assemblies; metabolism, membranes, and cell organization came later.

What driving force would underpin the progression from molecule to system? In a series of lucid and closely reasoned articles, the Israeli chemist Addy Pross argues that it is the kinetic power of replication.[12] All chemical reactions are governed by the laws of thermodynamics; a reaction can proceed only to an extent consistent with the chemical potential of its participants, as expressed in the change of free energy. This holds for replication too; molecular replication entails an input of substrates and energy (in the simplest instance, in the form of nucleoside triphosphates ready to be polymerized), and it can proceed only to the extent allowed by their chemical potential. But thermodynamics does not determine reaction *rates*. The autocatalytic nature of replication means that it will proceed much faster than any competing reaction, even one that would be thermodynamically favored if it were given time to go to completion. "Thus we can consider this kinetic phenomenon as an enormously powerful driving force responsible for the emergence and evolution of life. This force operates at every stage—from the primal replication reaction stage through to the single-cell stage and on to the multi-cell stage."[13]

Beginning with a basic molecular replicator, kinetic drive can support the exploration of complexity space by mechanisms that are wholly chemical yet guided by natural selection. Sequences tend to grow longer by accident during replication, increasing the amount of information they can potentially encode. For elongation to be favored despite its costs, increasing length must enhance the "effectiveness" of replication, defined by features that favor production of replicas over their decay. Such benefits may be purchased by incorporating accessory molecules into the replication pathway, with a premium on anything that enhances the supply of precursors and energy. So long as these can be found, increasing complexity can confer advantages that exceed its costs. The power of replication does not in itself explain how translation, coding, or membranes came to be, but it ensures that, if and when they are invented, they will be drawn into the service of replication. Pross stands on its head the traditional premise that "replication is a manifestation of life." On the contrary, from the chemical perspective, "life is a manifestation of replication." And the chemical, kinetic drive of replication suffices to explain life's teleonomic character, the undeniably purposeful behavior of living things. Initially, at the stage of plain molecular replication, that teleonomic character is rudimentary, expressed only in the "desire" to replicate. As evolution proceeds matters become more complicated, but "when all is said and done, all entities along that entire route, from replicating molecules through to human beings, in essence share the same 'dream'—to replicate."[14]

This is heady stuff, argued with conviction and style. I find the explosive power of replication very helpful in understanding how life has colonized every habitable nook and cranny, even the most inhospitable (a point also made by Richard Dawkins years ago). But there are at least two reasons to reserve judgment about the case Pross makes. First, the argument hinges on the chance appearance of a primal self-replicating molecule in the prebiotic broth. How likely is that? We know so little about the time, place, and circumstances that nothing can be ruled out. But the fact is that chemists have encountered insuperable difficulties in generating a working replicator, and many have expressed doubts about the project.[15] It is at least incumbent upon proponents of its spontaneous genesis to explain how the "correct" monomers could have been selected from the "prebiotic clutter," how a sufficient concentration of monomers was maintained, where the energy came from, and how the replicator evaded the tendency of polymers to break down by hydrolysis. Second, the gradual accretion of accessories implies that replication continued in a sustained manner over a protracted period, which calls for a very special locale. These misgivings would be somewhat allayed if, instead of a naked replicator, we envisage that replication was from the beginning a property of an organized system, sequestered in a compartment and sustained by a flux of energy.

ORGANIZED BY ENERGY FLOW. Nonconformists who doubt that life began with a molecular replicator and grew complex over time are apt instead to argue that life began in complexity, with self-organized networks of chemical reactions.[16] The basic premise is that the prebiotic environment supplied a rich diversity of organic substances, a variety of energy sources, and also potential catalysts, including clays, transition metals, and abiotic peptides. Chemistry was everywhere rampant, with reactions coalescing into networks of mounting complexity. A few will have attained "catalytic closure," the capacity to reproduce as a collective entity even though no single constituent replicates itself. Prebiotic reaction networks will have made up an ancestral version of metabolism, whose lineaments can still be traced in the metabolic machinery of contemporary cells. At the inception of life, the critical qualities were kinetic stability and error tolerance. Enclosure within a boundary must have been an early feature, but replicators, proteins, and structural order were later accretions.

These are stirring ideas. As Leslie Orgel put it, "The demonstration of the existence of a complex nonenzymatic metabolic cycle . . . would be a major step in research on the origin of life, while demonstration of an evolving family of such cycles would transform the subject."[17] But the proposal that metabolic organization can arise and persist of its own accord in the absence of both enzymes and genes seems to fly in the face of reason. There is no familiar precedent to lean on, and the spontaneous emergence of order out of chaos violates one's intuitive sense of how the world works. Indeed, if chemical networks could arise spontaneously, one would expect them to show up as a cause of spoilage in canned food. The case for self-organization becomes plausible only when the emergence of networks is strongly coupled to a flow of energy.

The cardinal point was made by Harold Morowitz forty years ago: the flow of energy through a system acts to organize that system. At the chemical level, coupled energy flux promotes the emergence of cyclic reaction pathways. At the physical level it causes the appearance of dissipative structures, including such familiar objects as flames, hurricanes, whirlpools, and those remarkable "Bènard cells" that appear when a pan filled with liquid is uniformly heated from below. There is nothing mystical about the emergence of large-scale patterns in systems kept far from equilibrium by the flux of energy; the phenomenon is well documented, and seems certain to play a major role in the inception of chemical and structural organization. I am inclined to go much further, in company with those who claim that the flux of energy is what called life into being.[18]

If this general proposition is to become a testable hypothesis for the

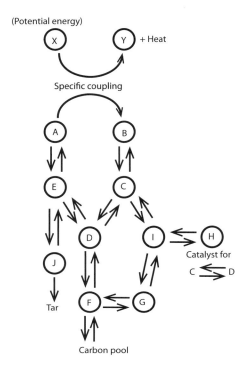

FIGURE 10.1. A metabolic network organized by a "driver" reaction. The exergonic (energy-yielding) transformation of X to Y, external to the cycle, is chemically coupled to the transformation of A to B (the "driver" reaction). This joint reaction moves the cycle represented by the letters A to E in the clockwise direction. The movement of material from side reactions (indicated by the equilibria associated with substances F to J), and from an external carbon source ("carbon pool") into the cycle is favored by the release of the potential energy in the transformation of X to Y. (From Shapiro 2006, with permission of the University of Chicago Press.)

origin of chemical complexity, it is necessary to specify both an energy source and a credible coupling reaction to funnel that energy into a reaction sequence. An abstract example, formulated by the late Robert Shapiro, is sketched in figure 10.1.[19] It postulates an external energy source, the conversion of X to Y, which is chemically coupled to the conversion of A to B, one reaction in a hypothetical metabolic cycle. Thanks to this coupling, the reaction sequence represented by A to E is moved in the clockwise direction, ultimately regenerating A; the cycle can expand by drawing in materials from external sources (reactions F through J) and generate a net yield of some product. Coupled reactions of this sort are a staple of today's biochemistry. Most cellular metabolites are made by re-actions that couple biosynthesis to the hydrolysis of ATP, which is regen-erated elsewhere (see chapter 5). For a network of reactions to operate, it must be sequestered in a compartment of some kind, if only to keep the reactants from diffusing away. There was no shortage of energy, with light and geochemical reactions being the most realistic.

All this sounds sensible enough in principle, at least to my mind, but a little reflection suffices to show how thin the ice is.[20] Modern biochemical cycles are catalyzed by specific and dedicated enzymes, manufactured by the cell for that purpose. But here we are thinking about a time before enzymes, when the only catalysts were minerals and metal ions. Is it really likely that all the requisite catalysts would occur together in one place, and would catalyze precisely the steps required to keep the cycle turning, rather than side-reactions that draw off the participants into byproducts (tar, in fig. 10.1)? It would be reassuring if one could point to a single known cycle, complete with energy source and coupling reaction, that op-erates in the absence of enzymes! Here, if anywhere, is an opportunity for experimentalists to make a contribution.

SELECTED FROM THE OUTSET.   There are, Richard Dawkins once observed, only two possible explanations for our existence: God and natural selec-tion. With the former excluded *a priori*, scientists must account for the transition from chemistry to biology by some sort of undirected evolution. The principles that drive biological evolution today are well understood: reproduction, heritable variation, competition, selection for reproductive advantage, and hence adaptation to a changing environment. The cen-tral issue is whether abiotic chemical systems, devoid of the apparatus for the expression and transmission of genetic information, could still exhibit some degree of heredity with variation and be subject to natural selection. If so, they might serve as a platform upon which evolution could construct more sophisticated mechanisms. On that premise, the origin of cells has much in common with subsequent evolution, even though the objects of selection, the criteria for success, and the mechanisms responsible have changed as the transition progressed.

On the premise that life began with a self-replicating molecule, conven-tionally taken to be RNA, the pattern of prebiotic evolution is comprehen-sible, at least in outline. In the absence of enzymes, replication was slow and subject to frequent errors that generated a cloud of mutants derived from the master sequence. It was natural selection that steered the kinetic

power of replication on a course that led to more accurate and stable replication and, later on, to complex assemblies culminating in cells.[21] The first
product of selection will have been RNA itself, selected out of a mixture
of abiotic molecules by virtue of its capacity to replicate. Selection would
have favored the acquisition of ancillary factors including amino acids,
abiotic peptides, and nucleic acids. At a later stage selection would have
favored replicators capable of specifying the structures of these auxiliary
factors; only then did replicators become primitive genes. Just how this
came about, and what were the criteria for selection, is not altogether
clear to anyone. But it seems fair to say that the evolution of replicators
cannot have gone very far in the absence of mechanisms to ensure a stable
milieu and a supply of energy and precursors. If ever there was an era of
free replicators, it must have been short-lived.

It's a grand leap from independent replicators to cells, because in a cell
replication is constrained for the benefit of the collective. The persistence
of a gene depends on the function it encodes and thus on the contribution
it makes to the survival of the community; it has little to do with the replication of that gene itself (there are exceptions, of course—parasitic replicators that make no contribution to the common welfare—but I shall here
set those aside). In fact, natural selection today does not see replicators at
all. It judges the phenotype of the cell, which may be quite remote from
the functions encoded by the genes. Just how those (hypothetical) free
replicators would have been herded into communities and then taught to
accept the collective discipline is one of the questions under debate.

On the alternative premise, that the first steps to life were represented
by self-organized metabolic cycles, evolution must be grounded in the
properties of systems. Individual cycles, like eddies, could "compete" for
resources, producing winners and losers. In theory, at least, they can undergo enlargement and a kind of propagation, even the transmission of
their compositional identity. I quite like this notion, but there was obviously a limit to the range of possible variations until the advent of digital
genes made unlimited heredity possible. Besides, a substantial baseline
level of functional organization seems to be a prerequisite for natural selection to favor progressively higher levels of adaptive order. The implication is that, even if nucleic acids were not first on the stage, they must
have made their entrance very early in the drama. Cellular organization,
founded on the collaboration of genes, catalysts, and membranes, was not
a late stage in the advance of either nucleic acids or metabolic cycles; it is
what the origin of life was all about.

How cellular organization came about is a great mystery, but it is not hard to see why natural selection would favor systems of this kind over unstructured ones. Enclosure in a pod, such as a membrane-bound vesicle, keeps multiple parts together and makes it possible to maintain an ionic environment that suits their operations. Membranes today are crucial to the generation of useful energy, and it is likely that this was true from the start. Free replicators compete for raw materials, but genes cooperate for the good of a larger unit that maintains autonomy from its environment. And with the advent of cells came the rudiments of individuality and a vast new range of variations for selection to sift.

Protocells

A decade ago, a hot topic for debate was which came first, replication or metabolism? That issue has not been resolved but has been largely super-seded by the recognition that neither of them, by itself, can take one far along the road to life. It is simply not credible to claim that anything be-yond the most rudimentary kind of replication or metabolism could have arisen in free solution. Instead, molecules and processes must have evolved concurrently and cooperatively in tandem with a basic mode of structural organization. This proposition comes in a variety of flavors,[22] all of which share the premise that life began to crystallize with the emergence of en-closed systems capable, to some degree, of harvesting matter and energy and directing these resources towards self-maintenance and reproduction. I shall refer to these as "protocells," entities that by definition lack nucleic acids, proteins, and the apparatus by which the former specify the latter (see also chapter 6). Just how such multimolecular assemblies arose is not well understood and is the subject of much research. Here we shall focus on the principles, beginning with the need for a physical boundary.

NO LIFE WITHOUT MEMBRANES. Cells as we know them are defined by their plasma membrane, which segregates the workplace from the envi-ronment. It is only thanks to the membrane that cells can maintain a dis-tinctive cytoplasmic milieu, which differs from all the rest of the world in its pH, ionic composition, and its complement of small molecules. The barrier function of the membrane stems from its structure, a phospholipid bilayer that is almost impermeable to ions and polar molecules. This fea-ture in turn imposes a requirement for selective portals and conduits to allow particular substrates to pass inward and out, mediated by a battery of specialized transport systems.

The functions of the plasma membrane go far beyond those of boundary and barrier. In prokaryotic cells (and to a lesser degree, in eukaryotic ones), the plasma membrane is the locus of energy transduction (see chapter 5); it is also deeply enmeshed in cell division and morphogenesis. And in today's world, membranes never arise *de novo*. Each membrane, complete with its complement of carriers and catalysts, is generated by growth of a preexisting membrane of the same kind. In principle, such membranes are as much a heritable feature of cells as the genome, passed continuously from one generation of cells to the next from the very dawn of life (see chapter 6).

All these sophisticated and purposeful activities depend today on protein catalysts; matters must have been far simpler and less specific prior to the invention of proteins encoded by genes. But it is noteworthy that proteins are not a prerequisite for the formation of compartments bounded by lipid membranes, which can arise spontaneously by self-assembly of diverse molecules that share a peculiar feature. Phospholipids, fatty acids, fatty alcohols, and many polycyclic molecules are "amphiphiles": one end repels water, the other attracts it, and that is sufficient to favor the formation of closed, membrane-bound vesicles. Since molecules of this kind form readily under abiotic conditions and occur in carbonaceous meteorites, it seems all but certain that vesicular structures were present early on and likely that they played a role in the genesis of life.[23]

Abiotic membranes were probably not made of phospholipids (which would have been unavailable) but of mixtures of medium-length fatty acids and alcohols. Jack Szostak and his associates showed that lipid vesicles can be made to incorporate molecules small and large, including polymerases that catalyze the replication of RNA and DNA with nucleotides from the medium as substrates (it is significant that vesicles made up of medium-length amphiphiles are permeable to nucleotides, whereas phospholipid vesicles are not). Though sensitive to perturbation by ions, the vesicles are remarkably stable at high temperatures. Under the right conditions, vesicles "grow," incorporating new molecules from the medium, and even undergo a kind of fission.[24] At first sight, such efforts appear to have more to do with the creation of artificial cells than with the evolution of ancestral ones. But given the likelihood that lipid vesicles were actually part of the prebiotic environment, this work represents the most promising approach to the genesis of cells.

The membrane biologist's dream, as recounted by David Deamer, runs something like this.[25] Four billion years ago, the young earth sweltered at 60°–70°C under a blanket of nitrogen and carbon dioxide, steaming and heaving with volcanic ruckus. Seas and lakes had become dilute stews of organic matter, covered with an oily surface film; and the incessant barrage of energy from lightning, heat and volcanic outgassing drove the mixture towards ever-greater complexity, including prototypes of today's macromolecules. These large molecules tended to aggregate on mineral surfaces and at the interface between water and atmosphere. Most of these deposits held no more potential than bathtub scum, but some took the form of bilayer vesicles, trapping samples of the stew. A few, perhaps, captured materials that could be driven up the gradient of complexity by persistent energy flux. And then, from the miasma of that earth, a cell precursor may have taken shape "when one or more of the assemblies found a way not only to grow, but also to reproduce by incorporating a cycle involving catalytic functions and genetic information."[26] A dream, for sure, but no more so than the molecular geneticist's dream of the ancestral replicator.

ENERGY FOR THE JOURNEY.   It's a long and arduous road from vesicles to the most primitive of cells, a passage that could never have been made without a plentiful and persistent supply of energy to do the work of staying organized and to drive the system towards greater autonomy. There was certainly no shortage of energy on the primitive earth; the difficult question is which sources fueled the ascent of primordial life and how they could have been coupled to the performance of work. Energy from the degradation of preformed organic compounds has been advocated in the past and remains a possibility, but the dilute abiotic broth cannot have sustained evolution for long. For energy supplies that last contemporary organisms turn to membranes, and it seems likely that this was true from the start. If this argument is correct, we are looking for ancient precursors to the machinery that today captures the energy of light and of redox reactions by transducing it into ion currents.

One such proposal, developed by Arthur Koch, is illustrated in figure 10.2.[27] In this primordial chemiosmotic energy transducer, the driving reaction is the oxidation of hydrogen sulfide by ferrous iron to produce pyrite, a reaction known to be highly exergonic thanks to the extreme insolubility of the product. Pyrite formation as a primordial energy donor was first proposed by Günter Wächtershäuser in the context of his theory of surface metabolism (see the following section), and has since morphed

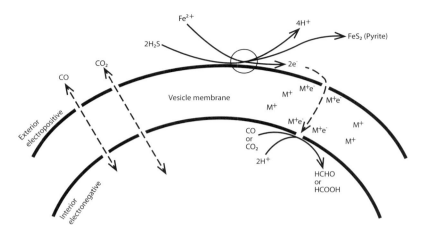

FIGURE 10.2. A model for the primordial generation of a proton potential. The energy source is the reduction of Fe^{2+} by H_2S to yield pyrite, assumed to take place on the outer surface of a membrane-bound vesicle. Protons remain outside, while the electrons are conducted into the vesicle's lumen by a chain of metal ions. CO or CO_2 are reduced at the inner surface to the level of formaldehyde or formic acid respectively. The consumption of protons inside, and their release outside, are equivalent to the translocation of protons from the vesicle interior to its exterior. (Adapted from Koch and Silver 2005, with permission.)

into several versions. In Koch's hypothesis, electrons generated by the oxidation of hydrogen sulfide are carried across the vesicle membrane by metal cations and reduce carbon dioxide or carbon monoxide on the inside surface. Since protons are consumed on the inside and produced on the outside, the effect is equivalent to translocating protons from the lumen to the outside. The proton potential thus generated could support useful work. Just how that happens, and just what the coupling reactions would be, remains to be worked out. A variation on this scheme, due to James Ferry and Christopher House, links pyrite production to the reductive fixation of carbon dioxide, generating acetyl Coenzyme A and eventually ATP.[28] If true, both of these proposals would constitute instantiations of Robert Shapiro's abstract scheme of ancestral energy coupling (fig. 10.1).

The alternative primordial energy source is light. In modern cells, light is captured by specialized pigments (chlorophyll or retinal), and the energy inherent therein is converted into a proton potential across the membrane by an elaborate cascade of redox carriers. These would not have been present at the beginning, but more primitive pigments and carriers might have been. In laboratory experiments, membrane vesicles doped with certain dyes respond to light with the generation of a pH gradient between the lumen and the medium.[29] Quinones, today's chief hydrogen carriers, have been found in lipid mixtures extracted from meteorites; they, together with carriers of electrons, are certainly molecules of great antiquity.[30] The notion that light was the first energy source has waxed and waned over the years; I would not count it out.

In modern cells, an extensive network of energy carriers, catalysts, and cofactors links the electrical potential generated by the membrane to a variety of cellular work functions. Being among the universals of biology, participants like ATP and ADP, NADH and NADPH, thioesters such as Coenzyme A, and also the ATPases that translocate protons and link the proton potential to cytoplasmic ATP are surely very ancient. Whether they go clear back to a time before proteins is quite unknown. Some authors have put forward inorganic polyphosphate as a relic of primordial bioenergetics.[31] Polyphosphates, long chains of phosphoryl groups that can be transferred to ADP and other acceptors, are widely distributed among microbes. They can be formed abiotically and occur as minerals, but a prebiotic ecosystem fueled by polyphosphates seems too pinched to take off and inherit the earth. Pyrophosphate is another possibility, the more so as proton-translocating pyrophosphatases are well known. The question of just how an early version of chemiosmotic energy transduction would be coupled to precellular metabolism, remains wide open.

Indeed, we have very little sense of that metabolism and the functions that it may have served. Did protocells rely on preformed organic molecules to construct their fabric, or synthesize their own? Some sort of protometabolism was surely required, and it seems reasonable to believe that its outlines foreshadowed the pathways used by contemporary cells.[32] Several authors have promoted the reductive TCA cycle or the Wood-Ljungdahl pathway as possible routes to capture carbon dioxide and distribute it to early biosynthesis.[33] Conceivably, these predate protein catalysts. But the truth is that we know nothing and can only speculate about the uses to which energy was put.

AN INORGANIC HATCHERY?   Readers will have noticed that many of the schemes and hypotheses sketched up to this point depend on the premise that the early earth supplied a generous menu of abiotic molecules; any protocells that emerged from the primordial broth will have been heterotrophs, relying on their environment for precursors and perhaps energy too. An alternative line of thought, which can be traced back to the reflections of A.G. Cairns-Smith and further still to the Russian biologist

Konstantin Mereschowsky a century ago, draws the beginnings of life from the earth itself.[34] Proponents envisage the earliest organisms as strict autotrophs, relying on inorganic reactions to generate all their constituents. An autotrophic genesis can call upon unlimited supplies of energy and seeks the origin of protocells in geochemistry.

The best known of these proposals has been developed in considerable detail by the German chemist Günter Wächtershäuser.[35] He locates the cradle of life in a volcanic setting of hot submarine vents, where simple inorganic molecules such as carbon dioxide, carbon monoxide, hydrogen sulfide, and nitrogen are exhaled from the deep. There they undergo chemical transformation, driven by chemical potentials such as the formation of pyrite and catalyzed by metals including nickel, cobalt, and iron. Organic molecules so produced would bind electrostatically to mineral surfaces to make a thin scum, within which further reactions took place that generated autocatalytic networks. Wächtershäuser designates this composite entity, made up of an inorganic substructure and an organic superstructure, as the "pioneer organism," a description that has not found favor with biologists. The chemical networks grow and become more elaborate when two previously independent catalytic cycles become so intertwined as to promote both and cement them together ("grafting"). Later stages call for the emergence of nucleic acids, proteins, lipid membranes, and cells but also require the reader to suspend disbelief. Serious critics found themselves unable to do that,[36] and neither can I. Experiments have shown that the conditions that prevail in hot volcanic vents can produce simple organic molecules, but the proposal that macromolecules and autocatalytic cycles can arise by surface metabolism will not stand in the absence of experimental verification.

Abundant energy is also a feature of a second proposal for autotrophic origins, developed over some two decades by Michael Russell and his collaborators,[37] which I find more appealing and to which I drew attention in chapter 5. This associates the beginnings of life with submarine vents of another kind, alkaline and warm rather than acid and hot (an example is the field-designated Lost City, off the Mid-Atlantic Ridge). Conditions in such vents sometimes favor the precipitation of iron sulfide into a honeycomb of minuscule compartments. These make plausible sequestered loci where organic substances might arise, accumulate, and undergo further reactions catalyzed by nickel and iron. The thermodynamic driving force to support such chemistry comes from the chemical potential of the gases discharged by the vent, which include carbon monoxide, methane, and most notably hydrogen gas generated by the geochemical reaction known as serpentinization.

The proposal has caught fire, and has now been taken a long step further by enthusiastic supporters.[38] Proponents see those cavities and crevices lined by catalytic minerals as the actual precursors of early cellular life, the wombs that nurtured metabolic pathways, self-replicating molecules, incipient genes, the trinity of DNA, RNA, and protein, and lastly, lipid membranes. Chemiosmotic energy transduction began when the protocells evolving in their stony matrix learned to draw upon the natural gradient of pH between alkaline vent effluents and the more acidic ocean outside the honeycomb. Over time, protocells acquired the means to harness this potential source of energy to their own purposes and subsequently generate a proton-motive force endogenously. Only then could primitive cells break free from their inorganic hatchery. In this view, today's membrane-bound, free-living cells represent quite a late stage in cell evolution: LUCA, the last common ancestor of the cellular domains, was a precellular entity with an RNA genome, sheltering in a labyrinth of interconnected micropores. Bacteria and Archaea emerged separately, each with its unique complement of lipids, to found the great prokaryotic domains.

Is the geochemist's dream credible? I am much intrigued by the idea that organic molecules may have been formed, concentrated, and melded in the cracks and fissures of submarine deposits, but balk at the later stages. Proteins, nucleic acids, the web of biosynthetic and catabolic reactions, all made in a hydrothermal vent? To the best of our knowledge, natural selection is required to drive evolution up the thermodynamic hill towards greater autonomy and complexity, and it is hard to see how selection can do this work on interconnected chambers in a mineral sponge. Could there be an alternative way to complexity that does not rely on competition among free-living units? Later on, there is also the problem of a precellular LUCA, which stands in flat contradiction to all the evidence that credits LUCA with membranes endowed with a basic set of redox pathways and energy-transducing ATPases (see chapter 4). But even skepticism can be carried to extremes. Those alkaline hydrothermal vents do make the most persuasive incubators for the beginnings of life, and there may be ways to overcome the objections. The urgent need now is to devise a working model of an alkaline vent, and to find out by experiment whether complex organic molecules can be produced under these very special conditions.

*   *   *

Did cellular organization come early? It must be so, but it is very hard to
see how it can be so. In truth, there is presently no persuasive hypothesis
to account for the emergence of protocells from the primal chaos. For the
time being, I must have recourse to a wizard's wand to forward us swiftly
to the point where the chemical phase of life's emergence can be said to
conclude. Koch and Silver identify that stage with the First Cell, a unique
historical entity that first brought together three indispensable chemical
features: a closed lipid membrane, chemiosmotic energy transduction,
and a self-replicating ribozyme.[39] One can quibble with the biochemical
particulars, and even with the postulate of a unique Cellular Eve, but the
biological point stands. Prebiotic evolution reached a turning point with
the advent of chemical systems that exhibited the rudiments of autonomy,
energy conversion, and reproduction. Explosive multiplication ensued,
out-competing all the neighbors, and both the mode and the tempo of
evolution were changed forever.

From Protocells to Progenotes

In all the constellation of mysteries that make up the genesis of cellular
life, nothing is more baffling than the origin of the principles and mechanisms of biological heredity. Life as we know it is ruled by the trinity of
DNA, RNA, and protein. The sequence of nucleotides in DNA encodes
genetic information, which specifies the sequence of amino acids in cellular proteins. The genetic information is perpetuated when DNA is replicated and copies are passed from one generation to the next. DNA serves
no other function. Like a book in the library, it is purely a repository of
information. Whatever cells do is done by proteins: they make up cellular structures, transport substances across membranes and within the
cytoplasm, and form the multitude of enzymes that catalyze all chemical
reactions. RNA serves as the intermediary that transmits sequence information from DNA to proteins. The sequence of nucleotides in DNA is
first transcribed into a corresponding RNA sequence (messenger RNA,
or mRNA), and that is subsequently translated into the (chemically quite
unrelated) sequence of amino acids in proteins. An intricate and sophisticated apparatus is required to carry out transcription and translation.
This includes an RNA polymerase to produce the messenger RNA, transfer RNAs to recognize particular amino acids and line them up on the

messenger RNA, activating enzymes to attach the amino acids to their cognate transfer RNAs, and ribosomes to link together amino acids into the finished proteins. Each amino acid is specified by a particular set of three nucleotides; the table of correspondences is known as the genetic code. Finally, information transfer has a direction: nucleic acids specify amino acid sequences, but proteins do not specify nucleotide sequences (the central dogma).

Whatever may have been the nature of the last common ancestor of cellular life, it clearly possessed the machinery of transcription and translation much as we know them today. This conclusion follows from the facts that the genetic code is, for all practical purposes, universal; and that Bacteria, Archaea, and Eukarya share the basic instrumentation of RNA polymerase, messenger RNA, transfer RNAs, and ribosomes. But this machinery cannot have been part of the endowment of protocells, a term loosely employed in the literature, which I restrict here to cell-like entities that lacked proteins specified by heritable information. The crucial step in the transfiguration of protocells into true cells will have been the invention of translation and the genetic code.

The more you reflect on this problem, the harder it becomes. Protein sequences are encoded in those of DNA, and proteins can only be produced with the help of the machinery of transcription and translation. But all this apparatus consists of proteins and relies on protein enzymes. Even the replication of DNA requires enzymes made of protein. DNA sequences specify the sequence of the replica but cannot produce that replica unaided. How, then, could genes have been made or expressed before there were proteins? How could proteins have been produced in the absence of genes to specify them? That seemingly random string of nucleotides has precise symbolic meaning, generated by the code that assigns to each triplet a corresponding amino acid; where did this semantic feat come from? Is it possible to disentangle all this intertwined complexity, and reconstruct it in the form of successive steps along an evolutionary pathway?

WHEN RNA RULED THE WORLD.   There appears to be but one way out of the impasse, beginning with the premise that protocells had neither proteins nor DNA but came to rely on RNA alone to bear the burden of both self-replication and catalysis. The idea germinated nearly forty years go in the writings of Francis Crick, Leslie Orgel, and Carl Woese, and was greatly reinforced by the unexpected discovery that certain chemical reactions in modern cells are catalyzed, not by conventional enzymes but by "ribozymes" made of RNA. At the time, RNA had only two known functions: as a major structural component of ribosomes and as the portable intermediary that ferried sequence information from DNA to ribosome. The discovery of ribozymes with a substantial repertoire of catalytic skills led Walter Gilbert and others to formulate the hypothesis commonly known as the RNA World.[40] At an early stage of cell evolution, prior to the existence of proteins and DNA, the functions of both genes and enzymes were performed by RNA. Contemporary mechanisms of transcription, translation, the genetic code and replication will all have begun to evolve in the RNA World, marking its culmination and also foreshadowing its extinction.

There is, of course, no direct evidence for an RNA-dominated stage of cell evolution, and there will probably never be any. The idea has been almost universally accepted thanks to its great explanatory power. It neatly evades the paradox of which came first, protein or DNA, by substituting RNA for both. And it supplies a reasonable explanation for the multiple roles of RNA in the physiology of contemporary cells, which can be seen as relics from the time when RNA did it all. Most important is the ribosome, which is now understood as an RNA-based device: it is RNA, not proteins, that performs the catalytic task of forging the peptide bonds that link one amino acid to the next. RNAse-P, a ubiquitous enzyme involved in RNA processing, is made up of both protein and RNA, and catalysis is a function of the latter. RNA serves as the genetic material of certain viruses (but never of cells), and is an essential component of several cellular organelles, including spliceosomes. The chemical structures of many coenzymes can be construed as modified nucleotides, which has been taken as evidence for a time when all of metabolism was catalyzed by ribozymes. Incidentally, DNA can also exhibit catalytic activity, but there appear to be no naturally occurring deoxyribozymes; this is consistent with the belief that natural ribozymes really are evolutionary relics.

How, then, should we imagine those hypothetical RNA-based protocells? On this there is little consensus. To paraphrase Carl Woese, everyone has his own take on that world. When the idea was first conceived, it was assumed that self-replicating RNA formed spontaneously in the prebiotic broth and clothed itself in the accoutrements that promoted the kinetic stability of such molecules; and some authors continue to hold to that view. However, as pointed out above, both the spontaneous appearance of RNA and its subsequent elaboration have lost much of their appeal. The preferred view today has the RNA World evolving within compartmentalized entities of some kind (lipid vesicles, mineral cavities, or on a surface), and supported by a significant metabolic base.[41] These protocells would have housed a diverse collection of ribozymes, which catalyzed all the reactions required for the replication of RNA, and perhaps for the expression of such "genetic" instructions as RNA came to hold. All other aspects are open to negotiation. Did ribozymes really catalyze energy transduction, biosynthesis of metabolites, or membrane transport? Ribozymes do not mediate such activities today, and it seems implausible that they did so in the past. More likely, metabolism was from the beginning the province of peptides. I shall not pursue these matters here, but must take a brief excursion into the primary historical task of that RNA-dominated era: the invention of proteins encoded by DNA. Proteins make better catalysts than RNA, faster and more specific by orders of magnitude; DNA makes a more stable platform for genes than RNA. How might they have evolved?

The short answer is that the origin of the principles that govern cellular operations today—genes specifying proteins and all the apparatus that this requires—remains quite unknown and points beyond the capacity of present-day biochemistry and biophysics.[42] A longer and more gracious statement takes account of a voluminous speculative literature that puts forward scenarios to generate these things without miracles, by small incremental steps, each advance building upon its predecessor.[43] In the beginning, given RNA, abiotic amino acids may have served as cofactors to early ribozymes, enhancing their activity or stability. Transfer RNAs may have begun as devices to accumulate and deliver amino acids. If peptides are superior to amino acids, selection would have favored catalysts for amino acid ligation. The precursors of the ribosome may have been involved in RNA replication and recruited to protein synthesis: the small subunit might have been a polymerase, the large one could have served as a ratchet (the triplet code, three nucleotides specifying each amino acid, finds a possible basis in this notion). There is more, much more, but it is hard to evaluate in the absence of testable predictions. As Yuri Wolf and Eugene Koonin rightly say, "The origin of the translation system is, arguably, the central and the hardest problem in the study of the origin of life."[44] Here, more than anywhere else on the journey through the murky realm of cell evolution, one can scarcely help wondering whether it will be necessary to step clear outside the box of conventional evolutionary mechanisms.

Once past this great hurdle, the familiar devices of cellular physiology appear to have evolved quickly and with mutual interactions.[45] There is evidence that ribosomes began with a small central core and underwent successive cycles of enlargement. Most of the basic protein families evolved early, apparently beginning with catalysts for metabolism; RNA came later, consistent with the idea that metabolism was from the beginning the province of peptides rather than ribozymes. Lipid membranes evolved in tandem with transport carriers and energy transduction. By the time of LUCA, life had passed beyond the formative stage and assumed its contemporary lineaments.

OF VIRUSES AND DNA.   As a substrate for genetic transactions, DNA is superior to RNA—chemically more stable and informationally less subject to ambiguity. The selective advantages of DNA are plain, but it is not at all obvious when and how DNA took over the genetic functions that (by hypothesis) were initially performed by RNA. There is general agreement that proteins specified by a code of some kind, and also the mechanisms of translation, came first. True enzymes may be required just to reduce ribonucleotides to deoxyribonucleotides, a step that ribozymes seem unable to perform. But contemporary biochemistry offers no indication as to the route by which DNA came to replace RNA, with one possible exception.

As mentioned in chapter 4, the distribution of DNA replication mechanisms found in contemporary cells, and of the genes that specify them, is unlike the pattern familiar from ribosomal proteins and RNAs. Instead of the three canonical versions of a common theme that spawned the conception of three domains diverging from a common ancestor, we have two kinds of DNA polymerase that are not homologous at all: one for the Bacteria and another for Archaea and Eukarya, the latter split into two branches that are but distantly related.[46] This suggests not only that DNA came on the scene later than did RNA and proteins, but that it came from outside the lineage of cellular life altogether. Patrick Forterre has championed the imaginative thesis that DNA and its replicases arose in the world of viruses, perhaps as a device to protect viral RNA genes from the defensive mechanisms devised by their hosts, who relied on RNA to make their own genomes. DNA would have been transferred to the protocellular hosts as a consequence of infection, and that may be how we came to have three domains of cellular life: three parallel and independent episodes of infection by DNA-bearing viruses. Viruses, it now appears, are as old as life itself.[47] Eugene Koonin, Jaana Bamford, and their associates have recently put forward well-articulated schemes that draw contemporary viruses from an Ancient Virus World reaching back to the RNA World, part of the biochemical turmoil at work in the mineral honeycombs deposited by alkaline submarine vents. Viruses were present at the creation, partners as well as plagues, exchanging genes and functions with all the neighbors and helping to shape progenotes and stem-group cells as they evolved.

From the invention of proteins and genes to true cells is still a long road. LUCA, if we read the evidence correctly, was not a primitive precellular assemblage of molecules but a cell or community endowed with many of the capabilities of today's cells: metabolism, energy transduction by ion currents, membranes with transport carriers, a cytoskeleton, cell division, and more. She may have floated free or skulked in a warm hydrothermal vent; her genome may have consisted of RNA or DNA, selfish replicators, or socialized genes. The clues are sparse and hard to read, the trail cold and confused. But with the advent of the trinity of RNA, proteins, and DNA, the darkness does begin to lift. Familiar mechanisms for variation and selection kick in, driving both ever-increasing autonomy from the environment and adaptation to its inevitable vicissitudes. Darwin's door opens, and the rest is cell history.

Enigma

"So far, no theory, no approach, no set of formulas, and no blackboard scheme has been found satisfactory in explaining the origin of life."[48] At the conclusion of a century of science, whose great glory is the discovery of how living things work, there is something downright disgraceful about this confession, an intimation that despite our vast knowledge and clever technology there may be questions that exceed our grasp. But its truth is indisputable. A survey of the literature devoted to the beginnings of life leaves one in no doubt that all the critical questions remain open. We know roughly when life first appeared on earth, but not where or how. There must have been a prebiotic soup of organic substances, but whether it was rich or thin (or even relevant) no one can say for sure. Was the emergence of life probable, even inevitable, or was it a unique event that will seldom (or never) happen elsewhere? Replicators or self-organizing networks, cells early or cells late, RNA first or peptides, and just how genes came to specify proteins by means of a universal symbolic language—theories come and theories go, but the questions remain.

Why has this problem proved so intractable? Some reasons quickly come to mind. Life arose unimaginably long ago in an unknown place and under conditions very different from those on earth today. Material relics of the most remote past have been lost beyond recall, and the experimental method can seldom be applied. The advent of genes and proteins brought down an opaque curtain, erasing all traces of what came before. Scientists are optimists by nature and can call history to witness. This, we say with pardonable self-assurance, is one of the hardest nuts we have ever sought to crack; it will take patience and a little luck, but a breakthrough is sure to come. Well, it may be so. But I cannot suppress the suspicion that the real reasons for our "abject failure," to borrow a phrase from Conway Morris, run deeper and that they have to do with those features that make living things special.[49]

Scientists hold as a central article of faith that all the phenomena of nature are consilient: organically interconnected, such that one grows naturally out of another. The spectacular advances of recent years have reinforced this conviction. Now that we know most of what there is to know about how living things work, we are more confident than ever that they obey all the laws of physics and chemistry. There is no vital force, no natural essence uniquely required to make molecules come to life. And yet, living things are obviously and undeniably special. While obeying the laws of physics and chemistry, life is not deducible from those laws. Nothing in all those weighty tomes would lead one to expect a world smothered in life, from bacteria to redwoods. The singularity of life is enshrined in some of biology's axioms: every cell comes from a preexisting cell, and there is no spontaneous generation, ever. The assertion that life did, long ago, germinate from the inanimate world forces us to confront the contradiction between life as part of the fabric of physics and chemistry, and life set apart. No wonder, then, that the origin of life has emerged as a deeply discomfiting issue, a black hole at the very heart of our science.

Understandably, biologists are reluctant to acknowledge that we face a problem that calls for unconventional, even radical ideas. Why can't we look to Darwin's theory to explain genesis, just as it has demystified the origin of eyes and flowers, and how fishes gave rise to amphibians, and reptiles to mammals? The sticking point is the first appearance of that startling novelty, entities that make themselves and act purposefully on their own behalf. There is no precedent for such a capacity in the inanimate world, no analog, no model. Once rudimentary living systems have come into existence, endowed with some ability to draw matter and energy into themselves, maintain their integrity, and reproduce their own kind, Darwin rides to the rescue. But it is not at all self-evident that the origin of autopoietic systems (or of autonomous agents) can be understood as a consequence of incremental variations sifted by natural selection.[50] On the face of it, their appearance seems to call for a large leap, a quantum jump into a new space of existence, whose nature passes comprehension. At the least, such musings imply that the origin of life may have called for some improbable event, effectively a miracle, that will have happened very seldom even in our enormous universe.

If we must look beyond the interplay of chance and necessity, where in the world of normal science might we find clues to the nature of an abrupt change of biological phase? Among the small number biologists to have confronted this issue head on is Stuart Kauffman, whom we met earlier in this chapter in connection with self-organizing reaction networks.[51] Kauffman rejects the implication of the gene-centered view of evolution, that cells and organisms are contraptions cobbled together by genes that happen to get on well together, and therefore lack intrinsic unity. That view fails to account for the integrity, the wholeness so evident in the operations of every cell. He does not deny the reality and importance of gene variation and selection, but insists that these operate against a background of order that grows spontaneously out of the organizing tendencies of complex systems. Integration and organization are not products of evolution but prerequisites to it; natural selection can only work when it complements that deeper level of order that arises spontaneously. In style as well as substance, Kauffman's writings can raise hackles, and self-organized order is still a far cry from the self-directed, purposeful objects that we instantaneously recognize as being living things, kin to ourselves. All the same, I believe that in "stalking new answers" behind the appearances he is taking first steps in the right direction. Others, no doubt, will follow his lead. The gap in the conventional framework is real and serious, and it is time to cast the net wider.

For the present, we are in limbo. The natural path from simple cosmic molecules to cells, from chemistry to biology, remains undiscovered. A path there must be, of that I am certain, else we would not be here seeking it. But the path is evidently well hidden, probably very narrow, and may require evolving life to have negotiated some tight corners by sheer rare good luck. I have a hunch that there is more, much more, to the origin of cells and of life than current philosophy knows. When—or better, if—we solve that mother of all conundrums it will change our perspective on life, the universe, and all that. But where we should look for illumination I cannot say.

CHAPTER ELEVEN

The Crooked Paths of Cell Evolution

Cell Evolution Is Special

Generalists are superficially mistaken about many things; specialists are profoundly mistaken about a few. — Walter Russell Mead, *Power, Terror, Peace and War*

Genes Hold the Limelight
Selected for Advantage?
Cell Evolution—Significantly Different
Those Evolvable Eukaryotes
Small Steps and Long Leaps
Do Variation and Selection Explain the Origin of Cells?

That the living world evolved by incremental steps over a very long period of time is not controversial—at least not in scientific circles— nor is the assertion that the underlying mechanism is the interplay of heredity, variation, and natural selection. Doubts, however, do persist over whether standard genetic mechanisms account for the entire spectrum of heredity and variation. The common understanding of evolution comes largely from studies on multicellular creatures: animals, plants, and fungi. Microorganisms, bacteria in particular, played little or no role in the formulation of the modern synthesis, and they remain somewhat marginal to this day. The question therefore arises whether anything more is required to explain the emergence and diversification of microbial life, particularly the origin of cells in the first place. Given that cells are a prerequisite for evolution by heredity with variation, and that for three-quarters of life's history all of life was microbial, it should not come as a surprise that our conception of evolution has been significantly skewed by the neglect of its earliest phases.

Incorporation of the microbial world into the framework of evolution has shaken the foundations of the modern synthesis.[1] Lateral gene transfer, rampant among prokaryotes, undermines the premise that organisms descend with modification from their parents, and also the claim that change must come slowly and gradually. The metaphor of the tree of life had to be replaced, particularly as applied to prokaryotes, with a web or net of interdigitating lineages. The inheritance of acquired characters, as exemplified by endosymbioisis, plays a major role in the history of life, especially in the genesis of eukaryotic cells. Mechanisms that explain the differences between species do not necessarily account for the origin of kingdoms. Variation is not always random, and even maladaptive changes can sometimes become fixed. And the admission of viruses into the biological universe seriously rattles our conceptual cage. Having said all that, it is important to keep the new developments in perspective.

For the past three decades, scholars of evolution have felt pinched by the modern synthesis and have called for its expansion.[2] So long as the history of life can be encapsulated as a tale of heredity, variation, and natural selection, microbial evolution nestles quite comfortably within a more flexible and generous edifice. In particular, the renovated doctrine retains the central role of genetic variation and selection as evolution's chief driver. The preeminence of genes extends to the crafting of life's essential devices, molecular machines such as ribosomes, ATPases, the cytoskeleton, and transport systems, all of which depend on proteins to carry out their functions and must have arisen subsequent to the invention of the basic mechanisms of gene variation, replication, and expression. To be sure, those mechanisms must themselves be products of evolution, and their origins reach back to a time before genes; but traces of that prehistory are too faint to read. Given genes and a substantial expansion of the conceptual framework, cell evolution requires no novel physical or biological principles, let alone purposeful design or intervention by a supernatural agency.

The state of affairs is much less satisfactory when we turn from the evolution of microbes, endowed with all the essentials of cellular life, to the origin of that organization itself. The evidence collected in the preceding chapters leaves one with the clear impression that the greatest of all the transitions in evolution was the advent of genes and of "unlimited heredity," the capacity to transmit an indefinitely large number of messages.[3] That requires replication by means of complementary templates, as in DNA, but it also requires mechanisms that link the information encoded in the sequence of bases to a cellular function, performed (as a rule) by a protein. Genes can only function in the context of cells, and evolution before there were proper cells must have been very different from the process we know.

Genes Hold the Limelight

The genome is not the cell's command center but a highly privileged data-bank, something like a recipe or a musical score (see chapter 1), yet for the purpose of parsing evolution, genes have a rightful claim to center stage. Genes are the primary carriers of heredity, passed from one generation to the next with extreme fidelity moderated by variation and rearrangement. Genes specify the primary structure (and at one remove, the function) of the proteins and RNAs, which together do most of the work that cells do. Most cellular traits are determined, directly or more often indirectly, either by a single gene or by many acting in concert; so it is genes acting through the products they encode that are primarily responsible for the phenotype upon which selection acts. The connections between genotype and phenotype are often cryptic and devious, and sometimes we do not know just what is in the black box. Besides, not all that is inherited is ge-netic and digital: certain chemical modifications of genes, and also some cellular structures, pass from one generation to the next by mechanisms that fall outside the standard rules, and the contribution of such "epige-netic" inheritance to evolution has yet to be worked out. All the same, stu-dents of evolution justifiably focus attention first and foremost on genes: their transmission and reassortment, the retooling of old genes, and the generation of new ones.

The sequence of bases in DNA can undergo manifold substitutions and rearrangements, all of which can potentially contribute to genetic novelty, but two appear to be particularly important to evolution.[4] One is the ef-fect of transposons, mobile genetic elements related to viruses that insert themselves at random locations in a genome and are replicated in tandem with it. When insertion happens to target a gene, the product may be al-tered or not made at all. It is generally believed that the majority of muta-tions of evolutionary significance are caused by transposons rather than local substitutions or deletions. The other prominent source of variation is the propensity of DNA sequences to undergo duplication, thanks to slip-page in the course of replication. Duplication may be restricted to a single gene or affect a string of them, even entire genomes. This is probably the chief source of new genes, required in the genesis of greater complexity: one copy of the duplicated gene carries out its original function while the other is free to vary and perhaps assume new duties. Keep in mind that an organism's phenotype is commonly the output not of a single gene but of a network of genes. Populations therefore hold more variation than once thought, offer more opportunity for change, and evolve more quickly than expected.

Why should you believe that the raw material of evolutionary history, which is ultimately displayed in the modification of organismal structures and functions, takes the form of accidental, undirected variations in the gene complement? Textbooks lay out a persuasive argument, buttressed by a massive body of evidence. The trouble is that evolution transpires over many generations. It can be inferred but seldom can be observed in action. This is where microbiologists get a chance to shine. A growing body of evidence from bacteria, yeast, and viruses whose generation times are measured in minutes provides solid support for the thesis that changes at the gene level are the grist for evolution's mill.[5]

A perfect case in point comes from the long-term experiments on *E. coli* evolution conducted in the laboratory of Richard Lenski.[6] *E. coli* does not normally use citrate under oxygenic conditions for lack of a suitable transport carrier, and mutants that have gained this capacity are exceedingly rare. Cells grown on a medium containing both a limiting amount of glucose and ample citrate were transferred to fresh medium daily for nearly twenty years—more than thirty thousand generations—before a citrate-utilizing mutant appeared in a single subculture. Since the evolving population had been sampled periodically, it was possible to reconstruct the chain of events that produced the successful mutant. Not only is that mutation uncommon, it had to be preceded by at least two prior potentiating mutations. All of these are contingent events that would probably never happen again in the same manner. Despite the great selective advantage conferred by the capacity to utilize citrate, its eventual acquisition was the consequence of several chance events.

Complex novelties that require multiple genetic changes, such as the generation of new cellular structures, are much harder to get at. The form and organization of microbial cells is quite strictly inherited and has always been taken to be genetically specified and subject to natural selection. Now we saw in chapter 6 that genes do not specify the higher levels of order; cells perpetuate their form and organization by building upon the preexisting structure. That architecture is composed of elements that the genes do specify—proteins in particular—and it is in this cryptic and devious fashion that cell forms and functions come under the genes' writ. The evidence comes from a large and diverse class of mutants that display changes in shape, thanks to alterations in the system that generates the form—the cytoskeleton, for example, or (still more circuitously) the cell wall. One can imagine new forms arising from rare advantageous mutations, reinforced by subsequent genetic changes that lock in the initial change in morphology. However, to the best of my knowledge, no instance has ever been documented at the molecular level; nor has any novel permanent structure or form ever arisen in the laboratory.

Of the many kinds of heritable variation, the ones we know best occur in DNA sequences that specify a useful function—a protein, perhaps, or a regulatory locus. So is this all of the story, or can heritable and selectable variations arise at other levels of biological organization? The answer is unclear and subject to much discussion under the heading of epigenetics.[7] One pertinent level is the chemical modification of genes, typically by the methylation of bases. Methylation does not alter the nucleotide sequence and is therefore not a mutation, but it is commonly passed from one generation to the next and alters the expression of the methylated gene. It is changes of this sort which ensure that, during the development of an embryo, some cells make bone and others brain, even though both contain the same genome. Do epigenetic changes contribute to the pool of variations that natural selection screens? The key to the answer is stability. Epigenetic changes tend to be easily reversed and are not well suited to long-term propagation. But one can imagine ways by which an advantageous epigenetic change may be converted into a genomic one, and thereby come under selection's scrutiny.

Selected for Advantage?

Heredity with variation and natural selection are the yin and yang of evolution: both are required, neither is sufficient by itself. Any population of entities subject to heritable variation will change over time, variations that promote reproduction will tend to be favored, and the population will become adapted to its environment. It's simple, comprehensible, and compatible with an enormous range of outcomes. Depending on the circumstances, which may include both the external and the internal environment, selection can favor a trend toward more complex entities or simplified ones, the gain of a function or its loss, changes in a phenotype or its stabilization, the elimination of species or their proliferation in endless variety. The great virtue of natural selection as an explanatory principle is also its chief defect: it can explain almost anything as long as the details are unknown, and they usually are. *Caveat emptor.*

Organisms from bacteria to whales make up the bulk of the biological world and are conventionally taken to be the objects upon which natural selection acts. Let us note in passing that organisms are not the only such objects. Viruses, peripheral components of the biological universe, vary and evolve all too quickly; in the laboratory, replicating molecules like RNA become targets of artificial selection, and some scientists maintain that species make up a level of selection over and above that of their constituent organisms. For purposes of the present inquiry into the origins of cellular organization and diversity, we can set these distractions to one side. But we cannot sidestep the question whether it is indeed organisms that are the objects of selection or their genes, as Richard Dawkins has insisted. The issue is subtle, and turns on the distinction between selection *of* versus selection *for*.[8] Selection does not normally "see" genes; it judges phenotypes (plasmids and some molecular parasites could be considered exceptions). Organisms are the entities that reproduce and compete, and natural selection assesses their "fitness," a measure of the capacity to leave offspring that are themselves capable of reproduction. Organisms must therefore be the objects upon which selection acts. However, to the extent that an organism's phenotype is the expression of its genotype, selection of a particular phenotype will favor reproduction of the genes that produced it and consequently increase their frequency in the population. Evolution is the consequence of selection *of* organisms *for* genomes that enhance their prospects of survival and reproduction.

Students of molecular evolution are usually safe to assume that two molecules with similar sequences descended from a common ancestor. Sequence space is so vast that a significant overlap between two molecules can seldom be blamed on convergence of separate lines of descent. No one doubts that, despite the differences among them, the ribosomal RNAs of Archaea, Bacteria, and Eukarya shared a common ancestry in the remote past. Cellular physiologists, by contrast, must keep in mind the reality of convergent evolution, which plays a large role in the history of the "higher" multicellular organisms and probably also in cell evolution.[9] Coccoid bacteria, for example, do not constitute a single line of descent; ancestral Bacteria were probably rod-shaped, and spherical forms arose on multiple occasions by loss of the apparatus that shapes the cylinder. Convergence also seems to be at the root of various aspects of cytoplasmic organization in protozoan cells.[10] Adaptations solve problems of existence, and the number of effective solutions is commonly quite limited. Convergent evolution demonstrates the power of variation and selection to explore all options and home in on the few that work well.

Anyone who looks upon the living world cannot fail to be impressed by its sheer variety; one of the persistent questions in biology is why there are so many kinds of organisms, or, for that matter, why organisms cluster into discrete species that often endure for millions of years. The conventional reply is that every species is adapted to its particular habitat, and the more complex an ecosystem becomes, the more niches will arise in it. This is particularly true of microorganisms: They can exploit tiny local differences in temperature or pH, swim or drift up and down to track changes in the level of light or oxygen, and find nooks and crannies among and within other organisms. This view implies that all—or at least most—features of organisms have been honed by selection; all favor survival or reproductive success and are by definition adaptive. Is it true that all traits of form and function represent adaptations, or could some be the result of factors and forces, including mere happenstance, that have nothing to do with selection for advantage?

This issue arose early in the context of protein evolution. Related proteins, such as the hemoglobin family, diverged over time thanks to mutations. As a result, amino acid sequences of hemoglobins from closely related organisms are nearly identical, while those from more distant species show increasing differences. Does each such alteration of sequence improve the performance of the molecule? Probably not. Most mutations are deleterious, either because they destabilize the protein's structure or impair its function and will be quickly eliminated by natural selection. Only those that improve the catalytic process or the stability of the protein as a whole will be positively favored. The great majority of mutations that show up in sequences appear to have little effect on either structure or function and are selectively neutral. The requirement that evolving proteins must maintain both physical stability and biological function at all times means that evolution treads a narrow path and change comes slowly and incrementally. After more than three billion years of variation and selection, ancient proteins preserve in their sequences clear traces of their ancestry, and only part of the global sequence space has been explored.

The same blending of features that are obviously useful, with others that seem immaterial, also applies to cell structure and physiology. Consider, for example, Bacterial flagella. In principle, organs that confer motility are surely valuable assets that confer selective advantage in most circumstances and support diverse lifestyles. It cannot be an accident that they are so widespread, and are in principle of "one kind" (i.e., homologous in general structure, function, and component proteins). But when examined in detail, flagella from different organisms differ with respect to some constituents and modes of operation (see chapter 6). It is presumably an instance of adaptation that in Spirochaetes flagella do not protrude into the medium but are tucked into the periplasmic space; their rotation causes the cell to move in a corkscrew fashion, which is thought to be advantageous when burrowing through soft mud. All right, but what does it mean that the flagella of *E. coli* randomly switch the direction of rotation, causing the cell to tumble, whereas those of *Rhodobacter sphaeroides* make sudden stops so that the cells halt in place? Is one mode superior to the other in the organism's niche, or is the choice a matter of chance?

By the same token, one wonders about the shapes of microbial cells. Are they wholly crafted by selection? Bacterial shapes are inherited from one generation to the next, well conserved over time, and often have discernible virtues.[11] For example, the distinctive form of *Caulobacter crescentus*, with its crescent-shaped cell body and long stalk, seems to be a clear case of form following function. The curved cell body may help resist predation by grazing protozoa, while the stalk absorbs nutrients and its tip serves as a holdfast. Whether the smaller differences in size and shape that sometimes distinguish one species from another likewise confer selective benefit is far less clear. The impression, that microbial forms reflect a combination of adaptation and "playfulness," comes to the fore when one muses on the baroque and bizarre architecture of many protists, notably the radiolaria. It is hard to believe that each and every one of their variations carries benefits in a particular habitat. More likely, the rules that govern the formation of these cells permit significant latitude so that many details of morphology and organization are products of neutral evolution and carry no selective benefit whatsoever.[12]

We are here looking at a very general principle, which Daniel McShea and Robert Brandon designate "biology's first law," by analogy with Newton's law of inertia: "In any evolutionary system in which there is variation and heredity, there is a tendency for diversity and complexity to increase, one that is always present but may be opposed or augmented by natural selection, other forces, or constraints acting on diversity or complexity."[13] The reason is simply that variation happens in biological systems, when heritable it accumulates, and in consequence organisms and their parts tend to become increasingly different. It bears repeating that this is a tendency, which in any particular set of circumstances may be suppressed or reinforced by other forces. The point is that the tendency persists: the wind still blows, and the results may crop up unexpectedly.

The mechanisms that effect variation differ from one level of the biological hierarchy to another. Thus, geneticists note that neutral or slightly deleterious mutations can become fixed by chance, particularly when population size is relatively small, and therefore increases in diversity and complexity do not necessarily bespeak adaptation.[14] This sort of undirected variation helps account for the observation that molecular machines often have many more parts than their function demands (eukaryotic ribosomes and spliceosomes, for example). Such gratuitous complexity may be the result of a ratchet-like process that tends to turn fortuitous molecular interactions into permanent mutual dependence even in the absence of positive selection.

Amidst the muddle of pressures, keep an eye on the compass. Undirected variation tends to increase diversity and complexity and supplies an abundance of raw materials that are potentially useful, but it does not explain the origin of adaptive functional structures such as ribosomes or ATPases. That remains the domain of natural selection, working against the prevailing wind. One lesson is that, whatever principles guide life's search for solutions, simplicity and rational design are not among them.

Cell Evolution—Significantly Different

Evolution in unicellular organisms, as in multicellular ones, revolves around heredity, variation, and adaptation. The fundamentals of genetics and natural selection are also much the same. But there are conspicuous differences with respect to the cellular mechanisms that mediate the acquisition and distribution of novelty. These center on lateral gene transfer, symbiosis, and various aspects of cell heredity; they strongly support the case for expansion of evolution's framework[15] and lend the field of cell evolution a distinctive tone.

LATERAL GENE TRANSFER.   In higher organisms, genes are almost always transmitted vertically from parents to offspring; exceptions have been recognized but are very uncommon. Perhaps the most conspicuous difference between microbial evolution and that of metazoa is the prevalence of lateral gene transfer in the former, particularly among prokaryotes but to a lesser degree also among unicellular eukaryotes (see chapter 2). As a rule of thumb, one may say that prokaryotes evolve by gene acquisition, eukaryotes by gene duplication and the expansion of gene families.

Lateral (or horizontal) gene transfer occurs in prokaryotic populations today, though it is hard to make out just how frequent it is. My sense is that it is common, especially among close relatives—frequent enough to erode lines of descent but not to erase them. Things may have been different in the early stages of cell evolution, prior to the emergence of the three domains, when lateral gene transfer may have been rampant (see chapter 4). Stem-group cells made up a community that freely exchanged genes and could therefore not generate discrete lineages. If that is true, early cell evolution is better described as a web or net, rather than a tree. In Woese's evocative imagery, the emergence of today's cellular domains from that fluid free-trade zone represents a phase change akin to crystallization, which occurred once protocells had attained a grade of organization such that genes imported from outside could no longer be easily assimilated. Only then could stable cell lineages emerge, and commence the mode of evolution with which we are familiar.

Is unbridled lateral gene transfer complementary to Darwinian evolution by variation and natural selection, or might it be the basis for an altogether different mode, as Woese and others have recently implied?[16] Lateral gene transfer can certainly supplement vertical inheritance and thus greatly enlarge the supply of selectable variations; the question is whether lateral gene transfer by itself can supply a basis for adaptive change. In my judgment, the answer must be no. Improvements in function require natural selection at the level of the organism as a whole. An imported gene that confers some benefit upon the recipient will become naturalized only if it persists long enough to be passed on to the next generation. Without vertical inheritance to stabilize the proceedings, lateral gene transfer alone provides no wellspring for adaptation or for increasing sophistication.[17] Therefore, vertical inheritance must have been the pivot of evolution at all times, enriched by other mechanisms but never supplanted.

CAN GENOMES REWRITE THEMSELVES?   The history of life revolves around adaptations, novel and useful features such as flagella, undulipodia, cell walls, and mitochondria. How do they arise? The standard reply invokes a static genome composed of discrete genes that are subject to variation by mutation and other random events. The variations are then culled by natural selection, and in consequence adaptation of the organism to its environment improves. Skeptics have questioned this scenario from the beginning, arguing that there never was time enough to bring forth the profusion of biological innovation, and in any event random mutations are more likely to degrade functions than to enhance them. A provocative recent book by James Shapiro spells out the thesis and offers an alternative vision based on contemporary knowledge of how genomes work.[18]

Briefly, the architecture of genomes turned out quite unlike what had been expected. In eukaryotes, individual genes are composed of both coding and noncoding portions, and transcripts must be spliced prior to translation. In metazoa, classical protein-coding sequences make up an astonishingly small proportion of the total, sometimes just a few percent. The remainder features regions that specify an intricate network of regulatory elements, many based on RNA rather than proteins. These are concerned, not only with the expression of individual genes, but with the fidelity and structure of the genome as a whole. Genomes also contain numerous stretches, long and short, of repetitious DNA whose functions are not obvious. Junk DNA? Not necessarily. Some segments can translocate from one locus to another (or, at least, they did so in the past), with the help of enzymes specified by the genome itself, and thereby reconfigure the instructions laid out in the recipe. The most conspicuous agents of change are the transposons, "jumping genes" that migrate from one place to another spontaneously or in response to stress. Transposons disrupt genes, but they often carry along genes or fragments of genes, which are thereby transferred from one neighborhood to another. Bacterial genomes are more streamlined than those of animals, but no more static. They are buffeted by a ceaseless rain of foreign genetic material, picked up from the environment or carried by viruses. Cells are forever restructuring their genomes, which are much more dynamic than we thought, more malleable and interactive, and correspondingly more evolvable.

One consequence of this continuous makeover is to supply more variations for selection to winnow, but Shapiro argues that there is much more to it than that. "Natural genetic engineering," as he terms it, generates different kinds of variations from those produced by classical mutations, one gene at a time. Rearrangements can take place at multiple locations at once and shuffle entire domains from one protein to another, producing novel combinations quickly and abruptly, perhaps even purposefully. Genomes, it seems, are built to evolve—not at the petty pace of classical genetics, but by leaps that entail rearrangement of the genetic architecture or the wholesale import of information from abroad. So could this be the way that organisms generate the multiple, coherent variations that seem required to manufacture complex organelles such as eyes or flagella? At the end of the day, that is a question that must be answered by experiment, and the proof is not yet there. But if it turns out that genomes can respond to environmental challenges by actively and productively rewriting their own content, then the process of evolution differs considerably from what the textbooks teach.

I have some reservations concerning this thesis. The malleability, even fluidity, of genomes seems to be established, and many of the underlying mechanisms are well understood. The question is whether those facts mandate a major shift in the evolutionary paradigm. First, we have already noted that conventional mutation and selection suffice to generate more complex functions, at least in bacteria. Second, the genomic modifications introduced by transposons and other agents are still "random," in the sense that their consequences are not directed toward the solution of specific needs or challenges; it is natural selection, not the variation, that supplies the "purpose." However, if future research uncovers such unanticipated capacities, James Shapiro will have earned the right to crow, "I told you so."

ENDOSYMBIOSIS.   Endosymbiosis is a prime example of genomic reconfiguration; its evolutionary significance in contemporary organisms is well understood, and it appears to have played an even grander role in the distant past. Symbiosis brings together not merely genes from disparate sources but unrelated organisms in their entirety. It sometimes generates novel organisms capable of exploiting niches not available to the partners living independently, such as lichens (products of symbiosis between a fungus and a unicellular alga) that thrive on bare rock. Many protists house symbionts on their surface or in the cytoplasm and could not do without them. But the role of symbiosis in cell evolution rises to a higher level, for there is good reason to believe that symbiosis was critical to the genesis of eukaryotic cells, and to the consequent ascent of biological complexity.

The story was reviewed in some detail in chapter 8. Briefly, one of the few certainties about cell evolution is that eukaryotic cells are products of endosymbiosis: mitochondria are descendants of proteobacteria that formed an intimate association with a host cell whose nature remains under debate. Since all known eukaryotic cells either harbor mitochondria or have done so in the past, it seems very likely that formation of this part-nership played a critical role in the genesis of eukaryotic organization; we shall return to this claim in the following segment. Another gift of endo-symbiosis is eukaryotic photosynthesis. We know that plastids derive from photosynthetic cyanobacteria, which took up residence in the cytoplasm of eukaryotic protistan cells. Subsequently, other heterotrophic protists gained access to this plentiful source of energy by secondary endosym-biosis, having captured a primary photosynthetic cell in its entirety and turned it into a plastid. Photosynthesis then spread widely across the eu-karyotic spectrum, testifying to the power of dissemination by symbiosis.

Symbiosis has become quite fashionable in recent years, invoked as a plausible explanation for the occurrence of seemingly disparate traits in a single cellular frame. The entity that hosted the mitochondrial precursor is a favorite candidate: it has been variously identified as a protoeukaryote, an archaebacterium, or something in between (see chapter 7), but some envisage it as the product of earlier mergers among prokaryotes.[19] Viruses may have contributed genes, possibly even the nucleus. And diderm pro-karyotes, cells with two exterior bilayer membranes (traditionally referred to as gram-negative), may (or may not) harbor traces of an early fusion between an ancient actinobacterium and an ancient clostridium.[20] In as-sessing the credibility of such proposals, keep in mind that while contem-porary eukaryotes commonly house endosymbionts, prokaryotes very seldom do. Only a few cases have been reported—just enough to sow doubt. The rule seems to be that symbiosis of one kind or another is not uncommon, but it seldom progresses to the point of fusion or merger of the partners. The rare occasions when formerly independent cell lines did merge identities mark major turning points in cell history, when new pat-terns of organization make their debut.

THE CONTINUITY OF CELLULAR ARCHITECTURE.   Every cell that has ever existed has been continuous with a previous cell, not only genetically but also structurally. The structural element is inconspicuous (albeit real) in the case of multicellular organisms, which tower over the gametes with which they began, but grows increasingly prominent as we delve into the world of unicellulars, both prokaryotic and eukaryotic. Chapter 1 sum-marized our understanding of the manner in which growing cells construct the architectural framework of the daughter upon that of the mother, by extension of elements including the plasma membrane, cell wall, and cytoskeleton (in eukaryotic cells the list includes the endoplasmic reticulum, centrioles, and chromosomes). A byproduct of architectural continuity is the phenomenon commonly known as structural heredity, the transmission of information that specifies location or orientation in cell space without the participation of genes.[21] The most spectacular example of the long-term perpetuation of structure is the persistence of biological membranes (chapter 6).

The question here is in what manner does architectural continuity contribute to cell evolution? Inheritance of structural changes unmediated by genes has been reported many times.[22] The most familiar examples come from ciliates, which construct the ciliary apparatus of a daughter cell upon that of its mother; surgically induced alterations in the orientation of cilia are transmitted from one generation to the next, even though no genetic change has taken place. Structural heredity of this sort is relatively labile and easily reversed; but if a change proves advantageous it can presumably be stabilized and reinforced by mutations in pertinent genes, just as envisaged by Conrad Waddington sixty years ago in his concept of "genetic assimilation." No doubt this happens, though I am not aware of any documented example. I suspect, however, that the real significance of architectural continuity lies elsewhere: it is not an alternative source of variations, but the stable core of cell organization whose transmission makes heredity, variation, and selection possible. A cell is not a heap of independent parts but an organization of interacting elements that operates and reproduces as a unit. Even if it were possible in theory to assemble it from the individual molecules, it is surely much more reliable and efficient to construct every cell by extension of the framework of its progenitor: by growth rather than self-assembly.

Seen from this vantage point, we can better appreciate the meaning of membrane heredity. Membranes are key players in the perpetuation of cell organization, far more so than genes (see chapter 6). The loss and acquisition of novel membranes are rare events in cell history, often associated with endosymbiosis. They signal the creation of new cellular compartments and commonly proclaim the birth of a new taxon. That most revolutionary event in cell evolution, the genesis of eukaryotic cells, is rooted in such an episode.

PRIONS. The annals of science hold no more romantic tale than the discovery of prions, infectious proteins that constitute a mode of cellular heredity quite independent of genes. It began with a fatal plague that afflicted tribal people living deep in the mountains of New Guinea. The disease was traced to the ritual consumption of human brains, which carried the causal agent from one victim to the next. The agent was first thought of as a "slow virus," but when isolated it turned out to contain no detectable nucleic acid whatsoever. After extensive investigation and furious debate, there is now general agreement that prions are a class of proteins that can fold into more than one configuration. In its normal state, the protein serves a physiological function; in the alternative fibrous (or "amyloid") configuration, it propagates the misfolded condition by associating with other molecules of the same protein and inducing them to adopt the misfolded configuration. Disease results from the accumulation of amyloid aggregates in the brain and other tissues. Note that prions are not self-replicating proteins; they are proper host-encoded proteins that undergo misfolding and act as templates that induce other protein molecules to adopt the same misfold.

A curious twist, to be sure, but what has it to do with cell evolution? It turned out that proteins that transmit their state of folding to other proteins are not uncommon and do not necessarily cause disease.[23] In unicellular organisms such as yeast, prions may be quite advantageous in helping to overcome environmental stress. A case in point is Sup35, a yeast protein that normally helps to terminate the translation of mRNA by the ribosomes. When Sup35 switches to its prion state, its capacity to terminate translation is diminished; readout continues with effects on the expression of many genes. This in turn alters characteristics such as cell adhesion, nutrient uptake, and sensitivity to various toxins, any of which may promote survival in particular circumstances. Prions of this kind are common in the wild; they enhance the plasticity of the genome and generate alternative phenotypes for selection to evaluate. Since the prion state is self-propagating, successful phenotypes are directly inherited; they represent a kind of Lamarckian inheritance, the transmission of an acquired character from one generation to the next without the participation of genes.

Do prions play a role in cell evolution? In yeast, at least, many of the known prions are regulatory proteins that control the expression of numerous genes at once. Consequently, a shift to the misfolded state can induce the sudden appearance of complex new traits. The propensity of proteins to misfold and adopt new configurations with altered functions is probably as old as protein-based life itself. Indeed, Halfmann and Lindquist suggest that transmissible folding states may have been more common in ancient cells whose proteins were shorter and simpler than those of contemporary cells.[24] Whether prions really made a significant contribution to cell evolution is quite unknown, but there is no doubt that they offer an altogether novel perspective on epigenetic variation and selection; we must keep them in mind when we reflect on just how cells evolve.

Those Evolvable Eukaryotes

The most conspicuous feature in the landscape of cell evolution is the profound rift that separates the eukaryotes from prokaryotes. This does not stand out at the level of ribosomal RNA or molecular biology in general, where the commonalities catch the eye, but is obvious to anyone concerned with morphology, physiology, and evolutionary potential. Prokaryotes are biochemically inventive and have found access to all the practicable energy sources our planet offers. It's a good rule of thumb that if a chemical reaction yields enough energy to support life, a prokaryote exists that exploits it. But when judged by their forms and organization, prokaryotes have advanced very little beyond their fossil progenitors of two to three billion years ago. Some, it is true, have attained structural and behavioral complexity beyond the norm: cyanobacteria and planctomycetes with their internal membranes come to mind, and so do the myxobacteria with their wolf-pack hunting habit and their elaborate fruiting bodies. Still, these pale in comparison with even the simplest eukaryotic protist. It almost seems as though prokaryotes made repeated starts up the ladder of complexity but always fell short. By contrast, eukaryotes, despite their meager metabolic repertoire, burst whatever constraints hamper prokaryotes to experiment with the opportunities afforded by larger cell size and more elaborate organization. Just what is it that makes the eukaryotic pattern of order so much more "evolvable" than the prokaryotic one?

Evolvability is an abstract term that refers to "the capacity to generate heritable, selectable phenotypic variation" and sometimes to the propensity to evolve novel structures.[25] Marc Kirschner and John Gerhart credit the conspicuous evolvability of metazoa to the flexibility and looseness of the developmental processes that mediate between genotype and phenotype. Loose reins also rank among the significant differences between prokaryotic cells and eukaryotic ones: as a rule the latter have many more genes and much extra DNA, more tolerant regulation of gene expression, a more fluid physiology and many more moving parts. These are surely important differences, but they look like consequences of the different evolutionary trajectories pursued by prokaryotes and eukaryotes rather than an explanation thereof. Nor does it seem plausible to attribute the difference to some singular and critical invention, such as nuclear membranes or linear chromosomes. Though characteristic of eukaryotic cells, both have been noted in some prokaryotes as well; even endocytosis has recently been documented in planctomycetes (see chapter 3).

A more convincing argument traces the gulf that divides prokaryotes from eukaryotes to their distinct ways of organizing the production of useful energy. In prokaryotes, energy generation is a function of the plasma membrane, whereas eukaryotes assign it to specialized compartments, the mitochondria and plastids. This greatly expands the membrane surface available for energy transduction by ion currents while freeing the plasma membrane for other tasks. However, as Nick Lane and William Martin explain in a stimulating paper, this is just the beginning of the story, for that anatomical difference has unexpected evolutionary implications.[26] In prokaryotes, the energy generated by each individual cell is used to support the output of its own genome; about three-quarters of that energy is required just to express the genomic information by way of protein synthesis. By contrast, in eukaryotic cells the energy produced by hundreds or even thousands of mitochondria, each one a prokaryotic power pack, is put at the service of a single central genome. In consequence, a eukaryotic genome governs far more energy than a prokaryotic one. This allowed eukaryotic genomes to expand, thus laying the foundations for the spectacular difference in cellular complexity. Lane and Martin argue that the spectacular rise of the eukaryotes was due to their acquisition of mitochondria at a very early stage in their evolution; and furthermore, that the only way structurally and functionally complex cells could make an appearance was by enslaving symbionts.

Let a few numbers taken from Lane and Martin bolster the thesis. When expressed on the basis of mass, the mean metabolic rates of aerobic bacteria and protozoa are not very different: 0.19 and 0.06 W/g, respectively (one watt corresponds to one joule per second). But their cellular masses are very different indeed, typically 2.6×10^{-12} g for the bacterium and $40,000 \times 10^{-12}$ g for the protozoan. In consequence, a protozoan cell has much more power at its command than a bacterial cell: 2,300 pw (picowatts) per cell, compared to 0.49 pw (a picowatt is equal to 10^{-12} watts). Eukaryotes made use of that surplus energy to expand their genomes by orders of magnitude: the mean haploid DNA content is only 6 megabases for a prokaryotic cell, compared to 3,000 megabases for a protozoan (mean gene numbers are 5,000 and 20,000, respectively). Even so, the energy available per gene is far greater for eukaryotes than for prokaryotes, 57,000 versus 30 pw per gene, respectively (other modes of calculation show even larger disparities). The mitochondria that power a eukaryotic cell still harbor a minimal genome, a remnant of the genome that came in with the ancestral endosymbiont; most of those genes were either lost or transferred to the host's nucleus. Taking organellar genomes into account makes little difference to the estimate of the power available per eukaryotic gene. Evidently, eukaryotic cells have energy to burn that is more than sufficient to evolve many novel protein families, explore sophisticated ways to regulate their production, and put them to work in elaborate physiological processes that would be far beyond the means of a prokaryotic cell.

Lane and Martin draw an instructive comparison between the energy budgets of a protozoan cell and a giant prokaryote, such as *Epulopiscium fishelsoni*, comparable in size to *Paramecium*. Every cell of *Epulopiscium* is a consortium of some 200,000 fully fledged nuclei, each reliant upon a share of the energy generated by the communal plasma membrane. Reproduction requires the cell to duplicate 760,000 megabases of DNA, compared to 6,000 megabases for a protozoan cell of similar size. *Epulopiscium* obviously generates sufficient energy to reproduce but would have none to spare on a quest for structural and functional elaboration.

So could one imagine some way for prokaryotes to compartmentalize their energy production in small pods and thereby augment the genome's power supply without going to all the bother of domesticating endosymbionts? Yes, one could, but bacteria have not followed that path, and Lane and Martin argue that in reality no such option was available. The reason is that the small genomes retained by both mitochondria and plastids perform an essential function. Maintaining a steady flux of electrons through the respiratory (and photosynthetic) redox chains requires the cell to make adjustments and repairs as needed, and that can only be done under the control of a genome localized to the particular organelle[27] (see also chapter 8). That kind of architecture can only be achieved by starting with endosymbionts enslaved and put to work as components of a larger community. Contrary to the conventional view, which has mitochondria coming late into a phagocytic cell already well on its way to eukaryotic organization, Lane and Martin would have the precursors of mitochondria present from the earliest stages and see them as a prerequisite to the rise of complexity. I shall reserve judgment as to whether the host was a standard-issue prokaryotic archaeon or already something rather different (see chapter 7). But the thesis that mitochondria were required for the evolution of the eukaryotic order has, to my ears, the ring of truth.

Let me not leave those evolvable eukaryotes without noting that the role of energetics in evolution is permissive, not prescriptive. A generous supply of useful energy made great things possible but mandated none of them. We are still obliged to specify the features of eukaryotic organization that natural selection favored, and the molecular inventions that marked the way. That task has not yet been fully accomplished, but the argument that predation based on phagocytosis is the key to the rise of the eukaryotes seems to me sound. Prokaryotes, by contrast, are constrained by limitations on the energy available to the genome, which bent their evolution into quite different channels. In their case, natural selection favored small and spare cells with streamlined genomes, little superfluous DNA, and tightly disciplined regulation of gene expression. Genes not required now are apt to be lost, but can be re-acquired by lateral gene transfer whenever they become useful. Quite unfazed by what we perceive as limitations, and marching under the banner "Small is Beautiful," prokaryotes flourished, multiplied, and kept most of the earth. In many ways, it's still a prokaryotic world.

Small Steps and Long Leaps

Geology is famously a matter of petty and mundane events continuing over long periods of time, punctuated by the occasional catastrophe. The same might be said of evolution; Darwin's emphasis on slow, gradual changes continuing over many generations surely owes a debt to Charles Lyell's doctrine of uniformitarianism. But in recent years evolutionists, too, have had to make room for rare, unique, or episodic events. Cell evolution in particular is sufficiently distant in time and circumstances that abrupt departures from the norm may have played a disproportionate role.

Forty years ago, Niles Eldredge and the late Stephen Jay Gould observed that the pattern of evolution as displayed in the fossil record does not conform to the expectation of gradual change, but is distinctly episodic. Species appear abruptly (by geological standards), endure for millions of years with little or no change, and then go extinct. Since an inch of sedimentary rock may represent a hundred thousand years, such "punctuated equilibrium" does not contradict the received mechanism of evolution, but it does indicate that the rate of evolution is not constant, and that speciation often corresponds to a period of rapid change.

The most familiar episode of instant innovation is the Cambrian explosion: the sudden appearance of animal body plans, seemingly out of nowhere and in the twinkling of a (geologist's) eye. This radiation of animal forms is captured in the wonderful fauna of the Burgess Shale of British Columbia, recently augmented by similar finds from China, which are currently dated to the very early Cambrian, 520 to 505 million years ago. What triggered their emergence remains uncertain. It may have been the realization of some critical invention (calcified skeletons, perhaps) or a permissive environmental change. Debate also continues as to whether the event really was as rapid as the term "explosion" suggests; remember that only 6 million years lie between modern humans and the ancestor we share with chimpanzees. But no one denies that a burst of evolutionary innovation changed the world.

It now appears that bursts of rapid and course-altering evolution are not exceptional but are normal features of the global pattern. That would not come as news to George Gaylord Simpson, who recognized this attribute nearly seventy years ago and coined the term "quantum evolution" to describe it. The emergence and proliferation of the three cellular domains—Bacteria, Archaea, and Eukarya—supplies striking examples. In each case the order in which the deep branches appeared is unresolved, perhaps unresolvable (fig. 2.3); this suggests a relatively brief interval of accelerated change, preceded and followed by more placid times. Some have argued that the bursts are artifactual, produced by the limited resolution of rRNA sequences and genomes in general, or by a fundamental misrooting of the tree of life. My own inclination is to take the pattern at face value and attribute such episodes to the opening of new niches or ways of making a living, which encouraged the quick proliferation of a successful design. The case seems especially convincing for the divergence of eukaryotic supergroups, which followed hard upon the emergence of the eukaryotic common ancestor (see chapter 7). The origin of viruses, protein folds, and perhaps cellular life itself seem to display such a pattern, interpreted by Eugene Koonin as transitions between evolutionary phases: complex novelties emerge in the "inflationary" episodes, and diversify in the calmer periods that follow.[28]

In principle there is no need to invoke special effects to account for the emergence of new forms, but singular events do lie at the root of some of evolution's most spectacular fireworks. None is more dramatic than the birth of eukaryotic cells by the symbiotic merger of two distinct cell types. Whichever scenario you favor (see chapter 7), it is indisputable that mitochondria derive from eubacterial symbionts; the partnership worked wonders and directed the course of evolution into a new channel. And then they did it again: another singular symbiosis gave rise to plastids, which then spread across the eukaryotic field; primary and secondary endosymbioses were pivotal events in the formation of major lineages (see chapter 8). It is debatable just how many such events are recorded in the history of the Eukarya and whether they also participated in shaping prokaryotic cells, but they are clearly both uncommon and consequential. Acquisition of endosymbionts introduces new membrane-bound compartments and lifts structural complexity to a higher level in one great leap.

Thomas Cavalier-Smith offers a radically different take on cell history from that developed here, but he, too, invokes rare events that accelerated the tempo of evolution and altered its direction (chapters 4 and 7). These include the loss of the outer membrane by the aboriginal diderms, or negibacteria, which initiated the evolution of the monoderms, or posibacteria. Some of these lost the cell wall as well, replacing it with a novel and more flexible surface coat; these neomura were the forerunners of both archaebacteria and eukaryotes. Both these clades, relatively recent branches of the tree of life, then underwent a period of quantum evolution, giving rise to the diversity of modern forms. Here again, major departures are said to stem not from the gradual accumulation of small change but from sudden and drastic reorganization.

A pattern of global evolution that features deep time dominated by microbes, saltatory events such as lateral gene transfer, symbiosis, periodic catastrophes, and bursts of innovation is not exactly what Darwin had in mind. In that vein, both Jan Sapp and Eugene Koonin have argued that the emerging vision of the evolutionary process represents a new beginning, a major departure from the conventional understanding of evolution.[29] Species are not necessarily discrete and objective entities genetically isolated from other gene pools. Gene mutation and recombination are not the chief fuels of evolution. Saltatory events dominate the big picture. And descent with modification from parents within the same taxon is not an adequate description of biological transformation. These claims are true and important. But even cell evolution is animated by the universal principles of heredity, variation and natural selection. Cell evolution is not the world Darwin knew, but it's Darwin's world all the same.

Do Variation and Selection Explain the Origin of Cells?

From Darwin's day to our own, skeptics have questioned whether evolution by small random steps guided by natural selection can really account for the generation of adaptive, functional organization. They point out that most mutations are deleterious—even destructive—and beneficial ones exceedingly rare. They doubt that there has been sufficient time for Darwin's mechanism to work and insist that the postulated intermediate stages have seldom been documented. Evolutionists respond with a barrage of evidence that leaves no reasonable doubt that the history of life is one of continuous change instigated by small heritable variations and firmly supports the role of natural selection in lending it direction. I have no inclination to enter into the rancorous and heavily politicized disputes that swirl around intelligent design but would insist that students of cell evolution cannot ignore the elephant in the room. Let us then consider, calmly and dispassionately, how sound are the reasons to believe that undirected variation, assisted by self-organization and guided by natural selection, created living cells in all their diversity, intricacy and wondrous beauty.

The same issue has recently been argued in a sprightly and engaging book by Michael Behe, a prominent figure in the intelligent design movement.[30] In the past Behe rejected Darwinism out of hand, but no longer; he now accepts as incontrovertible the evidence for the antiquity of life and for common descent, the role of natural selection in shaping forms and functions, and the capacity of random mutations to confer short-term selective advantage, such as resistance to antibiotics and parasites. What he doubts is that "Darwinian processes can take the multiple, coherent steps needed to build new molecular machinery, the kind of machinery that fills the cell." His thesis is grounded in a numerical argument drawn from the arms race between malaria parasites and their human hosts. The former evolved resistance to chloroquine, the latter threw up the sickle-cell mutation, but both have occurred very seldom despite ample time and large populations. So mutations happen and do influence biological events, but anything that requires more than one or two related mutations lies beyond the "edge of evolution." There is no way that the accumulation of independent random mutations can make two proteins come together, let alone craft intricate machines such as flagella or undulipodia. Instead, Behe invokes the widespread occurrence of nonrandom mutations: "That is, alterations to DNA over the course of the history of life on earth must have included many changes that we have no statistical right to expect, ones that were beneficial beyond the wildest reach of probability." These directed mutations Behe attributes to the intervention of a "designer," the historical alternative to Darwinian evolution, and really the only one that has been fully formulated.

In a scathing review of Behe's book, Kenneth Miller points out that the argument rests on a gross statistical fallacy and ignores the extensive evidence that multiple successive mutations did shape complex systems such as blood clotting and the immune response; and I see it so as well.[31] A spectacular recent example comes from the way bacteriophage lambda evades host resistance by multiple mutations in the tail-fiber protein J.[32] But doubts concerning the origin of biological organization continue to hang in the air. They go beyond statistics and mechanisms and should not get lost in the vicious dogfight between virtuous evolutionists and mendacious creationists. The central question is not whether variation and selection are real but whether they are sufficient; and if not, what forces supplement those discerned by Darwin. Whatever their nature, if forces over and above variation and selection need to be invoked, their elucidation becomes a major strand in the quest for cell history.

Those who seek the origin of cells—and thus of life—labor under a handicap that has no remedy. Everything important happened billions of years ago and left few traces. We have nothing so compelling as the fossils that abundantly document animal evolution, a record that is far richer today than it was in Darwin's time. The belief that the same forces that shaped the swallow's wing and the gazelle's leg also generated the forms and functions of cells necessarily rests on molecular evidence extracted from contemporary organisms, which we reviewed in the preceding chapters. That testimony includes the ready occurrence of mutations and other kinds of heritable variation, the production of supramolecular shapes by self-organization, and the demonstrable power of natural selection. It is indisputable that all living things are related as members of one grand family, and shared some sort of common ancestry in the extremely remote past. Many cellular structures and functions display molecular traces of standard evolutionary mechanisms (and of nonstandard ones also). Some complex machines, such as flagella and ribosomes, were apparently cobbled together from parts that originated elsewhere; one can make a plausible argument that they were recruited ("exapted," as Stephen Jay Gould would say) onto a preexistent root stock and evolved under the guidance of natural selection toward their present function. There is, of course, no proof or direct evidence, and there probably never will be. Understandably unease persists, and so it should: those eukaryotic undulipodia, made of several hundred proteins operating in concert, by random variation and selection, really? A more articulated explanation seems to be called for, here and elsewhere. Acceptance of the Darwinian thesis rests on its general plausibility, on the absence of positive evidence to the contrary, and on the lack of a credible alternative.

A word is in order here about the "randomness" of variation. When we claim that mutations occur at random, we do not mean that all genes and all sites have the same probability of mutation; it is well known that some loci are much more prone to vary than others. What we mean is that they are undirected. Mutations happen by chance when they happen, not necessarily when they would be useful. Two decades ago, the claim that in *E. coli* starvation preferentially induces mutations that allow the cells to mobilize alternative sources of energy sparked a flurry of excitement.[33] But it was a false alarm: those mutations switch on several generalized responses to metabolic stress, not a specific function, and they do not call into question the undirected nature of genetic variation. Indeed, there is no plausible molecular mechanism by which useful mutations could be specifically favored, and the more we appreciate the many steps that come between genes and their functions, the more far-fetched grows the notion of directed mutations. Behe's invocation of a "designer" who foresaw the ramifications of sequence changes in DNA or the far-flung consequences of assimilating a particular set of endosymbionts and directed them with a purpose in mind really boggles the imagination! For me, at least, this is the most powerful argument in favor of the thesis that life is the fruit of chance variation and natural selection: the alternative is just too much to swallow.

So far, so reasonable, but it leaves the heart of the matter untouched. Darwinian evolution is premised on the existence of mechanisms of heredity, variation, and natural selection. Implicit in this premise is the existence of an organized cell, at least a rudimentary one, complete with the capacity to couple heritable elements to functions visible to natural selection. That first cell must have possessed not only genes of some sort and the capacity to express their information content but a boundary, metabolism, energy transduction, and a degree of spatial organization. Variation and selection supply a convincing explanation for the elaboration, diversification and improvement of biological patterns; can they also account for the appearance of such patterns in the first place?

The sophisticated machinery of modern cells presumably arose by co-evolution of the primordial components within the context of reproducing protocells, but hardly any vestiges remain of what must have been a prolonged period of experimentation and evaluation. The ultimate riddle is the genesis of that first cell, the embryo of purposeful complexity, from the lifeless world of chemistry and physics, and that problem has stubbornly defied the best efforts of science for over three generations. Whatever alternatives exist that complement the standard Darwinian ratchet, here is where they came most forcefully into play. Nothing would advance our understanding of cell evolution so much as the discovery of persuasive and verifiable mechanisms to bring forth the beginnings of functional, adaptive organization before there were genes. The current ferment of experiment and debate was reviewed in chapter 10, and I have only praise and admiration for those who strive to crack the hardest nut of all. But the literature leaves me quite unconvinced that we are any closer to a solution than were the pioneers seventy years ago. One can argue that discoveries such as self-replicating ribozymes, self-organizing structures, and warm thermal vents show that life could in principle have originated here on earth; that there are no insuperable obstacles, only a lot of puzzles. But acceptance of this thesis still demands the suspension of much disbelief. There is, to be sure, pervasive continuity between the spheres of chemistry and biology; but I am equally impressed by the vast gulf that separates entities that are alive from those that are not.

As a boy I was fond of science fiction, and I still cherish a personal fantasy derived from that genre. It calls for the cosmic equivalent of Johnny Appleseed, a fabulous being that scoots about the galaxy in his starship; and when he (he, of course) chances upon a likely planet, he drops a packet of life-seeds. Most of them perish but some flourish, multiply, and evolve; we live on one of Johnny's successes. No, I don't take it seriously. But it is worth recalling that, thirty years ago, no less a thinker than Francis Crick published a book in which he considered the merits of directed panspermia versus autogenous evolution of life on earth and concluded that the evidence for the latter was no better and no worse than for the former.[34] Martin Line has more recently taken much the same position.[35] It is noteworthy, and a little disturbing, that thirty years on little has changed.

CHAPTER TWELVE

Summing Up

Journey without Maps

Old men have a weakness for generality, and a desire to see structures whole. That is why old scientists so often become philosophers. — Eugene Wigner, interview in *Science*

In Search of Cell History
Genesis
Progressive Evolution
Does Evolution Have a Purpose?

It's the best of science, it's the worst of science. In all of biology, possibly in all of science, there is no more profound mystery than the origin of cells and of life. Until we solve it, we cannot be sure that life grew naturally out of the lifeless world of chemistry and physics and must entertain the possibility that the gulf between life and nonlife was bridged with the assistance of forces that fall outside the reach of science as we understand it. At this point in time, no one can say with any confidence how life came to be. But we are, slowly and painfully, muddling toward a narrative history of cellular life, and we can already make out some of the major landmarks along the way. What we have learned is quite consistent with evolution by heredity, variation, and selection within an expanded framework that allows room for a variety of nontraditional mechanisms; there is no indication of guidance or intervention by any external directing agency. In sharp contrast, the fog lies as thick as ever over the beginning of things. The heart of the mystery is the genesis of the first self-sustaining and self-reproducing molecular systems, entities upon which natural selection could then act. In this final chapter, let me assess where, in my judgment, the inquiry presently stands and where it may be heading.

We began this journey with a small sheaf of general questions about cell history that clearly fall within the purview of natural science, all of which are currently foci of intense interest. How many kinds of cellular design does nature hold, and how are they related? How do viruses fit into our conception of the living world? Can all living things be mapped upon a unitary tree of life, or has this simile outlived its usefulness? Do all living things share a common ancestry, and if so, how should we envisage that entity? Why are eukaryotes so different from prokaryotes, and how did that cleavage come about? Can the origin and proliferation of life be understood within the general framework of evolutionary theory? Is the history of life just one damn thing after another, or can one discern a general trend toward mounting order, complexity, autonomy, and ultimately mind? What is life, anyway, and might there be a map for its journey after all? Multiple answers have been proposed for each of these questions, all entangled in controversy and obscured by clouds of ink. This makes the study of cell evolution frustrating as well as exhilarating, but also testifies to the irrepressible vitality of science for its own sake. After all, there can be few subjects less likely than cell evolution to contribute to the solution of mankind's manifold problems!

In Search of Cell History

Concerning the descent and diversification of cellular life, the literature presently offers two coherent yet sharply different accounts. The standard model, widely accepted among microbiologists and increasingly featured in textbooks, grew out of the late Carl Woese's discovery of the archaebacteria almost 40 years ago. It proposes three great evolutionary stems, or domains, that separated early in the history of life: Bacteria, Archaea and Eukarya (see fig. 2.1). The latter two are sister clades that share many molecular characteristics to the exclusion of Bacteria. Extensive lateral gene transfer among species, phyla and even domains has blurred the distinctive lineaments of the original stems, but the tripartite division of the living world remains discernible. The three domains were initially drawn from ribosomal RNA sequences but were soon reinforced by biochemical criteria; phylogenomic analyses commonly (but not always) recover three distinct stems. The root of the tree of life, the first bifurcation that separated the line that generated Bacteria from that which produced Archaea and Eukarya, represents LUCA, the last universal common ancestor of all life, whose nature has been much debated. LUCA was probably not a singular organism but a spectrum of related entities that readily swapped genes and evolved communally, rather than along discrete lineages. Current opinion holds that she was genetically and physiologically fairly advanced, endowed with all the essential capabilities of cellular life. The three domains emerged from LUCA by a process akin to "crystallization." In the case of the two prokaryotic domains, that entailed streamlining by selection for metabolic versatility, simple organization, and rapid reproduction; the more elaborate eukaryotes may have begun their career as humble scavengers. The chronology of cell evolution is uncertain, but judging by the fossil record prokaryotes originated more than 3.5 billion years ago and dominated the scene for the next 2 billion years. Convincing eukaryotic fossils are no older than 1.7 billion years, but molecular approaches suggest that the eukaryotic lineage may be much more ancient.

The alternative narrative has been articulated and tenaciously defended for the past 30 years by Thomas Cavalier-Smith, who rejects the standard model on grounds of both principle and observation. Cavalier-Smith recognizes only 2 evolutionary stems or empires, Bacteria and Eukaryota (see fig. 4.2b). Crucially, he sees the archaebacteria as a highly divergent and relatively recent subdomain of the Bacteria, not a separate domain of life. LUCA was a fairly well developed bacterium endowed with two lipid bilayer membranes (a diderm, or "negibacterium"), a peptidoglycan cell wall, chemiosmotic energy transduction, and probably photosynthesis. Its closest living relatives are found among the chlorobacteria, specifically *Chloroflexus*. Bacteria have been in existence for some 3.5 billion years, and their diversification into discrete phyla can be tracked by a combination of genomic and structural criteria. Lateral gene transfer is real but uncommon and is not a serious impediment to phylogeny. In Cavalier-Smith's view, a succession of innovations generated first the "posibacteria" (or monoderms, enclosed within a single bilayer membrane), and then the common ancestor of archaebacteria and eukaryotes. The latter two sister clades diverged quite recently, less than 1 billion years ago.

Persuasive arguments have been advanced in support of both basic narratives, but I can find no compelling objective criteria that would falsify either of them. For that matter, the same can be said of a clutch of variant hypotheses that have not yet attained the status of a comprehensive storyline. In the absence of proof, travelers in this wilderness have necessarily given much weight to plausibility, personal judgment, and aesthetics, and they have overwhelmingly come down in favor of the belief that the living world is divided into three great stems that reach back into the dawn of cell history. Three stems—not two—seem to make the best sense of the pattern of nature; the idea has become part of biology's conceptual framework and has guided the writing of this book from its inception. In the terminology made familiar by Thomas Kuhn, the three domains have become "normal science"; it would take a revolution to dethrone them. That's not to say that it could not happen, but I hear no rumbling of drums.

Bacteria and Archaea evolved directly from LUCA, that shadowy common ancestor of all cellular life; they are primary domains. But Eukarya did not; they are quite clearly chimeric, products of the fusion or merger of two or more cellular lines, and consequently, they represent a secondary or derived domain. The precise meaning of that statement remains open, for there is no thicket thornier than that which conceals the origin of eukaryotic cells. Everyone agrees that mitochondria and plastids derive from bacterial endosymbionts; what is at issue is the nature of the host, the timing of their association, and its role in shaping the structural and functional complexity of eukaryotic cells. Woese envisaged "urkaryotes," progenitors of the nuclear lineage, as one of the primary lines of cellular descent, more closely allied to Archaea than to Bacteria but distinct from both. Cavalier-Smith describes the host as a predatory protoeukaryote descended from the Bacterial line (specifically, from the "posibacteria" or monoderms). And many molecular biologists trace eukaryotic cells to a fusion or merger of Bacteria and Archaea, prokaryotic cells of modern kind. Fusion may have been a singular event, or it may have happened several times in succession. Depending on whom you read, the eukaryotic lineage is very ancient (more than two billion years), or relatively recent (a billion years or less); and mitochondria came early and drove the rise of cellular complexity, or else came late into a host cell that had already attained the essential eukaryotic features.

Recent results drawn from both genomics and cell physiology seem to converge on a resolution that combines features from several of the classical positions (fig. 12.1). This emerging synthesis places the elusive host very early, among the progenitors of the Archaea, most likely basal to the phylum called Thaumarchaeota. Everything suggests that it was already an organism of some structural and physiological sophistication. The renovated urkaryote made its living by scavenging or predation upon the prokaryotic primary producers, and acquired endosymbionts by a primitive version of phagocytosis. The first of these endosymbionts, relatives of today's α-proteobacteria, gave rise to mitochondria and made a substantial contribution to the eukaryotic genome. The acquisition of endosymbionts greatly expanded both energetic and genetic potentials and

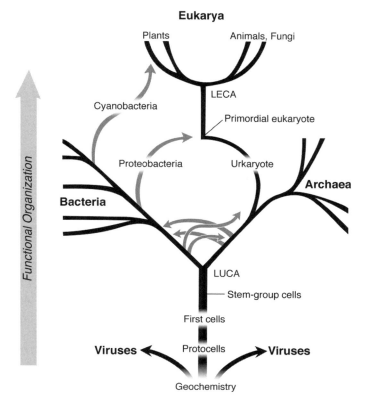

FIGURE 12.1   A brief history of life, a personal interpretation. In accord with much current literature, the diagram proposes that viruses originated in an ancient world, separate from but parallel to that of cells, and that LUCA (the last universal cellular ancestor) was itself the product of a lengthy evolution during which the basic cellular machinery took shape. The two primary domains, Bacteria and Archaea, descend from LUCA. Eukarya, a derived or secondary domain, originated in the fusion of an urkaryote (a scavenger/predator, basal to the Archaea) with an α-proteobacterium (progenitor of the mitochondrion). Another prolonged period intervened between the primordial eukaryote and LECA, the last eukaryotic common ancestor. Subsequent assimilation of a cyanobacterium in one lineage gave rise to plastids. Extensive interdigitation of lineages by lateral gene transfer has been omitted for clarity.

underpinned the emergence of eukaryotic organization. A subsequent, separate episode of endosymbiosis with a cyanobacterium initiated the evolution of plastids, which then dispersed photosynthesis across the eukaryotic universe.

Current narratives of evolution largely take for granted the potent simile of the tree of life: organisms descend from their ancestors in a linear

fashion with occasional divergences that generate new branches. The tree
has been battered by the storm of nontraditional hereditary mechanisms,
including lateral gene transfer and the acquisition of endosymbionts, to
the point where the very existence of a tree of life has been called into
question. That seems to me an extravagant reaction. The proper basis of
the universal tree of life is not genes but cells, and it remains true that
every cell derives from a preexisting cell. Nonlinear gene transfer undermines the assumption that one can use genes to track the history of cells
and may make it impossible to recover the correct tree, but that is no reason to deny that a history of cells exists in principle. As a practical matter,
traditional phylogeny is alive and well as applied to animals, plants, fungi,
and multicellular creatures in general. Gene transfer is a significant fac-
tor among eukaryotic microbes but not so prevalent as to seriously blur
the lines of descent. Prokaryotes, however, are different. Gene trafficking
is rampant, particularly among close relatives, and no single gene can be
considered altogether immune to transfer. Nonetheless, it appears that
even among prokaryotes vertical inheritance is the dominant mode and
the broad lines of phylogeny can be mapped by the use of selected genes.
When handled with caution, ribosomal RNAs retain their historical role
as the universal molecular chronometer. It is more alarming to note that
the tree of life has no sensible place for viruses. One can argue that viruses, being obligatory parasites, are not "alive" and do not belong on the
tree at all. Yet they clearly are part of the universe of living things, and not
a peripheral part either. The fact that we still do not know where to put
them points up a large lacuna in our view of life.

Genesis

All the uncertainties come to a head when we turn from the diversification and elaboration of cellular life to its origin, which I take to be synonymous with the advent of biological organization. Darwinian evolution
convincingly accounts for the burgeoning of life once cells had come into
existence, but it does not explain how cells came to exist in the first place.
One can speculate with some confidence on the doings of LUCA and even
her progenitors, drawing on evidence that these were already endowed
with genes and the means to express their instructions, not to mention
other cellular necessities. By hypothesis, progenotes (or stem-group cells)
were strands in a primordial tapestry of early life forms, possibly RNA-

based, and more makeshift and loosely built than contemporary cells, which probably evolved primarily by variation, competition and natural selection. Beyond this stage, the light quickly grows dim. Stem-group cells may have originated in a "First Cell," a speculative entity that first brought together the rudiments of metabolism, autonomy, and self-reproduction. That mythical ancestor, in turn, is separated by an opaque curtain from a previous era of protocells that preceded and produced the first cells. Conway Morris may be a little uncharitable, but he has it right: "The scientific annexe dedicated to the problem of the origin of life has not been marked by a series of sweeping and spectacular advances—the norm for science—but has lurched indecisively across a landscape dotted with stumbling blocks and crevasses."[1] The fact that so little of biological substance has come from so much labor raises the question whether the effort has been misdirected; at the least, it suggests that we are still missing a major piece of the puzzle.

No one still maintains that the first protocells assembled themselves directly from a menu of molecular precursors, obligingly supplied by a soup of abiotic chemicals. Instead, the hunt is afoot for some sort of self-sustaining network of chemical reactions that drew in matter and energy, constructed a boundary, achieved the capacity to reproduce itself and evolved toward increasing efficiency, complexity, and autonomy. This quest has again generated competing narratives, identified by the taglines "Genes First" and "Metabolism First." The former postulates that the climb began with the spontaneous advent of a self-replicating molecule, most probably RNA, that progressively acquired ancillary factors such as prebiotic peptides; the kinetic drive to replicate again and again is what underlies the genesis of cellular systems. Support for this viewpoint comes from the manifold roles of RNA in cellular operations, from the existence of ribozymes, as well as the recent laboratory synthesis of special kinds of self-reproducing RNA. The alternative storyline begins with metabolism—especially energy input—and looks to self-organization to generate mounting complexity. Proponents expect constituents of the primeval broth to undergo all sorts of chemical reactions, allowing the spontaneous emergence of cyclic loops; some of these will fix carbon and augment the proto-metabolic net. Persistence of the net constitutes a kind of heredity, and competition among networks for matter or energy could be the basis for a limited kind of natural selection. The argument relies heavily on theoretical models but is seriously deficient in material evidence.

There is a rich trove of chemistry in this endeavor, both experimental and theoretical. The devil lurks where he has always been, in the transition from chemistry to biology, and from this perch he taunts us conceited mortals with a barrage of riddles. One stumbling block is energy. To the best of my knowledge, no one has yet described a single reaction that couples a plausible prebiotic energy source to the production of organic compounds. Energy utilization always requires complex structures, not likely to have been available prior to the advent of proteins and membranes. No biopoietic scheme deserves serious consideration unless it specifies a realistic driving force! This is why some explorers have made the audacious proposal that life originated in the interstices of mineral mounds deposited by submarine hydrothermal vents, where geochemistry supplies both potential energy and raw materials. Membranes pose another fundamental problem. Biological membranes are very special and sophisticated constructs: an impermeable lipid fabric, studded with portals and pumps for nutrients and waste products, not to mention the machinery of energy transduction. Could working membranes have existed before there were proteins? And then there are genetic instructions encoded in nucleic acids but expressed as proteins: one symbolic language translated into another, with the aid of a universal apparatus of quite phenomenal sophistication. Proteins are indispensable because they perform tasks beneficial to the cellular system as a whole; they have functions. A polynucleotide may replicate itself or be a catalyst, but it is not a gene until it specifies a function. How could functions have come about in primordial, precellular systems? It all seems rather a lot to ask of a ball of cosmic dust, and it is hardly surprising that the answers have so far eluded our best efforts.

Something is going on here that transcends everyday chemistry, and this appears to have been true from the very beginning. To paraphrase an evocative statement by Stuart Kauffman, life wandered into realms of chemistry and physics that are never explored by inanimate matter.[2] We really have no clear idea how that happened—not even in principle—which is why the origin of life remains a mystery rather than a puzzle. I would like to believe that the answer has to do with an aspect of living things that has gone utterly out of fashion under the onslaught of molecular reductionism: the integrity, the wholeness displayed by every cell and organism, far exceeding comparable features of inanimate systems.[3] Life, it seems to me, would not have begun with simple molecules and grown complex; it began complex, perhaps even with a degree of "purposeful" organization, and grew more so over time. There will be no breakthrough, no leap of understanding, until we have real models to bring the incomprehensible within our ken. Laboratory exemplars of multimolecular, energy-driven metabolic systems would be a major step forward.

Whatever was the nature of the first protocells, and however they came into existence, they must have been pregnant with possibilities. Their eventual descendant, the common ancestor of the three domains, seems to have been a proper organism, endowed with the basic machinery of cellular life. Between protocells and LUCA, all the universals of cellular life took their present shape: DNA, transcription, translation and the genetic code, membrane transport and energy transduction, metabolic pathways, the cytoskeleton, spatial and functional organization, and much else besides. How much of this blizzard of innovation can be credited to the standard Darwinian ramp of gene-based variation winnowed by natural selection? There are convincing indications that descent with modification played a part in crafting intricate devices such as ribosomes and ATP synthases. But it is simply not clear whether the familiar processes suffice to account for the emergence of the biological machinery, let alone the emergence of spatial order or a genetic apparatus exquisitely honed for its specific purpose. Can natural selection really exert significant "pull" on chemical systems that possess neither individual identity nor the means for heredity and variation? It is customary at this point to invoke "self-organization," and indeed, there are physical as well as chemical mechanisms that generate ordered structures. The role of self-organization in the assembly of sophisticated cellular devices from their molecular constituents (the cytoskeleton, for example) is also well established. But biological self-organization begins with genetically specified building blocks. Can self-organization also explain the origin of functions, organization, and a rudimentary capacity to act on the system's own behalf? A demonstration that functional order can arise spontaneously by mechanisms that do not depend on Darwin's algorithm of variation and selection would transform the field. Until then, an open-minded but principled skepticism is the only defensible stance.

Progressive Evolution

For magnitude, duration, and sheer awesomeness, few tales match the history of life. Over the course of four billion years, life has unfolded from minute specks indistinguishable from contemporary bacteria to great and integrated creatures, mushrooms and pterodactyls, whales and oak trees. It has burgeoned to smother in organized biochemistry every habitable square inch of earth's surface, from the summits of the Himalayas to the depths of the Mariana Trench. Life's achievements include creatures intelligent enough to comprehend both their own nature and eventual mortality (though not, it seems, smart enough to manage their habitat), and parasites to prey on those creatures. Looking at life's immense journey as a whole, from meager and murky beginnings to swashbuckling present, what leaps to the eye is a spectacular progression in both quantity and quality, but neither uniformly nor universally. Life seems to advance by fits and starts, accumulation and reduction, all punctuated by long periods of stasis. That evolution is in some sense progressive is undeniable (fig. 12.1). But what, exactly, is the quality that has increased, expanded, and advanced over time? And granted that something has, is evolution by heredity, variation, natural selection, and adaptation sufficient to explain the blossoming of life, or must we postulate additional principles to lend direction to life's history?

Among the possible answers to the question "What quality has evolution maximized?" the fashionable one at present is "complexity." That is a wondrously flexible term, commonly employed in the literature (and in this book) in at least two senses. Daniel McShea and Robert Brandon, in their exposition of biology's first law, use the term rigorously as a trait measured by the number of distinct part types that make up the whole.[4] By this criterion, it is not unreasonable to assert that a yeast cell is more complex than a bacterial one (more kinds of molecules, arranged in more elaborate ways); a mouse, made up of several hundred cell types, is more complex than yeast; and a human being, whose brain alone consists of some hundred billion neurons (about as many as there are stars in our galaxy) stands on a still higher plane. Complexity in this technical sense has certainly increased, but the narrow usage fails to do justice to one's intuitive sense that what has unfolded over time is something of greater import than the number of parts. This is why those who try to make life intelligible generally have recourse to a colloquial usage of complexity, as a poorly differentiated mélange of multiplicity of parts and their interactions, functional adaptation, and sophistication of design. Complexity in this sense is little more than a label for the ineffable quality that has apparently increased over time. We must do better.

If the history of life has an overarching theme, it seems to be the overall expansion of adaptive, functional organization. From the literature on progress in evolution, two terms stand out. One is "autonomy," as in emancipation from the environment, which Bernd Rosslenbroich singles

out as the most relevant quality.[5] To be sure, organisms are not discon-
nected from their environment, far from it. But, as Rosslenbroich puts
it, in the course of evolution the direct influences of the environment
gradually diminished, while stability, flexibility, and self-referential functions intrinsic to the system came to the fore. Familiar examples include
membranes, homeostasis, and the displacement of physiological activities
from the surface to the interior of cells and organisms. The other term
that underscores an important aspect of the evolutionary progression is
"agency," the capacity of cells and organisms to act on their own behalf.[6]
Neither quality is easily measured, and there are bound to be examples of
autonomy and agency diminishing over time rather than increasing; but
taken together, autonomy and agency give me, at least, a better sense of
the benefit gained by maximizing adaptive, functional organization.

By now, even the most loyal reader will be wondering what these ruminations on evolutionary progress have to do with cell evolution. The
answer, in a word, is eukaryosis. The conspicuous advances during life's
history, the flowering of autonomy and agency and complexity, are all associated with eukaryotic cells and their descendants. Prokaryotes do display remarkable biochemical versatility and can lay claim to having first
perfected the very phenomenon of life. But for the past three billion years
they have remained minimally complex, not obviously more advanced
than their fossil progenitors in Archean rocks. The emergence of eukaryotic cells stands out, second only to the origin of life itself, as the great leap
forward that made all else possible. If the evolutionary progression owes
something to an inherent principle of directionality, let alone to guidance
by an external designer, one should expect to find their traces at the critical juncture represented by the origin of eukaryotic cells; but none declare
themselves. It is a little alarming that some of the leading investigators in
this field consider the birth of the eukaryotic cell to have been a special,
even a unique event. But the underlying mechanisms—endosymbionts
and their assimilation, lateral gene transfer, conversion into organelles,
and adaptation to changing environments—carry no overtones beyond
natural history, and the interplay of chance and necessity. If evolution is
God's way of creating the living world (a conciliatory formula once popular even among scientists), He gave the process free rein and refrained
from steering it.

Well then, does evolutionary doctrine explain the progression of life?
Progress, however defined, is not built into Darwin's theory; heredity, variation, and natural selection explain the adaptation of organisms to their

environment but do not predict a global increase in organismal complexity, agency, or autonomy. There is also no observational evidence for any inherent drive towards greater complexity and sophistication. Nevertheless, most biologists including myself are satisfied that standard mechanisms suffice to drive life up "Mount Improbable." Natural selection has clearly favored many inventions and transitions that made organisms more autonomous and better able to act on their own behalf, with a concomitant increase in complexity. Photosynthesis and peptidoglycan cell walls come to mind, and so do undulipodia and phagocytosis, not to mention legs, wings, and eyes to see with. In crafting its manifold adaptations, life drew on whatever sources of novelty came to hand. Mutation, gene transfer, symbiosis, and duplication were all grist for the mill, as were epigenetic and structural change. There is a universal tendency for things to fall apart and for variation and diversity to increase (in accord with biology's first law); that very process of creative destruction is also a source of novelty that selection builds on. Generally speaking, it seems that we need nothing more than heredity, variation and natural selection to explain the pattern of nature, progression and all.

Does Evolution Have a Purpose?

Most biologists appear to be content with this general picture, happy to focus their attention on more down-to-earth puzzles. But in some quarters dissatisfaction persists, which I share. Uneasiness probably stems in the first place from the realization that the conventional wisdom denies that tremendous history of life any meaning beyond itself: life happened, and that's all there is to it. To paraphrase a famous remark by the physicist Steven Weinberg, the more evolution seems comprehensible, the more it also seems pointless. Perhaps it is so, but I am unwilling to believe that life is merely some cosmic accident without significance; on the contrary, I surmise that life and its evolution are ultimately what the universe is all about. Reservations about the conventional wisdom are reinforced by the persistent uncertainty about the genesis of cells and their parts, and by the intractability of life's origin. To be sure, personal incredulity is not evidence. But it does supply motivation to ask whether there may be more to this story than current philosophy knows, and indeed hints that this may be so come from within the orbit of science itself. Some clues have been taken to suggest that, contrary to all appearances, there is a direction to life's journey that points towards self-awareness. Others come from peculiar features of the cosmos as a whole, which render it surprisingly friendly to the eventual appearance of life, and of mind.

Scientists who seek a larger meaning in the history of life tend to gravitate toward a marriage of evolution with intelligent design, though not necessarily of the theistic variety. One prominent figure is Simon Conway Morris, a paleontologist with impeccable evolutionary credentials, who is repelled by the reductionist fundamentalism that has come to dominate biological discourse. His thoughtful and provocative book, *Life's Solution* (2003), centers on the prevalence of convergence in evolution, and this leads him eventually to a set of biological qualities that seem to jar with the conventional account. First on his list is an unexpected simplicity to life, constructed from so small a number of basic building blocks that appear almost arbitrary (why tryptophan?). They can be combined in innumerable ways, but living things display an uncanny ability to navigate across the ocean of possibilities to the few tiny islands that actually work. Organisms end up perched on steep and narrow adaptive peaks, because all the alternatives are maladaptive. Conway Morris underscores that the conditions that would allow life to emerge are exceedingly stringent; earth, in providing such a habitat, will be one suitable planet in many millions, and may possibly be unique. But once life made a successful start, the emergence of ever more sophisticated forms is almost foreordained, culminating in intelligent beings whose appearance and physiology will be quite similar to our own. As he reads life's tea leaves, "salient facts of evolution are congruent with a creation,"[7] whose nature he is careful to leave unspecified. And Conway Morris is by no means alone; for instance, Michael Denton reached a similar conclusion from reflection on the remarkable "fitness" of life's solutions to the problems of existence.[8] But when it comes to explaining just why life exists in spite of the odds, few authors are as forthright as James Gardner, who argues that the universe was in fact designed by intelligent creatures, with the explicit objective of fostering the (Darwinian) evolution of subsequent generations of such creatures.[9] These perform an essential function in the life cycle of the universe, for they are the agents by which the universe reproduces itself.

Cosmologists have been much more receptive than biologists to the notion that beneath the contingency and randomness of events lies a predetermined course, perhaps even a purpose and a plan. The basis for this claim, all the more startling for the sobriety of its advocates, is the realization that the physical parameters of the universe are precisely those required to produce a setting in which life can evolve.[10] Sir Martin Rees, formerly astronomer royal of Great Britain, explains how it all turns on just six numbers that determine the key features of the universe: the extent and rate of its expansion, whether it can spawn galaxies, stars and planets, and whether the atoms of which molecules are made can be generated. The universe is astonishingly sensitive to the values of these parameters, which were fixed at the instant of the big bang, such that even minor deviations would have caused the universe to collapse, to fly apart, or to end up devoid of chemistry and barren of life. The findings seem solid and must somehow be incorporated into the fabric of reality, which thereby becomes ever weirder. Rees favors the interpretation that our observable universe is but one member of a huge ensemble of extant universes, only a few of which make abodes for life. Paul Davies inclines toward the idea that "the bio-friendliness of the universe arises from an overarching law or principle that constrains the universe/multiverse to evolve towards life and mind."[11] He takes life seriously, built into the workings of the cosmos at a fundamental level, and that is my inclination as well.

How might one imagine a universe fit for life without falling back on sentient creators? No one has traveled farther along this path than the venturesome cosmologist Lee Smolin, who begins by discarding some of the fundamental assumptions about the nature of the universe.[12] In his view, it is not true that the physical parameters of the universe were fixed for all time at the instance of the big bang. Instead, the cosmic constants are themselves products of a selective process that has much in common with biological evolution. Our universe is but one of a huge population of universes that propagate themselves by mechanisms centered on black holes. Universes do not compete among themselves for resources, nor do they harbor anything analogous to a genome. Nevertheless, since reproduction requires black holes, those universes whose parameters favor the formation of black holes will come to dominate the population of the multiverse. Such universes are also the ones most hospitable to the formation of galaxies, stars, the higher chemical elements, and ultimately the seeds of life. In Smolin's conception universes are dynamic and energetic entities, far from thermodynamic equilibrium and in no danger of running down. On the contrary, they generate structure and order at all levels by self-organization, and it is in this elusive but real process that we should seek the origin of life. Ours is a living universe, stunning in its beauty and variety, not because it was fashioned to be so, but precisely because it made itself. Evolution has no preset goal, no purpose; but the trend towards mounting complexity, autonomy, and mind is woven into the fabric of a self-organizing universe.

Science for the imagination—what should one make of it? In today's fractious climate it is necessary first of all to repeat that uncertainty about life's place in the universe does not call into question the evolution of life by natural processes, nor does it lend credence to the claims of biblical creationists. But it is clear that our perception of biology must be brought into harmony with a quite fantastic setting. The clockwork universe, made up of discrete particles moving and colliding at the behest of mechanical forces, and in which life is an anomaly, has long since been abandoned by physicists. In its place stands a strange and unintelligible construct, a universe that is expanding and making itself, composed of unfamiliar forms of matter and subject to "dark energy" whose nature is altogether obscure. Is this brave new world one in which life feels at home? I don't know. My mind keeps circling back to the famous remark of J. B. S. Haldane, one of the pioneers in the search for life's origin,[13] that the world is not only queerer than anyone has imagined, but queerer than anyone can imagine. For what it's worth, I suspect that of the beginnings of life we still know little and understand less. We are acquainted with but a single kind of life, a sample too small to warrant firm conclusions about the nature and genesis of this astonishing phenomenon. What we have learned about life in the cosmos comes from one tiny nook in the vastness of the universe, granting us little warrant for dogmatic pronouncements about what is and is not possible. The beginning of wisdom is to stay tuned, keep listening, relish uncertainty and never let go of the sense of wonder that drew most of us into science in the first place.

Let a counterculture have the last word. Richard Burton was an intrepid explorer of Queen Victoria's day; he knew the Muslim world intimately and had the chutzpah to write a *qasida*, a philosophical poem in that tradition:[14]

> As palace mirrored in the stream,
> As vapour mingled with the skies,
> So weaves the brain of mortal man
> The tangled web of Truth and Lies.

Notes

Preface

1. Beck 1957, 100.
2. Doolittle et al. 2002, 39.
3. Harold 2001.
4. Thomas Sprat, *The History of the Royal Society, London* (1722). The quotation is taken from *An Introduction to Comparative Biochemistry* by Ernest Baldwin (Cambridge University Press, 1949), iv, a small gem of a book that gave me my first glimpse of biochemistry sixty years ago and set my course for life.

Chapter One

1. For thoughtful discussions of how the cell theory was born and what it meant to naturalists of the time, see Mayr 1982 and Sapp 2003.
2. Stanier and Van Niel 1962.
3. Sapp 2005.
4. Stanier and Van Niel 1962.
5. Stanier, Doudoroff, and Adelberg 1963, cited in Sapp 2005.
6. Whittaker 1969.
7. Maynard Smith 1986; Harold 2001.
8. According to Kim, Klotchkova, and Kang (2001), certain giant algae can survive for a spell in the absence of a plasma membrane. Pollack (2001) accepts that membranes exist but denies their role as the indispensable permeability barrier.
9. The organization of bacterial cells is summarized in Harold 2001 and described in detail in any contemporary textbook of microbiology. On the meaning of complexity, see McShea and Brandon 2010.
10. *Collins English Dictionary*, 3rd ed., s.v. "system."
11. McShea and Brandon 2010.
12. *Collins English Dictionary*, 3rd ed., s.v. "biology."

13. Mayr 1982; Sapp 2003.

14. The doctrine that organisms represent nothing more than the execution of the instructions encoded in their genes has considerable historical depth. It may have been first stated seventy years ago by H. J. Muller, was given substance by the late George Williams, and came to full flower in the writings of Richard Dawkins (1976, 1982, 1995, 1996) and Daniel Dennett (1995). Love 'em or hate 'em, you cannot ignore their ideas which have largely shaped most scientists' understanding of life.

15. Lartigue et al. 2007; Gibson et al. 2010.

16. Jacob 1973.

17. Britten 1998.

18. Lartigue et al. 2007.

19. The idea that cells rather than genes are the proper level of biological abstraction also has deep roots. Tracy Sonneborn and Boris Ephrussi thought so sixty years ago. Contemporary authors broadly associated with this point of view include Cavalier-Smith 2006a, 2010; Goodwin 1994; Harold 2001, 2005; Jablonka and Lamb 2005; Kirschner and Gerhart 2005; Lewontin 2001; Noble 2006; Sapp 1987, 2003; and Shapiro 2011. For a recent reappraisal of the "organicist" conception of the living state, see the stimulating article by Denton, Kumaramanickavel, and Legge 2013.

20. All of the following, but especially Jablonka and Lamb 2005 and Shapiro 2011. Cavalier-Smith 2006a, 2010; Goodwin 1994; Harold 2001, 2005; Kirschner and Gerhart 2005; Lewontin 2001; Noble 2006; Sapp 1987, 2003; and Denton, Kumaramanickavel, and Legge 2013.

21. Spatial organization and morphogenesis in bacteria have become hot topics for research, but remain somewhat outside the main stream. Begin with reviews by Shapiro, McAdams, and Losick 2002, 2009; Carballido-Lopez and Errington 2003; Gitai 2005; Harold 2005; Margolin 2009.

22. See Sapp 1987, 2003; Harold 2001, 2005; Jablonka and Lamb 2005; Jablonka and Raz 2009.

23. Lartigue et al. 2007; Gibson et al. 2010.

24. Cavalier-Smith 2000, 2004.

25. Noble 2006, 13.

26. Darwin (1859) 1979.

27. Maynard Smith 1986, 4.

28. Maynard Smith 1986; Dennett 1995; Dawkins 1996, 2009; Coyne 2009; Lane 2009.

29. Eiseley 1975, 242.

Chapter Two

1. Darwin (1859) 1979, 171.

2. See Woese (2004, 2007) for personal recollections of the discovery of the

Archaea and how that shaped his understanding of evolution. Jan Sapp (2009) has contributed an invaluable historical account of that episode, and its significance for the study of evolution in general.

3. Woese 2004, 2007; Sapp 2009.

4. Gibson and Muse 2004. For current methods in microbial phylogenomics, see Delsuc, Brinkmann, and Philippe 2005; and Snel, Huynen, and Dutilh 2005.

5. See Woese 1987 for the foundations of modern bacterial taxonomy.

6. Woese, Kandler, and Wheelis 1990 is another classic of evolutionary science.

7. The root of the SSUrRNA tree cannot be inferred from RNA sequences alone, because of the lack of a suitable outgroup. Its position was located by the use of genes that underwent duplication prior to the deepest node on the tree. For a recent review of this methodology, see Zhaxybayeva, Lapierre, and Gogarten 2005.

8. Mayr 1998.

9. Ibid.

10. Woese 1998b. See also Wheelis, Kandler, and Woese 1992.

11. White 1968.

12. Should the term "prokaryote" be expunged from the literature? Yes, says Pace (2006). Whitman (2009) and Pace (2009a) conducted a formal debate.

13. See Brown and Doolittle 1997 for an early notice of the evidence for lateral gene transfer. Woese et al. (2000) analyze the complex case of aminoacyl t-RNA synthetases. The current status of genes that duplicated before the divergence of the two prokaryotic domains is considered by Zhaxybayeva, Lapierre, and Gogarten 2005.

14. Keeling et al. 2005, 674.

15. See Doolittle 1999; Doolittle et al. 2002; and Brown 2003. A striking example of a complex enzyme that is out of place thanks to lateral gene transfer comes from Lapierre, Shial, and Gogarten (2006), who found that certain strains of the Bacterium Deinococcus lack the usual F-ATP synthase, containing instead a synthase of the A/V type normally characteristic of Archaea. Sonea and Mathieu (2000) were among the first to consider the whole prokaryotic world as an extended "superorganism."

16. Keeling and Palmer (2008) survey the evidence for lateral gene transfer among eukaryotic microbes.

17. Reviewed in Doolittle, 1999; Doolittle et al. 2002; Boucher et al. 2003; Simonson et al. 2005; Fournier, Huang, and Gogarten 2009. For a recent example of large-scale transfer of genes, see Nelson-Sathi et al. 2012 on the origin of the Haloarchaea.

18. Nelson-Sathi et al. 2012.

19. Jain, Rivera, and Lake 1999; Simonson et al. 2005.

20. Woese 1987; Pace 1997, 2009b.

21. See Kurland, Canbäck, and Berg 2003, and Kurland 2005.

22. Doolittle 2009a, 2009b; Bapteste et al. 2009.

23. Keeling and Palmer 2008.

24. Amidst the din of discord, one appreciates calm and thoughtful voices. I have drawn on Walsh and Doolittle 2005; Beiko, Harlow, and Ragan 2005; Doolittle and Bapteste 2006; Doolittle 2009a, 2010; and McInerney, Cotton, and Pisani 2008.

25. Pace 2009b.

26. Recent examples include Cox et al. 2008 and Jun et al. 2010. Summaries of that burgeoning literature will be found in Snel, Huynen, and Dutilh 2005; Simonson et al. 2005; and McInerney, Cotton, and Pisani 2008.

27. Snel, Huynen, and Dutilh 2005.

28. Critical evaluations include Simonson et al. 2005; McInerney, Cotton, and Pisani 2008; Doolittle 2009a, 2009b; and especially Bapteste et al. 2009.

29. Feng, Cho, and Doolittle 1997.

30. Douzery et al. 2004; see also Sheridan, Freeman, and Brenchley 2003.

31. Battistucci, Feijao, and Hedges 2004.

32. Graur and Martin 2004.

Chapter Three

1. In the original quote, the word in brackets is "peoples."

2. Whitman, Coleman, and Wiebe 1998.

3. For recent discussions of this perennial topic, see Pepper and Herron 2008 and Queller and Strassman 2009.

4. Queller and Strassman 2009, 3144.

5. Ibid.

6. The term "undulipodia" was coined many years ago by the late Lynn Margulis. It failed to catch on but should have, for we do need a term to designate specifically the eukaryotic organs of motility, whose structure and operation is utterly unlike that of Bacterial and Archaeal flagella.

7. Koch 2003; Koch and Silver 2005.

8. The huge current literature on Archaea features noteworthy general reviews by Gribaldo and Brochier-Armanet (2006), Brochier-Armanet, Forterre, and Gribaldo (2011), Jarrell et al. (2011), and Cavicchioli (2011). There are also two comprehensive compendia, edited by Cavicchioli (2007), and by Garrett and Klenk (2007).

9. For Thaumarchaeota, see Brochier-Armanet et al. 2011 and Pester, Schleper, and Wagner 2011.

10. Elkins and colleagues (2008) report the first genome sequence from the Korarchaeota.

11. Valentine 2007. Possible adaptation to thermophily or antibiotics will be discussed in chapter 4, in conjunction with the views of Cavalier-Smith and Gupta.

12. The differences between Archaea and Bacteria were first spelled out in the writings of Carl Woese, Otto Kandler, and Wolfram Zillig twenty-five years ago; see Sapp 2009 for an account. For recent compilation and discussion, see Brown

and Doolittle 1997; Walsh and Doolittle 2005; Zhaxybayeva, Lapierre, and Gogarten 2005; Gribaldo and Brochier-Armanet 2006; and Jarrell et al. 2011.

13. Koch 1996; Trevors and Psenner 2001.

14. For the remarkable planctomycetes, see Fuerst 2005; Fuerst and Sagulenko 2011; and Lonhienne et al. 2010. Jun et al. (2010) report evidence that planctomycetes branched early off the Bacterial bush. Their microtubules are discussed by Pilhofer et al. 2011.

15. Fuerst 2005; Fuerst and Sagulenko 2011; Lonhienne et al. 2010; Jun et al. 2010; Pilhofer et al. 2011.

16. Just what is the relationship between planctomycetes and Eukarya? Reynaud and Devos (2011) and McInerney et al. (2011) debate the issue.

17. All these startling numbers come from Suttle 2005.

18. For an entry into this speculative but stimulating literature see Villareal 2005; Forterre 2006b; Koonin, Senkevich, and Dolja 2006; Brüssow 2009; and Krupovich et al. 2011.

19. Koonin et al. 2006, 4.

20. Forterre 2006b.

21. Claverie and Abergel 2010.

22. Claverie and Abergel (2010) argue that virus factories are living organisms; Moreira and Lopez-Garcia (2009) firmly exclude viruses from the tree of life.

Chapter Four

1. For the "unity of biochemistry," see Harold 2001 and Pace 2001.

2. See Sapp 2009 for a very satisfying account of the intellectual history of LUCA and the progenote. Sapp pays due attention to the contributions of Otto Kandler and Wolfram Zillig, which overlap with those of Woese. See also Darnell and Doolittle 1986.

3. Woese 1998a, 2000, 2002, 2004; Woese et al. 2000; Vetsigian, Woese, and Goldenfeld 2006.

4. Woese 1998a, 6856.

5. Woese et al. 2000; Vetsigian, Woese, and Goldenfeld 2006.

6. Vetsigian, Woese, and Goldenfeld 2006.

7. For critique of the suggestion that rampant gene transfer suffices to drive adaptive evolution see Kurland, Canbäck, and Berg 2003; De Duve 2005a, 2005b; and Penny 2005. Gray et al. (2010) make a case for constructive neutral evolution, a directional force that favors increasing complexity even in the absence of positive selection. The argument is plausible but seems insufficient to explain the emergence of functional adaptive structures.

8. For a formal statistical test of universal common ancestry see Theobald 2010. Koonin and Wolf (2010) reject the argument but agree that the conclusion is correct.

9. For the rootings advocated by Cavalier-Smith and Gupta, see below. Forterre

and Philippe (1999) place the root on the eukaryotic branch; Lake et al. (2009) between the monoderm bacteria and the diderms. Doolittle (2009b) explains lucidly why common ancestry does not necessarily imply a common ancestor.

10. Lake et al. 2009.

11. Forterre and Philippe 1999; Lake et al. 2009; Doolittle 2009b.

12. The literature holds diverse but overlapping perspectives on the nature of the last universal ancestor. I have drawn on Becerra et al. 2007; Glansdorff, Xu, and Labedan 2008; Zhaxybayeva and Gogarten 2004; Wang et al. 2010; and Di Giulio, 2011.

13. For the origin and implications of this important discovery, see Olsen and Woese 1997; Forterre 2002, 2006b.

14. Glansdorff, Xu, and Labedan (2008) spell out the argument in detail. Caetano-Anollès, Kim, and Mittenthal (2007) buttress the claim that LUCA was metabolically sophisticated. See Collins and Penny 2005 for spliceosomes; for introductions to the baffling distribution of lipids, see Peretò, Lòpez-Garcia, and Moreira 2004 and Lombard, Lòpez-Garcia, and Moreira 2012.

15. Glansdorff, Xu, and Labedan 2008.

16. Pace 1991; Stetter 2006.

17. Most authorities continue to be skeptical, but recent molecular evidence favors a thermophilic LUCA, in accordance with geochemical evidence for a warm ocean. See Di Giulio 2003; Stetter 2006; De La Rocha 2006; Gaucher, Govindarajan, and Ganesh 2008; Gouy and Chaussidon 2008; Boussau et al. 2008.

18. Cavalier-Smith's views have evolved over time, but remain consistent. Major presentations include his articles of 2002a, 2002b,; 2006a, 2006b; and 2010. Despite their length and unremitting detail they repay close reading, for they offer a fully articulated alternative to the conventional wisdom.

19. For a more positive take on the proposition that Archaea arose by transfiguration of one of the Bacteria, see Valas and Bourne 2011.

20. Gupta 2000, 2002.

21. Lake and Rivera 2004; Rivera and Lake 2004; Simonson et al. 2005, 6611.

22. Lake 2009.

23. Ibid.; for a devastating critique, see Swithers et al. 2011.

Chapter Five

1. For general introductions see Harold 1986; Nicholls and Ferguson 1992; and White 2007.

2. The case for glycolysis late has been spelled out by Martin and Russell, (2002) and by Lane, Allen, and Martin (2010).

3. Canfield, Rosing, and Bjerrum (2006) review the metabolic economics of the ancient anaerobic world. The argument that LUCA must have relied on inorganic energy donors is set forth by Martin and Russell (2002) and Lane, Allen,

and Martin (2010). Richardson (2000) has provided a useful broad-gauge review of the diversity of Bacterial respiration, while Croal et al. (2004) focus on pathways accessible from geochemistry.

4. For some recent examples see Biegel and Müller 2010 and Sapra, Bagramyan, and Adams 2003.

5. Stetter 2006.

6. For the metabolic patterns of acetogenesis see White 2007; recent reviews are due to Berg et al. 2010, Fuchs 2011, and Buckel and Thauer 2013. An excellent paper by Poehlein et al. (2012) discusses in some detail the metabolic economy of *Acetobacter woodii*, and clearly explains the reduction of CO_2, the generation and function of the sodium circulation, and how the near-magical mechanism called electron bifurcation makes it all hang together. The case for acetogenesis or methanogenesis as the primordial pathway of energy conservation is spelled out by Martin and Russell (2007) and Martin (2012).

7. Kelley et al. (2005) report on the Lost City hydrothermal field.

8. For recent discussions of the place of hydrothermal fields in cell evolution see Martin and Russell 2007; Martin et al., 2008; and Lane, Allen, and Martin 2010.

9. Lane, Allen, and Martin 2010.

10. H. Belloc, "The Microbe," 1912.

11. Woese 1987, 264.

12. Saier 2003; Therer and Saier 2009.

13. Several aspects of the redox construction kit are considered by Baymann et al. 2003; Martin and Russell 2007; Nitschke and Russell 2009; Martin 2012; and Schoepp-Cothenet et al. 2013.

14. For ATP synthase, see Boyer 1993; Capaldi and Aggeler 2002; and Mulkidjanian et al. 2007.

15. Mulkidjanian et al. 2007.

16. Martin and Russell 2007; Martin et al., 2008; and Lane, Allen, and Martin 2010.

17. Russell and Hall 1997; Martin and Russell 2002, 2007; Nitschke and Russell 2009.

18. For examples, see Berry 2002 and Leigh 2002. See also fig. 5.4.

19. Berg et al. 2010; Fuchs 2011; Buckel and Thauer 2013.

20. For a technical overview of photosynthesis, see Blankenship 2002. The evolution of photosynthesis has been considered by Xiong and Bauer 2002; Raymond et al. 2002; Mulkidjanian et al. 2006; Olson 2006; Allen and Martin 2007; Shi and Falkowski 2008; and by Hohmann-Marriott and Blankenship 2011.

21. Beja, Aravind, and Koonin 2000; Frigaard et al. 2006; Fuhrman, Schwalbach, and Stingl 2008.

22. The evidence concerning the evolution of respiratory chains is confusing and hard to interpret. For recent efforts, see Castresana 2001; Ducluzeau, Ouchane, and Nitschke 2008; and Brochier-Armanet, Talla, and Gribaldo 2009.

23. Ducluzeau, Ouchane, and Nitschke 2008; and Brochier-Armanet, Talla, and Gribaldo 2009.

Chapter Six

1. Darwin (1859) 1979, 217, 219.

2. Nilsson and Pelger 1994.

3. Behe 1996.

4. For a recent overview of prokaryotic motility mechanisms, see Jarrell and McBride 2008. The evolution of bacterial flagella is discussed in Pallen and Matzke 2006; Wong et al. 2007; and Snyder et al. 2008. The claim by Liu and Ochman (2007) that flagella arose from a small number of precursor molecules by repeated cycles of gene duplication, was immediately savaged by Doolittle and Zhaxybayeva (2007) and appears to be incorrect.

5. For Archaeal flagella and their designation as archaella see Jarrell and Mc-Bride 2008 and Jarrell and Albers 2012.

6. Jarrell and McBride 2008; Jarrell and Albers 2012.

7. Gould 1997.

8. Maezawa et al. 2006.

9. Bray 2009.

10. Lenski et al. 2003.

11. The literature on the spatial organization of prokaryotic cells is growing by leaps and bounds. Recent reviews include Shapiro, McAdams, and Losick 2009; Margolin 2009; Cabeen and Jacobs-Wagner 2010; Vendeville, Larivière, and Fourmentin 2011; and Ingerson-Mahar and Gitai 2012.

12. Harold 2005 lays out the case.

13. Blobel 1980.

14. Sapp 1998, 224.

15. Cavalier-Smith 2000, 2004, 2010. Quotation from Cavalier-Smith 2000, 175, 176.

16. See, for example, Wächstershäuser, 2003; Peretò, Lòpez-Garcia, and Moreira 2004; Martin and Russell 2002; Lane, Allen, and Martin 2010; and Lombard, Lòpez-Garcia, and Moreira 2012.

17. Shapiro, McAdams, and Losick 2009; Margolin 2009; Cabeen and Jacobs-Wagner 2010; Vendeville, Larivière, and Fourmentin 2011; and Ingerson-Mahar and Gitai 2012.

18. Erickson 2007.

19. Recent contributions to our knowledge of the archaeal cytoskeleton include Makarova et al. 2010; Ettema, Lindås, and Bernander 2011; and Guy and Ettema 2011.

20. Erickson 2007; Makarova et al. 2010; Barry and Gitai 2011.

21. Barry and Gitai 2011.

22. Erickson 2007; Makarova et al. 2010; Barry and Gitai 2011.

23. See Koch and Silver 2005 and Woese 1987, 1998a, 2004. For a historical overview, see Sapp 2009.

24. Woese 1987, 1998a, 2004.

25. Woese 1998a, 6856.

Chapter Seven

1. Sogin 1994; Pace 1997. In a more recent study of SSU rRNA sequences, Dawson and Pace (2002) report a somewhat different order of branching, but the general pattern remains the same.

2. Keeling 1998; Roger 1999; Cavalier-Smith 2002b; Embley and Martin 2006; Shiflett and Johnson 2010; Müller et al. 2012. By contrast, Margulis et al. (2006) defend the archaezoan hypothesis and insist that many anaerobic protists will prove to be ancestrally devoid of mitochondria.

3. Hirt et al. 1999; Embley and Martin 2006.

4. The new tree of the eukaryotes is presented by Cavalier-Smith 2002; Baldauf 2003; Keeling et al. 2005; and Adl et al. 2005.

5. For picoeukaryotes, see Baldauf 2003 and Massana 2011. The genomes of the two record holders have been reported by Matsuzaki et al. 2004 and Derelle et al. 2006.

6. Derelle et al. 2006.

7. Matsuzaki et al. 2004.

8. Cavalier-Smith 2002b, 2009; Stechman and Cavalier-Smith 2002, and Richards and Cavalier-Smith 2005.

9. Roger and Hug (2006) and Roger and Simpson (2009) cast a critical eye on the state of play.

10. Keeling and Palmer 2008; Bapteste et al. 2009.

11. Margulis 1993; Margulis and Sagan 2002.

12. Gast, Sanders, and Caron 2009.

13. Cavalier-Smith 2002b, 2009.

14. Hartman and Fedorov 2002; Kurland, Collins, and Penny 2006; Poole and Penny 2006.

15. Margulis 1970; Stanier 1970; Cavalier-Smith 1975.

16. Davis 2003, 73.

17. The literature on the origin of eukaryotic cells is enormous, and growing by leaps and bounds. I have drawn especially on review articles by Cavalier-Smith (2002b, 2009), Martin (2005), Embley and Martin (2006), Kurland, Collins, and Penny (2006), Poole and Penny (2006), De Duve (2007), and Gribaldo et al. (2009).

18. Margulis et al. 2006.

19. Lòpez-Garcia and Moreira 1999, 2006.

20. Martin and Müller 1998; Müller et al. 2012.

21. Rivera and Lake 2004; Simonson et al. 2005.

22. Pisani, Cotton, and McInerney 2007; Cox et al. 2008; Yutin et al. 2008; Davidov and Jurkevitch 2009; Guy and Ettema 2011.

23. Davidov and Jurkevitch 2009.

24. Cavalier-Smith 2002b, 2009. See also Kurland, Collins, and Penny 2006, De Duve, 2007.

25. Kurland, Collins, and Penny 2006; Poole and Penny 2006.

26. Kurland, Collins, and Penny 2006; Poole and Penny, 2006. For a slashing critique see Martin et al. 2007, and its rebuttal.

27. Jèkely (2007) and De Nooijer, Holland, and Penny (2009) consider the social context in which eukaryotes emerged.

28. Hartman and Fedorov 2002; Kurland, Collins, and Penny 2006; Collins and Penny 2005.

29. Dacks and Field 2007; Dacks, Poon, and Field 2008.

30. Concerning undulipodia and their origin, see Cavalier-Smith 2002b, 2009; Chapman, Dolan, and Margulis 2000; Margulis et al. 2006; Jèkely and Arendt 2006.

31. For nuclear membranes, see Cavalier-Smith 2002b, 2009; Martin and Koonin 2006; Lòpez-Garcia and Moreira 2006.

32. Martin and Koonin 2006; Lòpez-Garcia and Moreira 2006.

33. Kurland, Collins, and Penny 2006; Poole and Penny 2006. Jèkely, 2007, and De Nooijer, Holland, and Penny 2009.

34. Kurland, Collins, and Penny 2006.

35. The most pertinent genomic analyses come from Cox et al. 2008; Williams et al. 2012; Yutin et al. 2008; Guy and Ettema 2011; and Yutin and Koonin 2012. Odd, is it not, that the A-ATPase of Archaeal plasma membranes is homologous to the V-ATPase of eukaryotic endoplasm (rather than to an ATPase of the latter's plasma membrane), and that the apparatus of cell division in some Crenarchaeota is homologous to a eukaryotic system for endoplasmic vesicle formation? There must be a lesson here, but I cannot read it.

36. Lester, Meade, and Pagel 2005; Embley and Martin 2006; Timmis et al. 2004; Pisani, Cotton, and McInerny 2007.

37. Lane and Martin 2010; Lane 2011.

38. Ibid. See also chapter 11.

39. Embley and Martin 2006, 623.

Chapter Eight

1. Margulis 1970.

2. Thomas 1974, 82.

3. Gray and Doolittle 1982.

4. Woese 1987.

5. The literature on the evolution of organelles is large and intimidating, but well covered by current reviews. Among the most helpful, see Gray, Burger, and Lang 1999; Andersson et al. 2003; Dyall, Brown, and Johnson 2004; Embley and Martin 2006; Reyes-Prieto, Weber, and Bhattacharya 2007; Gould, Waller, and McFadden 2008; Archibald 2009; and Keeling 2010. For a general introduction to mitochondria and their place in the world see Lane 2005.

6. Gast et al. 2009; Nowack and Melkonian 2010.

7. Gray, Burger, and Lang 1999; Andersson et al. 2003; Embley and Martin 2006; Cavalier-Smith 2006a; Shiflett and Johnson 2010; Müller et al. 2012.

8. John and Whatley 1975.

9. Davidov and Jurkevitch 2009.

10. Cavalier-Smith 2006c.

11. Fenchel and Bernard 1993.

12. Andersson et al. 2003.

13. Martin and Müller 1998.

14. Gray, Burger, and Lang 1999; Burger, Gray, and Lang 2003; Gray, Lang, and Burger 2004.

15. Gibbs 1981, 2006.

16. The evidence that plastids are monophyletic is reviewed in articles by Palmer 2003; Rodriguez-Ezpelata et al. 2005; Reyes-Prieto, Weber, and Bhattacharya 2007; Archibald 2009; and Keeling 2010. Confirmation comes from the genome of the glaucophyte alga *Cyanophora*, reported by Price et al. (2012).

17. Deschamps et al. 2008.

18. Criscuolo and Gribaldo 2011.

19. Yoon et al. 2004.

20. Cavalier-Smith 2006c; Keeling 2010.

21. For a sample of the extensive literature on secondary plastids see Falkowski et al. 2004; Bhattacharya et al. 2007; Lane and Archibald 2008; Archibald 2009; Elias and Archibald 2009; Keeling 2010.

22. See Gray, Burger, and Lang 1999; Andersson et al. 2003; Dyall, Brown, and Johnson 2004; Embley and Martin 2006; Reyes-Prieto, Weber, and Bhattacharya 2007; Gould, Waller, and McFadden 2008; Archibald 2009; Keeling 2010.

23. For diverse views on the origin of metabolite exchange and protein import in mitochondria, see Andersson et al. 2003; Dyall, Brown, and Johnson 2004; Cavalier-Smith 2006c; Dolezal et al. 2006; Gross and Bhattacharya 2009; Clements et al. 2009; Lithgow and Schneider 2010.

24. Ibid.

25. Gray, Burger, and Lang 1999; Burger, Gray, and Lang 2003; Gray, Lang, and Burger 2004; Timmis et al. 2004.

26. Ibid.

27. Gray, Burger, and Lang 1999; Allen 2003.

28. Ibid.

29. Lane and Martin, 2010, 2012.

30. Osteryoung and Nunnari 2003; Hoppins, Lackner, and Nunnari 2007.

31. The molecular mechanisms of primary and secondary plastid formation have been discussed at length. See Cavalier-Smith 2003; Reyes-Prieto, Weber, and Bhattacharya 2007; Bhattacharya et al. 2007; Gould, Waller, and McFadden 2008; Lane and Archibald 2008; Archibald 2009; Gross and Bhattacharya 2009; Keeling 2010.

32. Allen 2003.

33. Archibald 2009, R81.

34. Marin, Nowack, and Melkonian 2005; Nowack, Melkonian, and Glöckner 2008; Keeling 2010; Nowack and Grossman 2012.

35. Ibid.

36. Müller 1993.

37. Embley et al. 2003; Lithgow and Schneider 2009; Hjort et al. 2010; Shiflett and Johnson 2010; Müller et al. 2012.

38. Ibid.

39. Gould, Waller, and McFadden 2008; Elias and Archibald 2009; Keeling 2010.

40. Ibid.

Chapter Nine

1. Confusingly, there is no obligatory connection between the Archean eon and the prokaryotic Archaea, which may or may not have thrived in that distant day. It does not help matters that the eon Americans spell Archean is designated in the United Kingdom as the Archaean.

2. Knoll 2003; the indispensable introduction to early life on earth, and a delightful read.

3. The symposia include Cavalier-Smith, Brasier and Embley 2006; Nisbet et al. 2008; the textbook is Sullivan and Baross 2007.

4. Lane 2002.

5. Knoll 2003.

6. Hoffman 1976; Knoll 2003; Awramik and McNamara 2007.

7. Brocks et al.1999; Summons et al. 2006; Brocks and Banfield 2009.

8. Canfield, Glazer, and Falkowski 2010.

9. For the molecular phylogeny of cyanobacteria, see Tomitani et al. 2006; Schirrmeister, Antonelli, and Bagheri 2011.

10. Kasting and Siefert 2002; Catling and Kasting 2007.

11. Canfield, Rosing, and Bjerrum 2006.

12. Stolper, Revsbech, and Canfield 2010.

13. Moorbath, 2005; Fedo, Whitehouse, and Kamber 2006; Eiler 2007.

14. For reviews of this controversial subject, see especially the magisterial survey by Buick 2007; also Altermann and Kamierczak 2003; Schopf, 2006; Westall

2005, 2009. Fossil bacteria from the deep Archean, 3300 to 3400 million years ago, have been described by Westall et al. 2006 and Wacey et al. 2011.

15. Awramik 2006; Buick 2007; Allwood et al. 2009; Arndt and Nisbet 2012.

16. Garcia-Ruiz et al. 2003; Brasier et al. 2006.

17. Schopf 1993, 2006.

18. Dalton 2002; Brasier et al. 2006.

19. Ueno et al. 2006; and commentary by Canfield 2006.

20. Javaux, Marshall, and Mekker 2010.

21. Buick 2010.

22. Brocks et al. 1999.

23. Rasmussen et al. 2008. Waldbauer et al. (2009) reaffirm the biomarker evidence for cyanobacteria, oxygenic photosynthesis, respiration and eukaryotes about 2600 million years ago. See also George et al. (2008), who found biomarker molecules in droplets of oil extracted from somewhat younger rocks.

24. Brocks et al. 1999; Rasmussen et al. 2008; Waldbauer et al. 2009; George et al. 2008.

25. Falkowski et al. 2004; Knoll et al. 2006; Berney and Pawlowski 2006; Brocks and Banfield 2009.

26. Bengtson et al. 2009.

27. Javaux, Knoll, and Walter 2004; Knoll et al. 2006.

28. Javaux, Marshall, and Mekker 2010.

29. Knoll et al. 2006.

30. Cavalier-Smith 2002a, 2006a, 2006b, 2010.

31. Kasting and Siefert 2002; Catling 2005; Kopp et al. 2005; Falkowski 2006; Holland 2006; Falkowski and Isozaki 2008; Buick 2008.

32. Knoll 2003; Holland 2006.

33. Kasting and Siefert 2002; Catling 2005; Kopp et al. 2005; Falkowski 2006; Holland 2006; Falkowski and Isozaki 2008; Buick 2008.

34. For the relationship of oxygen to steranes, see Summons et al. 2006, and Waldbauer, Newman, and Summons 2011.

35. Kasting and Ohno (2006) consider the evidence for a methane-rich atmosphere on the early earth.

36. Ueno et al. 2006; and commentary by Canfield 2006.

37. For recent efforts to date LECA, see Berney and Pawlowski 2006; Chernikova et al. 2011; and Parfrey et al. 2011.

38. Cavalier-Smith 2010.

Chapter Ten

1. Books that survey the origin of life include Eigen 1992; Morowitz 1992; Maynard Smith and Szathmàry 1995; Davies 1999; Fry 2000; Popa 2004; De Duve 2005a; Hazen 2005; and Koonin 2012.

2. Haldane 1929; Oparin 1938.

3. Miller 1953.

4. A sample of this literature includes Lazcano and Miller 1996; Orgel 1998; Deamer et al. 2002; Chen et al. 2006; Powner, Gerland, and Sutherland 2009; Cooper et al. 2011; and Saladino et al. 2012.

5. Ibid.

6. Ibid.

7. The quest for a self-replicating RNA has been reviewed with varying degrees of skepticism by Joyce 2002; Orgel 2004; Joyce and Orgel 2006; and Shapiro 2006, 2007.

8. Lincoln and Joyce 2009; Wochner et al. 2011.

9. See, for example, Fry 2000, 2011; De Duve 2005a; Hazen 2005; Penny 2005; Joyce and Orgel 2006; Shapiro 2006, 2007.

10. Hazen 2005, 131.

11. Eigen 1992.

12. Pross 2003, 2004, 2005a, 2005b. For a current and highly readable presentation see Pross 2012.

13. Pross 2003, 2004, 2005a, 2005b, 2012. Quotation from Pross 2003, 400.

14. Pross 2003, 403, 404.

15. See, for example, Fry, 2000, 2011; De Duve 2005a; Hazen 2005; Penny 2005; Joyce and Orgel 2006; Shapiro 2006, 2007.

16. For examples, see Dyson 1985; Kauffman 1993, 2000; Shapiro 2006, 2007; Trefil, Morowitz, and Smith 2009.

17. Lifson 1997; Fry 2000; Pross 2003; Orgel 2000, 2008. Quotation from Orgel 2008, 11.

18. Wicken 1987; Schneider and Sagan 2005.

19. Shapiro 2006, 2007.

20. Orgel 2000, 2008. Also see Pross 2003.

21. Concerning the role of natural selection in prebiotic evolution, see Maynard Smith and Szathmàry 1995; De Duve, 2005a, 2005b; Penny 2005; Fry 2000, 2011; and Moreno and Ruiz-Mirazo 2009.

22. The presumption that biological molecules and processes evolved together in membrane-bound spaces has become commonplace, and features in many of the books and articles cited above. For a recent collection of articles on protocells, see Rasmussen et al., 2009. Incidentally, while most authors envisage protocells to have had the same topology as contemporary cells, a few imagine them to have been everted ("obcells"); see Cavalier-Smith 2001; Griffiths 2007.

23. For the role of membranes in prebiotic evolution see Morowitz 1992; Szostak, Bartel, and Luisi 2001; Deamer et al. 2002; Koch and Silver 2005; Luisi, Ferri, and Stano 2006; Chen et al. 2006.

24. Hanczyk, Fujikawa, and Szostak 2003; Chen et al. 2006; Mansy et al. 2008.

25. Deamer 2008.

26. Ibid., 37.

27. Koch and Schmidt 1991; Koch and Silver 2005.

28. Ferry and House 2006.

29. Steinberg-Yfrach et al. 1997.

30. Berry 2002; Deamer et al. 2002; Baymann et al. 2003

31. Kornberg, Rao, and Ault-Richè 1999.

32. De Duve 2005a; Smith and Morowitz 2004; Trefil, Morowitz, and Smith 2009.

33. Among the proponents of either the reductive TCA cycle or the Wood-Ljungdahl pathway, see Wächtershäuser 1992, 2006; Lindahl 2004; Smith and Morowitz 2004; Ferry and House 2006; Martin and Russell 2007; Trefil, Morowitz, and Smith 2009.

34. Cairns-Smith 1985.

35. Wächtershäuser 1992, 2006; Lindahl 2004; Cody 2004.

36. Lifson 1997; Orgel 2000, 2008.

37. Russell and Hall 1997; Martin and Russell 2002, 2007.

38. See Koonin and Martin 2005; Martin et al. 2008; Nitschke and Russell 2009; Lane, Allen, and Martin 2010; Martin 2011; Lane and Martin 2012. In a recent variation on this theme, Mulkidjanian et al. (2012) draw attention to the fact that cytoplasm is rich in K+ but poor in Na+, the very opposite of seawater. This and other arguments leads them to seek the origin of life in inland geothermal fields, some of which generate fluids whose ionic composition resembles that of cytoplasm.

39. Koch and Silver 2005.

40. Gilbert 1986. For reviews of what the RNA World stands for today see Joyce, 2002; Penny, 2005; and Gesteland, Czech, and Atkins 2006, particularly the chapter by Joyce and Orgel. Copley, Smith, and Morowitz 2007, consider the coevolution of RNA genes with metabolism, and Kurland 2010, raises serious questions about the whole conception of a pure RNA World.

41. Gilbert 1986; Joyce 2002; Penny 2005; Gesteland, Czech, and Atkins 2006; Copley, Smith, and Morowitz 2007.

42. For a radical critique of current ideas, see Trevors and Abel 2004; Abel and Trevors 2005, 2006.

43. For examples, see Szathmàry 1999; Penny 2005; Wolf and Koonin 2007; Yarus 2010.

44. Wolf and Koonin 2007, 1.

45. Bokov and Steinberg 2009; Caetano-Anollès, Kim, and Mittenthal 2007, Caetano-Anollès et al. 2011, Caetano-Anollès, Kim, and Caetano-Anollès 2012. Mulkidjanian et al. 2009.

46. Villareal 2005; Forterre 2006a, 2006b; Koonin and Dolja 2006; Jalasvuori and Bamford 2008; and Koonin 2012.

47. Ibid.

48. Popa 2004, 138.

49. Conway Morris 2003.

50. "Autopoietic" means self-making. The proposal that living organisms should

be defined as entities that make themselves was put forward more than thirty years ago by F. G. Varela and H. R. Maturana, and has proven both practical and illuminating; see Luisi 2003, for a review. "Autonomous agent," as used by Kauffman (2000), is not exactly the same but the meanings overlap.

51. Kauffman 1993, 2000.

Chapter Eleven

1. The thesis that recent discoveries supersede the modern synthesis has been vigorously argued by Sapp (2009), Shapiro (2011), and most notably by Koonin (2012).

2. Scholars differ on the shortcomings of the modern synthesis, and on what expansion should entail. For examples, see Gould 1982; Margulis and Sagan 2007; Kutschera and Niklas 2004; Jablonka and Lamb 2005; and Pigliucci 2007. Cell evolution demands a still more radical transformation of the framework.

3. Maynard Smith and Szathmáry 1995.

4. For some perspectives on the genomic basis of evolution see Long et al. 2003; Taylor and Raes 2004; Frost et al. 2005; and Bard 2010.

5. Buckling et al. 2009; Elena and Lenski 2003.

6. Blount, Borland, and Lenski (2008) and Blount et al. (2012) analyze the origin of mutants that can utilize citrate.

7. Jablonka and Lamb 2005; Bird 2007; Jablonka and Raz 2009; Huang 2011.

8. Mayr 1997; Griesemer 2000.

9. Conway Morris, 2003.

10. Leander 2008; Lukeš, Leander, and Keeling 2009.

11. Young 2006.

12. Webster and Goodwin 1996; Lukeš, Leander, and Keeling 2009.

13. McShea and Brandon 2010, 4. It should be noted that these authors use the term "complexity" in a rigorous fashion, to refer exclusively to the number and disparity of constituent parts. In colloquial usage, widespread in the literature, the term implies much more including functionality, adaptation, sophistication and integration. There seems to be no generally accepted term for this all-important quality of living things, and so I will sometimes employ "complexity" to designate it. We shall revisit this matter in chapter 12.

14. See Lynch 2007. Gray et al. (2010) make a strong case for neutral evolution as a constructive force in the generation of complexity, and Finnigan et al. (2012) present a striking example of its operation. Speijer (2011) is not convinced and defends the importance of natural selection in explaining the sophisticated complexity of biological form and function.

15. Gould 1982; Margulis and Sagan 2007; Kutschera and Niklas 2004; Jablonka and Lamb 2005; Pigliucci 2007.

16. Vetsigian, Woese, and Goldenfeld 2006; Goldenfeld and Woese 2007; Koonin 2007.

17. For defense of the traditional position see De Duve 2005a, 2005b; Kurland 2005; and Kurland, Canbäck, and Berg 2003.

18. Shapiro 2011.

19. For prokaryotic mergers at the base of the dubious host, see Margulis et al. 2006; Horiike et al. 2001; and Simonson et al. 2005. Claverie and Abergel (2010) touch on the possible role of viruses in cell evolution.

20. Lake (2009) argues that diderm prokaryotes are products of an early symbiosis, but his case has been vigorously challenged by Swithers et al. (2011).

21. Sapp 1987, 2003; Harold 2005.

22. Ibid.

23. Halfmann and Lindquist 2010; Halfmann et al. 2012.

24. Ibid.

25. Kirschner and Gerhart 1998, 8420; Pigliucci 2008.

26. Lane and Martin 2010.

27. Allen 2003.

28. Koonin 2007, 2012, with reservations concerning the cosmological parallels.

29. Sapp 2009; Koonin 2007, 2012.

30. Behe 2007 (quotes, 162, 165).

31. Miller 2007.

32. Meyer et al. 2012.

33. Foster 2000.

34. Crick 1981.

35. Line 2002.

Chapter Twelve

1. Conway Morris 2003, 23; Denton 1998.

2. Kauffman 2000.

3. Wholeness is a conspicuous feature of living things that pertains to the integration of disparate parts, but is not easily pinned down, let alone measured. A preoccupation with wholeness is historically linked to the philosophical stance called Organicism or Holism, traceable to the German philosopher Immanuel Kant, which emphasizes the emergent autonomous activity of whole organisms. For discussion, see Mayr 1982 and Goodwin, 1994. The subject remains highly relevant, as documented in an excellent recent article by Denton, Kumaramanickavel, and Legge 2013.

4. McShea and Brandon 2010.

5. Rosslenbroich 2006. A thoughtful and readable survey of suppositions and controversies surrounding the idea of evolutionary progress.

6. Kauffman 2000.

7. Conway Morris 2003, 329.

8. Denton 1998.

9. Gardner 2003. This author also provides a lucid and current summary of the findings and speculations that have brought us to this pass.

10. Rees 1999.

11. Davies 2006, 300.

12. Smolin 1997.

13. For a biography of this intellectual giant who helped to shape both biochemistry and genetics early in the twentieth century see Clark 1969.

14. Burton (1880) 1994.

Glossary

actin: major protein of the eukaryotic cytoskeleton; the chief component of microfilaments

adaptation: in evolution, any change in the structure or functioning of an organism that makes it better suited to its environment

adenosine diphosphate (ADP): see adenosine triphosphate

adenosine triphosphate (ATP): the universal energy currency of living organisms; the molecule consists of adenine, ribose, and three phosphoryl groups; successive release of phosphate gives rise to adenosine diphosphate (ADP) and adenosine monophosphate (AMP)

ADP: see adenosine triphosphate

algae: in this book, used generically to designate photosynthetic protists

allele: one of two or more alternate states of a gene that typically arise by mutation

amino acid: a small, water-soluble, organic compound that possesses both a carboxyl group (-COOH) and an amino group ($-NH_2$); proteins are polymers made up of a characteristic set of twenty amino acids

antiport: the coupled transport of two substrates by a single carrier, such that the substrates move in opposite directions (contrast with *symport*)

Archaea: microbial clade originally defined by ribosomal RNA sequences and associated with extreme environments, now known to be widespread; Archaea constitute one of the three domains of life

archaeon: an organism classified in the domain Archaea

Archean eon: the period from about 4 to 2.5 billion years ago

Archezoa: a proposed kingdom-level division of the protists whose members lack mitochondria and chloroplasts and which may have diverged from the eukaryotic stem prior to the acquisition of endosymbionts

ATP: see adenosine triphosphate

ATP synthase: a class of complex enzymes pivotal to biological energy transduction, that link ATP chemistry in the cytoplasm to the translocation of protons (sometimes

Na$^+$) across membranes; in bacteria, mitochondria and plastids, ATP synthases catalyze ATP synthesis during oxidative and photosynthetic phosphorylation

autopoiesis: the capacity of living organisms to make and reproduce themselves au-
tonomously

autotroph: an organism that can derive all its carbon from CO_2

Bacteria: diverse microbial domain that includes most of the familiar lineages, as well
as the ancestors of both mitochondria and plastids; when spelled with a lower-case
"b," refers to the traditional informal term for prokaryotes

bacteriophage (-phage): a virus that is parasitic on bacteria

biopoiesis: the development of life from nonliving matter; synonymous with the origin
of life

catalyst: a substance that increases the rate of a chemical reaction without itself under-
going permanent chemical change; *enzymes* are the chief catalysts in biochemical
reactions

cell: the structural and functional unit of living organisms; cell size varies, but most are
microscopic (0.001 to 0.1 mm); many organisms consist of but a single cell (bacteria
and most protists), others are multicellular (fungi, plants, and animals)

cell heredity: the transmission of cellular traits by structural continuity or other
means, without the participation of genes (see also under *structural heredity*)

cell wall: the strong and relatively rigid envelope of many cells external to the plasma
membrane; cells of plants, fungi, many protists, and most bacteria are walled; animal
cells are not

cellulose: a polymer of glucose characteristically found in eukaryotic cell walls, especially those of plants and certain protists

chemoautotroph: a bacterium that lives by extracting energy from inorganic chemical
reactions

chemolithotroph: a bacterium that lives by extracting energy from inorganic chemical
reactions; sometimes called chemoautotroph

chloroplast: a membrane-bound organelle in cells of plants and green algae that contains chlorophyll and is the locus of photosynthesis (see also *plastid*).

chromosome: a threadlike structure found in cell nuclei, which carries genes in linear
array

ciliate: a protozoan whose surface is studded with cilia, which serve as the organs of
motility

cilium: surface appendage of eukaryotic cells that typically serves as an organ of motility; in the literature of cell biology, the relatively short cilia are distinguished from
their longer cousins, called flagella (see also under *flagellum* and *undulipodia*)

clade: a group of organisms comprised of a common ancestor and all its descendants
(see also *monophyletic group*)

coding: in biology, the procedure by which the sequence of nucleotides in a gene determines the sequence of amino acids in a protein (see also codon)

codon: a triplet of nucleotide bases within a molecule of DNA or mRNA that specifies a particular amino acid

complexity: the condition of a system made up of multiple interacting parts whose behavior is not easily deduced from the properties of those components

cyanobacteria: a phylum of Bacteria characterized by photosynthesis of the same kind as that of plastids

cytokinesis: the process by which a cell divides into two (or more) separate daughter cells

cytoplasm: the jellylike material that fills cellular space exclusive of the nucleus and other organelles

cytoskeleton: a network of microscopic filaments and tubules that pervades the cytoplasm of both eukaryotic and prokaryotic cells; it's functions include motility, cell division, and secretion

cytosol: the fluid phase of cytoplasm exclusive of organelles

deoxyribonucleic acid (DNA): a nucleic acid composed of two intertwined polynucleotide chains; the sugar is deoxyribose; DNA is the genetic material of all cells and of many viruses

diderm: a prokaryotic cell whose envelope features two lipid membranes, the plasma membrane and an outer membrane; it typically stains gram negative (contrast with monoderm; see also under *gram-negative, gram-positive*)

DNA polymerase: an enzyme that catalyzes the synthesis of DNA from its nucleotide precursors, typically using an existing strand of DNA as the template; the agent of gene replication

domain: the highest level of biological classification; the three domains are Archaea, Bacteria, and Eukarya

Escherichia coli: a normal inhabitant of the human gut classified in the phylum Proteobacteria; of all living organisms, this one is most fully understood

emergence: the appearance of new properties in a system that were not present nor easily predictable from the properties of the components

endergonic: a biochemical reaction that requires energy and will not proceed unless an external energy source is present (e.g., most biosynthetic processes)

endocytosis: a mode of nutrient uptake in which the substrate is enclosed within a vesicle derived from the plasma membrane

endomembrane: a generic term for a membrane within the cytoplasm (e.g., the endoplasmic reticulum and the Golgi apparatus of eukaryotic cells)

endosymbiosis: a symbiosis in which one of the partners (the endosymbiont) resides within the cytoplasm of the other

energy: the capacity to do work; in biology, the capacity to drive processes that do not occur spontaneously, such as the synthesis of complex molecules

entropy: a portion of the energy of a system that is unavailable for work; in a wider sense, a measure of the system's disorder; as disorder increases, so does the entropy

enzyme: a protein that acts as a catalyst in biochemical reactions

epigenesis: the approximately stepwise process by which genetic information, modified by environmental factors, is translated into the substance and behavior of an organism

epigenetic inheritance: refers to heritable features not encoded in the nucleotide sequence of genes

Eukarya: the domain of life that contains all eukaryotes, whether unicellular or multicellular (also spelled *Eucarya*)

eukaryote: an organism whose genetic material is enclosed in a true, membrane-bound nucleus; Eukaryotic cells also typically have a cytoskeleton, internal membranes, and organelles (also spelled *eucaryote*)

exergonic: a biochemical process that releases energy (e.g., respiration)

exocytosis: a mode of secretion in which substances are enclosed within a vesicle that fuses with the plasma membrane and discharges its contents to the outside

exon: a nucleotide sequence in a gene that codes for part or all of the gene product; exons are sometimes separated by noncoding sequences called introns

fermentation: a biochemical pathway that breaks down organic substances in the absence of oxygen, with production of metabolites and usable energy

fimbria: hairlike projections commonly seen on the surface of bacterial cells

Firmicutes: a large phylum of monoderm (gram-positive) Bacteria; the most familiar example is *Bacillus subtilis*

fitness: the condition of an organism that is well adapted to its environment, as measured by the ability to reproduce its kind

flagellate: a motile protist equipped with undulipodia (eukaryotic flagella)

flagellum: a relatively long, whiplike structure present on the surface of many cells and serving as an organ of motility; in this book, flagellum refers specifically to the organelles of prokaryotes; eukaryotic flagella (and their shorter cousins, the cilia) have an entirely different structure and are designated "undulipodia"

foraminifer: a kind of protozoan whose elaborate silica shells are often found fossilized

free energy: a measure of a system's ability to do work; in biochemistry, the fraction of the energy released by a chemical reaction that is available to do work (the unavailable fraction is the entropy)

gene: a unit of heredity, usually consisting of a stretch of DNA that codes for some biological function

genome: the total gene complement of an organism; more precisely, all the genes carried on a single set of chromosomes

glycolysis: the sequence of reactions by which sugar is converted to ethanol in the absence of oxygen

Golgi apparatus: assembly of vesicles and membranous tubules found in the cytoplasm of eukaryotic cells that participates in the secretion of cell products

gram-negative, gram-positive: two classes of prokaryotes that respond differentially to a certain staining procedure, thanks to a difference in envelope structure; gram-negative prokaryotes, or diderms, have both a plasma membrane and an outer membrane; gram-positive prokaryotes, or monoderms, lack the outer membrane

halophile: an organism that lives in environments rich in salt.

heterotroph: an organism that derives carbon and energy from preformed organic substances

Histone: a class of proteins that occur in chromosomes in association with DNA

holism: the doctrine that a biological system is more than the sum of its components; to understand the whole one must know its parts, but that dissection always leaves an unexplained residue that turns on relationships among those parts

homeostasis: the regulation by an organism of its chemical composition and other aspects of its internal environment

homology (homologous): similarity of structure or sequence due to common ancestry

horizontal gene transfer: transfer of genes by mechanisms other than cell reproduction; in this book, the term "lateral gene transfer" is used instead

hydrogenosome: the organelle of energy production in many protists that inhabit environments devoid of oxygen; its characteristic feature is the production of hydrogen gas

intron: nucleotide sequence within a gene that does not code for the gene product (also called an "intervening sequence")

kingdom: in taxonomy, a high level of classification; traditional kingdoms include animals, fungi, and plants

last (universal) common ancestor (LUCA): a hypothetical organism ancestral to all three domains of life

lateral gene transfer: transfer of genes by mechanisms other than cell reproduction (also called *horizontal gene transfer*)

lipid: any of a diverse group of organic compounds found in living organisms that are insoluble in water but soluble in organic solvents such as chloroform or benzene (e.g., fats, oils, steroids, and terpenes)

lipopolysaccharide: a complex molecule consisting of both lipid and carbohydrate moieties that makes up the outermost envelope of gram-negative Bacteria

liposome: an artificial vesicle bounded by a lipid bilayer membrane

lysosome: an organelle found in the cytoplasm of eukaryotic cells that carries out the breakdown of food particles and sometimes of cell constituents

membrane: in this book, the thin sheet made up of lipids and proteins that forms the boundaries of cells, organelles, and intracellular compartments.

messenger RNA (mRNA): the product of transcription of structural genes; a type of RNA that carries the genetic instructions from DNA into the cytoplasm

metabolism: the sum of the chemical reactions that take place in a living organism; compounds that take part in, or are formed by, these reactions are called "metabolites"; a sequence of reactions that generates (or degrades) a metabolite is considered a "metabolic pathway"

Metazoa: animals whose body is composed of many cells grouped into tissues and organized by a nervous system. The term includes all vertebrates and invertebrates, excluding protists

methanogenesis: a metabolic sequence by which CO_2 is reduced to methane, with the production of useful energy and cell constituents; all methanogens belong to the domain Archaea

microfilament: an element of the cytoskeleton of eukaryotic cells, a helical thread, 6–7 nm in diameter whose chief constituent is the protein actin

microtubule: an element of the cytoskeleton of eukaryotic cells; a cylindrical filament about 25 nm in diameter whose chief constituent is the protein tubulin

mitochondrion: an organelle of eukaryotic cells that is the locus of respiration and oxidative phosphorylation

mitosis: the division of a cell to form two daughter cells, each of which has a nucleus containing the same number and kind of chromosomes as the mother cell

modern synthesis: the theory of evolution as reformulated in the mid-twentieth century

monoderm: a prokaryotic cell whose envelope features but a single lipid membrane; it typically stains gram positive (contrast with *diderm*; see also under *gram-negative, gram-positive*)

monophyletic group: a group of organisms comprising a common ancestor and all its descendants (see also *clade*)

morphogenesis: the development of form and structure in an organism

murein: the characteristic wall polymer of Bacteria; it consists of repeating units of muramic acid and N-acetyl glucosamine, cross-linked by short chains of amino acids; also called *peptidoglycan*, the term preferred in this book

mutant: an organism or a gene that has undergone a mutation

mutation: a heritable change in the genetic material of a cell that may cause it and its descendants to differ from the normal in appearance or behavior

NADH: the reduced form of nicotinamide adenine dinucleotide; a major donor of electrons to the respiratory chain

Nicotinamide adenine dinucleotide (NAD): a coenzyme made up of two linked nucleotides that takes part in numerous reactions of cell metabolism; its reduced form, NADH, is a major donor of electrons to the respiratory chain

nuclear membrane: the membrane that delimits the nucleus and regulates the flow of materials between nucleus and cytoplasm; by definition, eukaryotic nuclei are bounded by a membrane, and prokaryotic ones are not

nucleic acid: a large and complex biological molecule consisting of a chain of nucleotides; there are two types: DNA and RNA

nucleoid: a defined region in the cytoplasm of prokaryotic cells that contains the genetic material; unlike the nuclei of eukaryotic cells, prokaryotic nucleoids are not enclosed in a membrane

nucleotide: the basic building block of nucleic acids; Each nucleotide consists of a nitrogenous base, a sugar, and one or more phosphate groups

nucleus: the large, membrane-bounded body embedded in the cytoplasm of eukaryotic cells that contains the genetic material

operon: a functionally integrated genetic unit consisting of adjoining genes that encode proteins, together with loci that control their expression

order: a state in which the components are arranged in a regular, comprehensible, or predictable manner; various degrees of order are represented by the letters of the alphabet, the arrangement of a keyboard, and the sequence of amino acids in a protein

organelle: a minute structure within a cell that has a particular function (e.g., nucleus, mitochondria, and flagella)

organism: an individual living creature, either unicellular or multicellular

organization: purposeful or functional order, as in the arrangement of the parts of a bicycle or a skeleton

oxidative phosphorylation: the chief mechanism of ATP production in aerobic organisms; the enzymatic generation of ATP coupled to the transfer of electrons from a substrate to oxygen

peptide: one of a large class of organic compounds consisting of two or more amino acids linked by peptide bonds

peptidoglycan: the characteristic wall polymer of Bacteria; it consists of repeating units of muramic acid and N-acetyl glucosamine, cross-linked by short chains of amino acids

phagocytosis: the process by which cells engulf and digest minute food particles

phagotroph: an organism that lives by engulfing food particles

phospholipid: one of a group of lipids that contain both a phosphate group and one or more fatty acids

photosynthesis: the chemical process by which green plants (and many other organisms) synthesize organic compounds from CO_2 and water, by use of the energy of light

phototroph: an organism that lives by photosynthesis

phylogenomics: in a phylogenetic context, large-scale reconstruction of a phylogeny using genomic data (usually ten genes or more)

phylogeny: evolutionary relationship among organisms or their parts (e.g., genes)

phylum: a category used in the classification of animals, such as the phyla Protozoa or Mollusca: in plant classification, the term "division" is preferred; both terms are used to designate the major categories of Bacteria and Archaea.

plasma membrane: The membrane that forms the outer limit of a cell and regulates the flow of material into the cell and out (also called the *cytoplasmic membrane*)

plasmid: a unit of DNA within the cytoplasm that can exist and replicate independently of the chromosomes

plastid: an organelle of photosynthetic organisms or their derivatives, generally involved in photosynthesis and certain metabolic pathways; plastids include chloroplasts

polymer: a molecule or complex composed of repeating elements (monomers) (e.g., proteins are polymers of amino acids)

polynucleotide: any polymer of nucleotides, including DNA and RNA

polypeptide: a peptide containing more than ten amino acids, commonly a hundred or more; all proteins are polypeptides

progenote: a hypothetical stage of cell evolution in which gene transcription and translation are still plastic and loosely coupled (see also *stem-group cell*)

prokaryote: an organism whose genetic material is not separated from the cytoplasm by a nuclear membrane; more generally, a grade of cellular organization lacking a true nucleus and most organelles; both Bacteria and Archaea belong to this grade (also spelled *procaryote*).

protein: a large molecule consisting of one or more polypeptide chains; the molecular mass ranges from 6,000 Da to the millions

proteinoid: polypeptides and related structures formed by subjecting a mixture of amino acids to dry heat

proteobacteria: a large and diverse phylum of gram-negative bacteria; the most familiar example is *Escherichia coli*.

Proterozoic eon: the period from the close of the Archean to the beginning of the Cambrian, approximately 2.5 to 0.5 billion years ago

protist: a member of the kingdom Protista (or Protoctista), which includes the unicellular eukaryotic organisms and some multicellular lineages derived from them (also called protoctist)

protocell: in this book, a hypothetical precellular entity that lacks genes and encoded gene products such as proteins

protoplasm: the jellylike material that fills cellular space, exclusive of the nucleus and other organelles; in the contemporary literature, the term *cytoplasm* is preferred

protozoa: in this book, used generically to designate nonphotosynthetic protists, such as amoebas or ciliates

pseudopod: a temporary outgrowth of certain cells (especially amoebas) that serves as an organ of locomotion or feeding

punctuated equilibrium: a model of the pattern of evolution that proposes that long periods of relative constancy are punctuated by episodes of rapid change

receptor: a molecule on the surface of a cell whose function is to detect a particular stimulus and initiate a response

redox chain: a cascade of enzymes that carry out successive steps of reduction and oxidation; the respiratory chain is an example

reductionism: the doctrine that the properties of a complex system can be largely (or even wholly) understood in terms of its simpler parts or components

replication: the production of an exact copy; usually employed in reference to the replication of DNA, in which one strand provides a template for the formation of a complementary strand, which is then copied once more to reproduce the original

respiration: the utilization of oxygen; in cell biology, the oxidative degradation of organic substances with the production of metabolites and energy; oxygen usually serves as the oxidant, but sulfate or nitrate sometimes take its place in "anaerobic respiration"

respiratory chain: the biochemical basis of respiration; a cascade of proteins and other metabolites that carries electrons from substrates to oxygen in aerobic cells

ribonucleic acid (RNA): a nucleic acid made up of polynucleotide chains whose sugar is ribose (e.g., ribosomal RNA, transfer RNA, and messenger RNA)

ribosome: an intracellular organelle that carries out protein synthesis and is found in all cells; it is composed of several species of RNA and some fifty proteins

ribozyme: a biological catalyst composed of RNA (unlike enzymes, which are proteins)

RNA polymerase: an enzyme that catalyzes the synthesis of RNA from its nucleotide precursors, typically using an existing strand of DNA or RNA as the template; the agent of gene transcription

self-assembly: a mode of self-organization in which supramolecular order emerges from the association of molecules without input of either external information or energy (e.g., the polymerization of tubulin into microtubules)

self-construction: this author's term for a mode of self-organization in which supramolecular order arises by the association of preformed molecules in a manner that requires the input of energy but not of information; contrast with *self-assembly*

self-organization: in this book, the emergence of supramolecular order from the interaction of many molecules that obey only local rules, without reference to an external template or global plan

sequence: the order of amino acids in a protein or of nucleotides in a nucleic acid; the determination of that order is referred to as "sequencing"

small-subunit ribosomal RNA (SSU rRNA): a species of RNA found in the small subunit of all ribosomes; extensively used in phylogenetic studies because its sequence is well conserved across all domains of life

species: in taxonomy, a group of similar and closely related individuals that can usually breed among themselves

spliceosome: an organelle of eukaryotic cells that excises introns and splices the remaining transcript into a continuous chain

Stem-group cell: in this book, a hypothetical stage of cell evolution in which basic cellular mechanisms are in the process of formation (see also *progenote*)

Steroid: Any of a group of lipids derived from a certain phenanthrene derivative that has a nucleus of four rings. Examples include cholesterol and steroid hormones.

stromatolite: a rocky, cushionlike mass, produced by the growth of large populations of cyanobacteria and algae; they are commonly found fossilized, though living ones still occur in favorable locales

structural heredity: the transmission of cellular traits by structural continuity without participation of genes (also called *cell heredity*)

symbiosis: living together; an interaction between individuals of different species that is beneficial to both partners

symport/antiport: the coupled transport of two substrates by a single carrier; we speak of symport when both substrates move in the same direction and of antiport when they move in opposite directions

syntrophy: metabolic symbiosis; association of two or more organisms in which the end products of one partner serve as substrates for the other

system: an entity composed of elements that interact, or are related to one another, in some definite manner; a bicycle and a cell are systems; a lump of granite is not

taxonomy: the practice or science of classification

teleonomy: the appearance of purpose in biological organization, due to evolution by heredity, variation and natural selection; Contrast with teleology, the doctrine that living systems are products of purposeful design

thermophile: an organism that can live at high temperatures

transcription: assembly of an RNA molecule complementary to a stretch of DNA; this is the first step in protein synthesis, and represents the transfer of sequence information from DNA to RNA.

transfer RNA (tRNA): a small RNA molecule that conducts activated amino acids to the ribosome during protein synthesis

translation: assembly of a protein by a ribosome, using messenger RNA to specify the order of the amino acids.

transposon: a mobile genetic element that can become integrated at many different sites along a chromosome

tubulin: a globular protein that is the chief constituent of microtubules

undulipodium: eukaryotic flagellum (see *flagellum*)

urkaryote: a hypothetical ancestor of eukaryotic cells, prior to the acquisition of the endosymbionts that gave rise to mitochondria

vacuole: A membrane-bounded space within the cytoplasm of a living cell that is filled with fluid or gas; in plants and some protists, vacuoles occupy much of the cells' volume

vectorial: having a direction in space

vesicle: a small, membrane-bounded sac (usually filled with fluid) within the cytoplasm of a living cell

virus: a particle too small to be seen with a light microscope or to be trapped by filters but capable of reproduction within a living cell; viruses have very limited metabolic capacities and are obligatory intracellular parasites

work: a process that runs counter to the spontaneous direction of events (e.g., protein synthesis represents work, but their degradation does not)

References

Abel, D. L., and Trevors, J. T. 2005. Three subsets of sequence complexity and their relevance to biopolymeric information. *Theoretical Biology and Medical Modelling* 2:29.

———. 2006. Self-organization vs. self-ordering events in life-origin models. *Physics of Life Reviews*. doi:10.1016/j.plrev.2006.07.003.

Adl, S. M., Simpson, A. B. G., Farmer, M. A., Andersen, R. A., Anderson, O. R., Barta, J. R., et al. 2005. The new higher-level classification of eukaryotes with emphasis on the taxonomy of protists. *Journal of Eukaryotic Microbiology* 52:399–451.

Allen, J. F. 2003. The function of genomes in bioenergetic organelles. *Philosophical Transactions of the Royal Society of London, Series B* 358:19–38.

Allen, J. F., and Martin, W. 2007. Out of thin air. *Nature* 445:610–12.

Allwood, A. C., Grotzinger, J. P., Knoll, A. H., Burch, I. W., Anderson, M. S., Coleman, M. L., and Kanik, I. 2009. Controls on development and diversity of early Archean stromatolites. *Proceedings of the National Academy of Sciences U S A* 106:9548–52.

Altermann, W., and Kamierczak, J. 2003. Archean microfossils: a reappraisal of early life on earth. *Research in Microbiology* 154:611–17.

Andersson, S. G. E., Karlberg, O., Canbäck, B., and Kurland, C. G. 2003. On the origin of mitochondria: a genomics perspective. *Philosophical Transactions of the Royal Society of London, Series B* 358:165–79.

Archibald, J. M. 2009. The puzzle of plastid evolution. *Current Biology* 19:R81–88.

Arndt, N. T., and Nisbet, E. G. 2012. Processes on the young earth and the habitats of early life. *Annual Review of Earth and Planetary Science* 40: 521–49.

Ascherson, N. 1995. *Black sea.* New York: Hill and Wang.

Awramik, S. M. 2006. Respect for stromatolites. *Nature* 441:700–1.

Awramik, S. M., and McNamara, K. J. 2007. The evolution and diversification of life. In Sullivan and Baross 2007, 313–34.

Baldauf, S. L. 2003. The deep roots of eukaryotes. *Science* 300:1703–6.

Bapteste, E., O'Malley, M. A., Beiko, R. G., Ereshefsky, M., Gogarten, P. J., Franklin-Hall, L., et al. 2009. Prokaryotic evolution and the tree of life are two different things. *Biology Direct* 4:34.

Bard, J. 2010. A systems biology view of evolutionary genetics. *BioEssays* 32:559–63.

Barry, R., and Gitai, Z. 2011. Self-assembling enzymes and the origins of the cyto-skeleton. *Current Opinion in Microbiology* 14:704–11.

Battistucci, F. U., Feijao, A., and Hedges, S. B. 2004. A genomic timescale of pro-karyote evolution: insights into the origin of methanogenesis, phototrophy, and the colonization of land. *BMC Evolutionary Biology* 4:44.

Baymann, F., Lebrun, E., Brugna, M., Schoepp-Cothenet, B., Guidici-Orticoni, M.-T., and Nitschke, W. 2003. The redox protein construction kit: pre-last uni-versal common ancestor evolution of energy-conserving enzymes. *Philosophi-cal Transactions of the Royal Society of London, Series B* 358:267–74.

Becerra, A., Delaye, L., Islas, S, and Lazcano, A. 2007. The very early stages of bio-logical evolution and the nature of the last common ancestor of the three major cell domains. *Annual Review of Ecology, Evolution and Systematics* 38:361–79.

Beck, W. S. 1957. *Modern science and the nature of life*. New York: Harcourt, Brace & Co.

Behe, M. J. 1996. *Darwin's black box: the biochemical challenge to evolution*. New York: Free Press.

Behe, M. J. 2007. *The edge of evolution: the search for the limits of Darwinism*. New York: Free Press.

Beiko, R.G., Harlow, T.J., and Ragan, M. A. 2005. Highways of gene sharing in prokaryotes. *Proceedings of the National Academy of Sciences U S A* 102:14332–37.

Beja, O., Aravind, L., and Koonin, E. V. 2000. Bacterial rhodopsin: evidence for a new type of phototrophy in the sea. *Science* 289:1902–6.

Belloc, H. 1912. *More Beasts for Worse Children*. London: Duckworth.

Bengtson, S., Belivanova, V., Rasmussen, B., and Whitehouse, M. 2009. The con-troversial "Cambrian" fossils of the Vindhyan are real but more than a billion years older. *Proceedings of the National Academy of Sciences U S A* 106:7729–34.

Berg, I. A., Kockelhorn, D., Ramos-Vera, W. H., Say, R. F., Zarzyski, J., Hugler, M., et al. 2010. Autotrophic carbon fixation in Archaea. *Nature Reviews Microbiol-ogy* 8:447–60.

Berney, C., and Pawlowski, J. 2006. A molecular timescale for eukaryote evolution recalibrated with the continuous microfossil record. *Proceedings of the Royal Society of London, Series B* 273:1867–72.

Berry, S., 2002. The chemical basis of membrane bioenergetics. *Journal of Molecu-lar Evolution* 54:595–613.

Bhattacharya, D., Archibald, J. M., Weber, A. P. M., and Reyes-Prieto, A. 2007. How do endosymbionts become organelles? Understanding early events in plastid evolution. *BioEssays* 29:1239–46.

Biegel, E., and Müller, V. 2010. Bacterial Na^+-translocating ferredoxin-NAD^+ oxidoreductase. *Proceedings of the National Academy of Sciences U S A* 107:18138–42.

Bird, A. 2007. Perceptions of epigenetics. *Nature* 447:396–98.

Blankenship, R. E. 2002. *Molecular mechanisms of photosynthesis*. Oxford: Blackwell Science.

Blobel, G. 1980. Intracellular membrane topogenesis. *Proceedings of the National Academy of Sciences U S A* 77:1496–500.

Blount, Z. D., Barrick, J.E., Davidson, C. J., and Lenski, R. E. 2012. Genomic analysis of a key innovation in an experimental *Escherichia coli* population. *Nature* 489:513–18.

Blount, Z. D., Borland, C. Z., and Lenski, R. E. 2008. Historical contingency and the evolution of a key innovation in an experimental population of *Escherichia coli*. *Proceedings of the National Academy of Sciences U S A* 105:7899–906.

Bokov, K., and Steinberg, S. V. 2009. A hierarchical model for evolution of 23S ribosomal RNA. *Nature* 457:977–80.

Boucher, Y., Douady, C. J., Papke, R. T., Walsh, D. A., Boudreau, M. D., Nesbø, C. L., et al. 2003. Lateral gene transfer and the origins of prokaryotic groups. *Annual Review of Genetics* 37:283–28.

Boussau, B., Blanquart, S., Necsulea, A., Lartillot, N., and Gouy, M. 2008. Parallel adaptations to high temperatures in the Archaean eon. *Nature* 456:942–45.

Boyer, P. 1993. The binding-change mechanism for ATP synthase—some probabilities and possibilities. *Biochimica et Biophysica Acta* 1140:215–50.

Brasier, M., McLoughlin, N., Green, O., and Wacey, D. 2006. A fresh look at the fossil evidence for early Archaean cellular life. *Philosophical Transactions of the Royal Society, Series B* 361:887–902.

Bray, D. 2009. *Wetware: A computer in every living cell*. New Haven, CT: Yale University Press.

Britten, R. J. 1998. Underlying assumptions of developmental models. *Proceedings of the National Academy of Sciences U S A* 95:9372–77.

Brochier-Armanet, C., Forterre, P., and Gribaldo, S. 2011. Phylogeny and evolution of the Archaea: one hundred genomes later. *Current Opinion in Microbiology* 14:274–81.

Brochier-Armanet, C., Talla, E., and Gribaldo, S. 2009. The multiple evolutionary histories of dioxygen reductases: Implications for the origin and evolution of aerobic respiration. *Molecular Biology and Evolution* 26:285–97.

Brocks, J. J., and Banfield, J. 2009. Unraveling ancient microbial history with community proteogenomics and lipid geochemistry. *Nature Reviews Microbiology* 7:601–9.

Brocks, J. J., Logan, G. A., Buick, R., and Summons, R. E. 1999. Archean molecular fossils and the early rise of eukaryotes. *Science* 285:1033–36.

Brown, J. R. 2003. Ancient horizontal gene transfer. *Nature Reviews Genetics* 4:121–31.

Brown, J. R., and Doolittle, W. F. 1997. Archaea and the prokaryote-eukaryote transition. *Microbiological Reviews* 61:456–502.

Brüssow, H. 2009. The not so universal tree of life, or the place of viruses in the living world. *Philosophical Transactions of the Royal Society of London, Series B* 364:2263–74.

Buckel, W., and Thauer, R. K. 2013. Energy conservation via electron bifurcating ferredoxin reduction and proton/Na^+ translocating ferredoxin oxidation. *Biochimica et Biophysica Acta* 1827: 94–113.

Buckling, A., Maclean, R. C., Brockhurst, M. A., and Colegrave, N. 2009. The *Beagle* in a bottle. *Nature* 457:824–29.

Buick, R. 2007. The earliest records of life on earth. In Sullivan and Baross 2007, 237–64.

———. 2008. When did oxygenic photosynthesis evolve? *Philosophical Transactions of the Royal Society of London, Series B* 363:2731–34.

———. 2010. Ancient acritarchs. *Nature* 463:885–86.

Burger, G., Gray, M. W., and Lang, B. F. 2003. Mitochondrial genomes: anything goes. *Trends in Genetics* 19:709–16.

Burton, R. F. (1880) 1944. *The Kasidah of Haji Abdu El-Yezdi*. New York: Willey Books.

Cabeen, M. T., and Jacobs-Wagner, C. 2010. The bacterial cytoskeleton. *Annual Review of Genetics* 44:365–92.

Caetano-Annollés, G., Kim, K. M., and Caetano-Anollés, D. 2012. The phylogenomic roots of modern biochemistry. *Journal of Molecular Evolution* 74: 1–34.

Caetano-Anollés, G., Kim H. S., and Mittenthal, J. E. 2007. The origin of modern metabolic networks inferred from phylogenomic analysis of protein architecture. *Proceedings of the National Academy of Sciences U S A* 104:9358–63.

Caetano-Anollés, D., Kim, K. M., Mittenthal, J. E., and Caetano-Anollés, G., 2011. Proteome evolution and the metabolic origins of translation and cellular life. *Journal of Molecular Evolution* 72:14–33.

Cairns-Smith, A. G. 1985. *Seven clues to the origin of life*. Cambridge: Cambridge University Press.

Canfield, D. E. 2006. Gas with an ancient history. *Nature* 440:426–27.

Canfield, D. E., Glazer, A. N., and Falkowski, P. G. 2010. The evolution and future of earth's nitrogen cycle. *Science* 330:192–96.

Canfield, D. E., Rosing, M. T., and Bjerrum, C. 2006. Early anaerobic metabolisms. *Philosophical Transactions of the Royal Society of London, Series B* 361:1819–36.

Capaldi, R. A., and Aggeler, R. 2002. Mechanism of the F_1F_0-type ATP synthase, a biological rotary motor. *Trends in Biochemical Sciences* 27:154–60.

Carballido-Lopez, R., and Errington, J. 2003. A dynamic bacterial cytoskeleton. *Trends in Cell Biology* 13:577–83.

Castresana, J. 2001. Comparative genomics and bioenergetics. *Biochimica et Biophysica Acta* 1506:147–62.

Catling, D. 2005. The story of O_2. *Science* 308:1730–32.

Catling, D., and Kasting, J. F. 2007. Planetary atmospheres and life. In Sullivan and Baross 2007, 91–116.

Cavalier-Smith, T. 1975. The origin of nuclei and of eukaryotic cells. *Nature* 256: 463–68.

———. 2000. Membrane heredity and early chloroplast evolution. *Trends in Plant Science* 5:174–82.

———. 2001. Obcells as proto-organisms: membrane heredity, lithophosphorylation, and the origins of the genetic code, the first cells and photosynthesis. *Journal of Molecular Evolution* 53:555–95.

———. 2002a. The neomuran origin of archaebacteria, the negibacterial root of the universal tree and bacterial megaclassification. *International Journal of Systematic and Evolutionary Microbiology* 52:7–76.

———. 2002b. The phagotrophic origin of eukarytotes and phylogenetic classification of protozoa. *International Journal of Systematic and Evolutionary Microbiology* 52:297–35.

———. 2003. Genomic reduction and evolution of novel genetic membranes and protein-targeting machinery in eukaryote-eukaryote chimaeras (meta-algae). *Philosophical Transactions of the Royal Society of London, Series B* 358:109–34.

———. 2004. The membranome and membrane heredity in development and evolution. In Hirt, P. R., and Horner, D. S., eds., *Organelles, genomes and eukaryotic phylogeny* (Boca Raton, FL: CRC Press), 335–51.

———. 2006a. Cell evolution and earth history: stasis and revolution. *Philosophical Transactions of the Royal Society of London, Series B* 361:969–1006.

———. 2006b. Rooting the tree of life by transition analysis. *Biology Direct* 1:19.

———. 2006c. Origin of mitochondria by intracellular enslavement of a photosynthetic purple bacterium. *Proceedings of the Royal Society of London, Series B* 273:1943–52.

———. 2009. Predation and eukaryote cell origins: a coevolutionary perspective. *International Journal of Biochemistry & Cell Biology* 41:307–22.

———. 2010. Deep phylogeny, ancestral groups and the four ages of life. *Philosophical Transactions of the Royal Society of London, Series B* 365:111–32.

Cavalier-Smith, T., Brasier, M., and Embley, T. M. 2006. Major steps in cell evolution, proceedings of a discussion meeting held at the Royal Society. *Philosophical Transactions of the Royal Society of London, Series B* 361, no 1470.

Cavicchioli, R., ed. 2007. *Archaea: Molecular and cellular biology*. Washington, DC: American Society for Microbiology Press.

———. 2011. Archaea—timeline of the third domain. *Nature Reviews Microbiology* 9:51–61.

Chapman, M. J., Dolan, M. F., and Margulis, L. 2000. Centrioles and kinetosomes: form, function and evolution. *Quarterly Review of Biology* 75:409–29.

Chargaff, E. 1978. *Heraclitean fire*. New York: Rockefeller University Press.

Chen, I. A., Hanczyc, M. M., Sazani, P. L., and Szostak, J. W. 2006. Protocells: genetic polymers inside membrane vesicles. In Gesteland, Czech, and Atkins 2006, 57–88.

Chernikova, D., Motamedi, S., Csürös, M., Koonin, E. V., and Rogozin, I. B. 2011. A late origin of extant eukaryotic diversity: divergence times estimated using rare genomic changes. *Biology Direct* 6:26.

Clark, R. W. 1969. *The life and work of J.B.S. Haldane*. New York: Coward-McCann.

Claverie, J.-M., and Abergel, C. 2010. Mimivirus: the emerging paradox of quasi-autonomous viruses. *Trends in Genetics* 26:431–37.

Clements, A., Bursac, D., Gatsos, X., Perry, A. J., Civciristov, S., Celik, A. N., et al. 2009. The reducible complexity of a mitochondrial molecular machine. *Proceedings of the National Academy of Sciences U S A* 106:15791–95.

Cody, G. D. 2004. Transition metal sulfides and the origins of metabolism. *Annual Review of Earth and Planetary Sciences* 32:569–99.

Collins, L., and Penny, D. 2005. Complex spliceosomal organization ancestral to extant eukaryotes. *Molecular Biology and Evolution* 22:1053–66.

Conway Morris, S. 2003. *Life's solution: inevitable humans in a lonely universe*. Cambridge: Cambridge University Press.

Cooper, G., Reed, C., Nguyen, D., Carter, M. and Wang, Y. 2011. Detection and formation scenario of citric acid, pyruvic acid and other possible metabolism precursors in carbonaceous meteorites. *Proceedings of the National Academy of Sciences U S A* 108: 14015–20.

Copley, S. D., Smith, E., and Morowitz, H. J. 2007. The origin of the RNA world: co-evolution of genes and metabolism. *Bioorganic Chemistry* 35:430–43.

Cox, C. J., Foster, P. G., Hirt, R. P., Harris, S. R., and Embley, T. M. 2008. The archaebacterial origin of eukaryotes. *Proceedings of the National Academy of Sciences U S A* 105:20356–61.

Coyne, J. A. 2009. *Why evolution is true*. Oxford: Oxford University Press.

Crick, F. H. C. 1981. *Life itself: its origin and nature*. New York: Simon and Schuster.

Criscuolo, A., and Gribaldo, S. 2011. Large-scale phylogenomic analyses indicate a deep origin of plastids within cyanobacteria. *Molecular Biology and Evolution* 28:3019–32.

Croal, L. R., Grainick, J.A., Malasam, D., and Newman, D. K. 2004. The genetics of geochemistry. *Annual Review of Genetics* 38:175–202.

Dacks, J. B., and Field, M. C. 2007. Evolution of the eukaryotic membrane-trafficking system: origin, tempo and mode. *Journal of Cell Science* 120:2977–85.

Dacks, J. B., Poon, P. P., and Field, M. C. 2008. Phylogeny of endocytic components yields insight into the process of nonendosymbiotic organelle evolution. *Proceedings of the National Academy of Sciences U S A* 105:588–93.

Dalton, R. 2002. Squaring up over ancient life. *Nature* 417:782–84.

Darnell, J. E., and Doolittle, W. F. 1986. Speculations on the early course of evolution. *Proceedings of the National Academy of Sciences U S A* 83:1271–75.

Darwin, C. (1859) 1979. *The origin of species by means of natural selection.* New York: Avenel Books.

Davidov, Y., and Jurkevitch, E. 2009. Predation between prokaryotes and the origin of eukaryotes. *BioEssays* 31:748–57.

Davies, P. 1999. *The fifth miracle.* New York: Simon and Schuster.

———. 2006. *The Goldilocks enigma: why is the universe just right for life?* London: Allen Lane.

Davis, R. H. 2003. *The microbial models of molecular biology: from genes to genomes.* Oxford: Oxford University Press.

Dawkins, R. 1976. *The selfish gene.* Oxford: Oxford University Press.

———. 1982. *The extended phenotype.* Oxford: Oxford University Press.

———. 1995. *River out of Eden.* New York: Harper Collins.

———. 1996. *Climbing Mount Improbable.* New York: W. W. Norton.

———. 2009. *The greatest show on earth! The evidence for evolution.* New York: Free Press.

Dawson, S. C., and Pace, N. R. 2002. Novel kingdom-level eukaryotic diversity in anoxic environments. *Proceedings of the National Academy of Sciences U S A* 99:8324–29.

Deamer, D., Dworkin, J. P., Sanford, S. A., Bernstein, M. P., and Allamandola, L. J. 2002. The first cell membranes. *Astrobiology* 2:371–81.

Deamer, D. W. 2008. How leaky were primitive cells? *Nature* 454:37–38.

De Duve, C. 2005a. *Singularites—landmarks on the road to life.* London: Cambridge University Press.

———. 2005b. The onset of selection. *Nature* 433:581–82.

———. 2007. The origin of eukaryotes: a reappraisal. *Nature Reviews Genetics* 8:395–403.

De La Rocha, C. 2006. In hot water. *Nature* 443:920–21.

Delsuc, F., Brinkmann, H., and Philippe, H. 2005. Phylogenomics and the reconstruction of the tree of life. *Nature Reviews Genetics* 6:361–75.

Dennett, D. C. 1995. Darwin's dangerous idea: evolution and the meanings of life. New York: Simon and Schuster.

De Nooijer, S., Holland, B. R., and Penny, D. 2009. The emergence of predators in early life: there was no Garden of Eden. *PLoS One* (46): e5507. doi: 10.1371.

Denton, M. J. 1998. *Nature's destiny: how the laws of biology reveal purpose in the universe.* New York: Free Press.

Denton, M. J., Kumaramanickavel, G., and Legge, M. 2013. Cells as irreducible wholes: the failure of mechanism and the possibility of an organicist revival. *Biology and Philosophy* 28:31–52.

Derelle, E., Ferraz, C., Rombauts, S., Rouze, P., Worden, A. Z., Robben, S., et al. 2006. Genome analysis of the smallest free-living eukaryote unveils many unique features. *Proceedings of the National Academy of Sciences U S A* 103:11647–52.

Deschamps, P., C. Colleoni, Y. Nakamura, E. Suzuki, L.-L. Putaux, A. Buleon, et al. 2008. Metabolic symbiosis and the birth of the plant kingdom. *Molecular Biology and Evolution* 25:536–48.

Di Giulio, M. 2003. The universal ancestor and the ancestor of Bacteria were hyperthermophiles. *Journal of Molecular Evolution* 57:721–30.

———. 2011. The last universal common ancestor (LUCA) and the ancestors of archaea and bacteria were progenotes. *Journal of Molecular Evolution* 72:119–26.

Dolezal, P., Likic, V., Tachezy, J., and Lithgow, T. 2006. Evolution of the molecular machines for protein import into mitochondria. *Science* 313:314–18.

Doolittle, W. F. 1999. Phylogenetic classification and the universal tree. *Science* 284:2124–28.

———. 2009a. Eradicating typological thinking in prokaryotic systematics and evolution. *Cold Spring Harbor Symposia on Quantitative Biology* 74:197–204.

———. 2009b. The practice of classification and the theory of evolution, and what the demise of Charles Darwin's tree of life hypothesis means for both of them. *Philosophical Transactions of the Royal Society of London, Series B* 364:2221–28.

———. 2010. The attempt on the life of the Tree of Life: science, philosophy and politics. *Biology and Philosophy* 26:455–73.

Doolittle, W. F., and Bapteste, E. 2006. Pattern pluralism and the tree of life hypothesis. *Proceedings of the National Academy of Sciences U S A* 104:2043–49.

Doolittle, W. F., Boucher, Y., Nesbø, C. L., Douady, C. J., Andersson, J. O., and Roger, A. J. 2002. How big is the iceberg of which organellar genes in nuclear genomes are but the tip? *Philosophical Transactions of the Royal Society of London, Series B* 358:39–58.

Doolittle, W. F., and Zhaxybayeva, O. 2007. Reducible complexity: the case of the bacterial flagellum. *Current Biology* 17:R510–12.

Douzery, E. J. P., Snell, E. A., Bapteste, E., Delsuc, F., and Philippe, H. 2004. The timing of eukaryotic evolution: does a relaxed molecular clock reconcile proteins and fossils? *Proceedings of the National Academy of Sciences U S A* 101:15386–91.

Ducluzeau, A.-L., Ouchane, S., and Nitschke, W. 2008. The cbb3 oxidases are an ancient innovation of the domain Bacteria. *Molecular Biology and Evolution* 25:1158–66.

Dyall, S. D., Brown, M. T., and Johnson, P. J. 2004. Ancient invasions: from endosymbionts to organelles. *Science* 304:253–57.

Dyson, F. 1985. *Origins of life*. London: Cambridge University Press.

Eigen, M. 1992. *Steps towards life: a perspective on evolution*. New York: Oxford University Press.

Eiler, J. M., 2007. The oldest fossil or just another rock? *Science* 317:1046–47.

Eiseley, L. (1946) 1959. *The immense journey*. New York: Vintage Books.

————. 1975. *All the Strange Hours—The Excavation of a Life*. New York: Charles Scribner's Sons.

Elena, S. F., and Lenski, R. E. 2003. Evolution experiments with microorganisms: the dynamics and genetic basis of adaptation. *Nature Reviews Genetics* 4:457–69.

Elias, M., and Archibald, J. M. 2009. Sizing up the genomic footprint of endosymbiosis. *BioEssays* 31:1273–79.

Elkins, J. G., Podar, M., Graham, D. E., Makarova, K. S., Wolf, Y., Randau, L., et al. 2008. A korarchaeal genome reveals insights into the evolution of the Archaea. *Proceedings of the National Academy of Sciences U S A* 105:8102–7

Embley, T. M., and Martin, W. 2006. Eukaryotic evolution, changes and challenges. *Nature* 440:623–30.

Embley, T. M., Van der Giezen, M., Horner, D. S., Dyal, P. L., and Foster, P. 2003. Mitochondria and hydrogenosomes are two forms of the same fundamental organelle. *Philosophical Transactions of the Royal Society of London, Series B* 358:191–203.

Erickson, H. P. 2007. Evolution of the cytoskeleton. *BioEssays* 29:668–77.

Ettema, T. J. G., Lindås, A-C., and R. Bernander 2011. An actin-based cytoskeleton in archaea. *Molecular Microbiology* 80:1052–61.

Falkowski, P. G. 2006. Tracing oxygen's imprint on earth's metabolic evolution. *Science* 311:1724–25.

Falkowski, P. G., and Isozaki, Y. 2008. The story of O_2. *Science* 322:540–42.

Falkowski, P. G., Katz, M. E., Knoll, A. H., Quigg, A., Raven, J. A., Schofield, O, and Taylor, F. J. R. 2004. The evolution of modern eukaryotic phytoplankton. *Science* 305:354–60.

Fedo, C. M., Whitehouse, M. J., and Kamber, B. S. 2006. Geological constraints on detecting the earliest life on earth: perspective from the early Archaean (older than 3.7 gyr) of southwest Greenland. *Philosophical Transactions of the Royal Society of London, Series B* 361:851–68.

Fenchel, T., and Bernard, C. 1993. A purple protist. *Nature* 362:300.

Feng, D. D., Cho, G., and Doolittle, R. F. 1997. Determining divergence times with a protein clock: update and reevaluation. *Proceedings of the National Academy of Sciences U S A* 94:13028–33.

Ferry, J. G., and House, C. H. 2006. The stepwise evolution of early life driven by energy conservation. *Molecular Biology and Evolution* 23:1286–92.

Finnigan, G. C., Hanson-Smith, V., Stevens, T. H., and Thornton, J. W. 2012. Evolution of increased complexity in a molecular machine. *Nature* 481:360–64.

Forterre, P. 2002. The origin of DNA genomes and DNA replication proteins. *Current Opinion in Microbiology* 5:525–32.

————. 2006a. The origin of viruses and their possible roles in major evolutionary transitions. *Virus Research* 117:5–16.

————. 2006b. Three RNA cells for ribosomal lineages and three DNA viruses to replicate their genomes: a hypothesis for the origin of cellular domain. *Proceedings of the National Academy of Sciences U S A* 103:3669–74.

Forterre, P., and Philippe, H. 1999. Where is the root of the universal tree of life? *BioEssays* 21:871–79.

Foster, P. L. 2000. Adaptive mutations: implications for evolution. *BioEssays* 22:1067–74.

Fournier, G. P., Huang, J., and Gogarten, P. J. 2009. Horizontal gene transfer from extinct and extant lineages: biological innovation and the coral of life. *Philosophical Transactions of the Royal Society of London Series B* 364:2229–39.

Frigaard, N.-U., Martinez, A., Mincer, T. J., and DeLong, E. F. 2006. Proteorhodopsin lateral gene transfer between marine planktonic Bacteria and Archaea. *Nature* 439:847–50.

Frost, L. S., Leplae, R., Summers, A. O., and Toussaint, A. 2005. Mobile genetic elements: the agents of open source evolution. *Nature Reviews Microbiology* 3:722–32.

Fry, I. 2000. *The emergence of life on earth: a historical and scientific overview*. New Brunswick, NJ: Rutgers University Press.

————. 2011. The role of natural selection in the origin of life. *Origins of Life and Evolution of the Biosphere* 41:3–16.

Fuchs, G. 2011. Alternative pathways of carbon dioxide fixation: Insights into the early evolution of life? *Annual Review of Microbiology* 65:631–58.

Fuerst, J. A. 2005. Intracellular compartmentation in Planctomycetes. *Annual Review of Microbiology* 59:299–328.

Fuerst, J. A., and Sagulenko, E. 2011. Beyond the bacterium: planctomycetes challenge our concepts of microbial structure and function. *Nature Reviews Microbiology* 9:403–13.

Fuhrman, J. A., Schwalbach, M. S., and Stingl, U. 2008. Proteorhodopsins: an array of physiological roles? *Nature Reviews Microbiology* 6:488–94.

Garcia-Ruiz, J. M., Hyde, S. T., Carnerup, A. M., Christy, A. G., Van Kranendonk, M. J., and Welham, N. J. 2003. Self-assembled silica-carbonate structures and detection of ancient microfossils. *Science* 302:1194–97.

Gardner, J. N. 2003. *Biocosm: the new scientific theory of evolution—intelligent life is the architect of the universe*. Makawao, Hawaii: Inner Ocean Publishing.

Garrett, R., and Klenk, H.-P., eds. 2007. *Archaea: evolution, physiology and molecular biology*. Oxford: Blackwell.

Gast, R. J., Sanders, R. W., and Caron, D. A. 2009. Ecological strategies of protists and their symbiotic relationships with prokaryotic microbes. *Trends in Microbiology* 17:563–69.

Gaucher, E. A., Govindarajan, S., and Ganesh, O. K. 2008. Palaeotemperature trend for Precambrian life inferred from resurrected proteins. *Nature* 451:704–7.

George, S. C., Volk, H., Dutkiewicz, A., Ridley, J., and Buick, R. 2008. Preservation of hydrocarbons and biomarkers in oil trapped inside fluid inclusions for >2 billion years. *Geochimica et Cosmochimica Acta* 72:844–70.

Gesteland, R. F., Czech, T. R., and Atkins, J. F., eds. 2006. *The RNA world: the nature of modern RNA suggests a prebiotic RNA world*. 3rd ed.. Cold Spring Harbor, NY: Cold Spring Harbor Laboratory Press.

Gibbs, S. P. 1981. The chloroplast endoplasmic reticulum: structure, function, and evolutionary significance. *International Review of Cytology* 72:49–99.

———. 2006. Looking at life: from binoculars to the electron microscope. *Annual Review of Plant Biology* 57:1–17.

Gibson, D. G., Glass, J. I., Lartigue, C., Noskov, V. N., Chuang, R.-Y., Algire, M. A., et al. 2010. Creation of a bacterial cell controlled by a chemically synthesized genome. *Science* 329:52–56.

Gibson, G., and Muse, S. V. 2004. *A primer of genome science*. 2nd ed. Sunderland, MA: Sinauer Associates.

Gilbert, W. 1986. The RNA world. *Nature* 319:618.

Gitai, Z., 2005. The new bacterial cell biology—moving parts and subcellular architecture. *Cell* 120:577–86.

Glansdorff, N., Xu, Y., and Labedan, B. 2008. The last universal common ancestor: emergence, constitution and genetic legacy of an elusive forerunner. *Biology Direct* 3:29.

Goldenfeld, N., and Woese, C. 2007. Biology's next revolution. *Nature* 445:369.

Goodwin, B., 1994. *How the leopard changed its spots: the evolution of complexity*. New York: Simon and Schuster.

Gould, S. B., Waller, R. F., and McFadden, G. I. 2008. Plastid evolution. *Annual Review of Plant Biology* 59:491–517.

Gould, S. J. 1982. Darwinism and the expansion of evolutionary theory. *Science* 216:380–87.

———. 1997. The exaptive excellence of spandrels as a term and prototype. *Proceedings of the National Academy of Sciences U S A* 94:10750–55.

Gouy, M., and Chaussidon, M. 2008. Ancient bacteria liked it hot. *Nature* 451:635–36.

Graur, D., and Martin, W. 2004. Reading the entrails of chickens: molecular timescales of evolution and the illusion of precision. *Trends in Genetics* 20:80–86.

Gray, M. W., Burger, G., and Lang, B. F. 1999. Mitochondrial evolution. *Science* 283:1476–81.

Gray, M. W., and Doolittle, W. F. 1982. Has the endosymbiont hypothesis been proven? *Microbiological Reviews* 46:1–42.

Gray, M. W., Lang, B. F., and Burger, G. 2004. Mitochondria of protists. *Annual Review of Genetics* 38:477–524.

Gray, M. W., Lukeš, J., Archibald, J. M., Keeling, P. J., and Doolittle, W. F. 2010. Irremediable complexity? *Science* 330:920–21.

Gribaldo, S., and Brochier-Armanet, C. 2006. The origin and evolution of

Archaea: a state of the art. *Philosophical Transactions of the Royal Society, Series B* 361:1007–22.

Gribaldo, S., Poole, A. M., Daubin, V., Forterre, P. and Brochier-Armanet, C. 2009. The origin of the eukaryotes and their relationship with the Archaea: are we at a phylogenomic impasse? *Nature Reviews Microbiology* 8:743–52.

Griesemer, J. 2000. The units of evolutionary transitions. *Selection* 1–3:67–80.

Griffiths, G. 2007. Cell evolution and the problem of membrane topology. *Nature Reviews Molecular Cell Biology* 8:1018–24.

Gross, J., and Bhattacharya, D. 2009. Mitochondrial and plastid evolution in eukaryotes: an outsiders' perspective. *Nature Reviews Genetics* 10:495–505.

Gupta, R. 2000. The natural evolutionary relationships among prokaryotes. *Critical Reviews in Microbiology* 26:111–31.

———. 2002. Phylogeny of Bacteria: Are we now close to understanding it? *ASM News* 68:284–91.

Guy, L., and Ettema, T. J. G. 2011. The archaeal "TACK" superphylum and the origin of eukaryotes. *Trends in Microbiology* 19:580–87.

Haldane, J. B. S. 1929. The origin of life. Repr. in Bernal, J. D., ed., *The origins of life*. London: Weidenfeld and Nicholson, 1967.

Halfmann, R., Jarosz, D. F., Jones, S. K., Chang, A., Lancaster, A. K., and Lindquist, S. 2012. Prions are a common mechanism for phenotypic inheritance in wild yeasts. *Nature* 482:363–68.

Halfmann, R., and Lindquist, S. 2010. Epigenetics in the extreme: prions and the inheritance of environmentally acquired traits. *Science* 330:629–32.

Hanczyc, M. M., Fujikawa, S. M., and Szostak, J. W. 2003. Experimental models of primitive cellular compartments: encapsulation, growth, and division. *Science* 302:618–22.

Harold, F. M. 1986. *The vital force—a study of bioenergetics*. New York: W. H. Freeman.

———. 2001. *The way of the cell: molecules, organisms and the order of life*. Oxford: Oxford University Press.

———. 2005. Molecules into cells: specifying spatial architecture. *Microbiology and Molecular Biology Reviews* 69:544–64.

Hartman, H., and Fedorov, A. 2002. The origin of the eukaryotic cell: a genomic investigation. *Proceedings of the National Academy of Sciences U S A* 99:1420–25.

Hazen, R. M. 2005. *Gen.e.sis*. Washington, DC: Joseph Henry Press

Hirt, R. P., Longsdon, J. M., Healy, B., Dorey, M. W., Doolittle, W. F and Embley, T. M. 1999. Microsporidia are related to fungi: evidence from the largest subunit of RNA polymerase II and other proteins. *Proceedings of the National Academy of Sciences U S A* 96:580–85.

Hjort, K., Goldberg, A. V., Tsaousis, A. D., Hirt, R. P., and Embley, T. M. 2010. Diversity and reductive evolution of mitochondria among microbial eukaryotes.

Philosophical Transactions of the Royal Society of London, Series B 365:713–27.

Hoffman, H. J. 1976. Precambrian microflora, Belcher Islands, Canada: significance and systematics. *Journal of Palaeontology* 50:1040–73.

Hohmann-Marriott, M. F., and Blankenship, R. E. 2011. Evolution of photosynthesis. *Annual Review of Plant Biology* 62:515–48.

Holland, H. D. 2006. The oxygenation of the atmosphere and oceans. *Philosophical Transaction of the Royal Society, Series B* 361:903–15.

Hoppins, S., Lackner, L., and Nunnari, J. 2007. The machines that divide and fuse mitochondria. *Annual Review of Biochemistry* 76:751–821.

Horiike, T., Hamada, K., Kanaya, S., and Shinozawa, T. 2001. Origin of eukaryotic cell nuclei by symbiosis of Archaea in Bacteria is revealed by homology-hit analysis. *Nature Cell Biology* 3:210–14.

Huang, S. 2011. The molecular and mathematical basis of Waddington's epigenetic landscape: A framework for post-Darwinian biology? *BioEssays* 34:149–57.

Ingerson-Mahar, M., and Gitai, Z. 2012. A growing family: the expanding universe of the bacterial cytoskeleton. *FEMS Microbiology Reviews* 36:256–66.

Jablonka, E., and Lamb, M. 2005. *Evolution in four dimensions: genetic, epigenetic, behavioral and symbolic variation in the history of life.* Cambridge, MA: MIT Press.

Jablonka, E., and Raz, G., 2009. Transgenerational epigenetic inheritance: prevalent mechanisms, and implications for the study of heredity and evolution. *Quarterly Review of Biology* 84:131–76.

Jacob, F., 1973. *The Logic of life.* New York: Pantheon Books.

Jain, R., Rivera, M. C., and Lake, J. A. 1999. Horizontal gene transfer among genomes: the complexity hypothesis. *Proceedings of the National Academy of Sciences U S A* 96:3801–6.

Jalasvuori, M., and Bamford, J. K. H. 2008. Structural co-evolution of viruses and cells in the primordial world. *Origins of Life and Evolution of the Biosphere* 38:165–81.

Jarrell, K. F., and Albers, S.-V. 2012. The archaellum: an old motility structure with a new name. *Trends in Microbiology* 20:307–12.

Jarrell, K. F., and McBride, M. J. 2008. The surprisingly diverse ways that prokaryotes move. *Nature Reviews Microbiology* 6:466–76.

Jarrell, K. F., Walters, A. D., Bochiwal, C., Borgia, J. M., Dickinson, T., and Chong, J.P. J. 2011. Major players on the microbial stage: why archaea are important. *Microbiology* 157:919–36.

Javaux, E. J., Knoll, A. H., and Walter, M. 2004. Recognizing and interpreting the fossils of early eukaryotes. *Origins of Life and Evolution of the Biosphere* 33:75–94.

Javaux, E. J., Marshall, C. P., and Bekker, A. 2010. Organic-walled microfossils in 3.2-billion-year-old shallow-marine siliciclastic deposits. *Nature* 463:934–38.

Jèkely, G. 2007. Origin of phagotrophic eukaryotes as social cheaters in microbial biofilms. *Biology Direct* 2:3.

Jèkely, G., and Arendt, D. 2006. Evolution of intraflagellar transport from coated vesicles and autogenous oigin of the eukaryotic cilium. *BioEssays* 28:191–98.

John, P., and Whatley, F. R. 1975. Paracoccus denitrificans Davis (Micrococcus denitrificans Beijerinck) as a mitochondrion. *Nature* 254:495–98.

Joyce, G. F. 2002. The antiquity of RNA-based evolution. *Nature* 418:214–21.

Joyce, G. F., and Orgel, L. E. 2006. Progress towards understanding the origin of the RNA world. In Gesteland, Czech, and Atkins 2006, 23–56.

Jun, S.-R, Sims, G. E., Wu, G. A., and Kim, S. H. 2010. Whole genome phylogeny of prokaryotes by feature frequency profiles: an alignment-free method with optimal resolution. *Proceedings of the National Academy of Sciences U S A* 107:133–38.

Kasting, J. F., and Ohno, S. 2006. Palaeoclimates: the first two billion years. *Philosophical Transactions of the Royal Society of London, Series B* 361:917–29.

Kasting, J. F., and Siefert, J. L. 2002. Life and the evolution of earth's atmosphere. *Science* 296:1066–68.

Kauffman, S. 1993. *The origins of order: self-organization and selection in evolution*. New York: Oxford University Press.

———. 2000. *Investigations*. New York: Oxford University Press.

Keeling, P. J. 1998. A kingdom's progress: Archezoa and the origin of eukaryotes. *BioEssays* 20:87–95.

———. 2010. The endosymbiotic origin, diversification and fate of plastids. *Philosophical Transactions of the Royal Society of London, Series B* 365:729–48.

Keeling, P. J., Burger, G., Durnford, D. G., Lang, B. F., Lee, R. W., Pearlman, R. E., et al. 2005. The tree of the eukaryotes. *Trends in Ecology and Evolution* 20:670–76.

Keeling, P. J., and Palmer, J. D. 2008. Horizontal gene transfer in eukaryotic evolution. *Nature Reviews Genetics* 9:605–18.

Kelley, D. S., Karson, J. A., Fruh-Green, D. L., Yoerger, D. R., Shank, T. M., Butterfield, D. A., et al. 2005. A serpentinite-hosted ecosystem: the Lost City hydrothermal field. *Science* 307:1428–34.

Kim, G. H., Klotchkova, T. A., and Kang, Y.-M. 2001. Life without a cell membrane: regeneration of protoplasts from disintegrated cells of the marine green alga *Bryopsis plumosa*. *Journal of Cell Science* 114:2009–11.

Kirschner, M., and Gerhart, J. 1998. Evolvability. *Proceedings of the National Academy of Sciences U S A* 95:8420–27.

———. 2005. *The plausibility of life: resolving Darwin's dilemma*. New Haven, CT: Yale University Press.

Knoll, A. H. 2003. *Life on a young planet*. Princeton, NJ: Princeton University Press.

Knoll, A. H., Javaux, E. J., Hewitt, D., and Cohen, P. 2006. Eukaryotic organisms in proterozoic oceans. *Philosophical Transactions of the Royal Society of London, Series B* 361:1023–38.

Koch A. L., 1996. What size should a bacterium be? A question of scale. *Annual Review of Microbiology* 50:317–48.

———. 2003. Development of the sacculus marked emergence of the Bacteria. *ASM News* 69:229–33.

Koch, A. L., and Schmidt, T. M. 1991. The first cellular bioenergetic process: primitive generation of a protonmotive force. *Journal of Molecular Evolution* 33:297–304.

Koch, A. L., and Silver, S. 2005. The first cell. *Advances in Microbial Physiology* 50:227–59.

Koonin, E. V. 2007. The biological big bang model for the major transitions in evolution. *Biology Direct* 2:21.

———. 2012. *The logic of chance: the nature and origin of biological evolution.* Upper Saddle River, NJ: Pearson Education.

Koonin, E. V., and Dolja, V. V. 2006. Evolution of complexity in the viral world: the dawn of a new vision. *Virus Research* 117: 1–4.

Koonin, E. V., and Martin, W. 2005. On the origin of genomes and cells within inorganic compartments. *Trends in Genetics* 21:647–54.

Koonin, E. V., Senkevich, T. G., and Dolja, V. V. 2006. The ancient virus world and evolution of cells. *Biology Direct* 1:29.

Koonin, E. V., and Wolf, Y. 2010. The common ancestry of life. *Biology Direct* 5:64.

Kopp, R. E., Kirschvink, J. L., Hilburn, I. A., and Nash, C. Z. 2005. The palaeozoic snowball earth: a climate disaster triggered by the evolution of oxygenic photosynthesis. *Proceedings of the National Academy of Sciences U S A* 102:1131–36.

Kornberg A., Rao, N. N., and Ault-Richè, D. 1999. Inorganic polyphosphate: a molecule of many functions. *Annual Review of Biochemistry* 68:89–125.

Krupovich, M., Prangishvili, D., Hendrix, R. W., and Bamford, D. H. 2011. Genomics of bacterial and archaeal viruses: dynamics within the prokaryotic virosphere. *Microbiology and Molecular Biology Reviews* 75:610–35.

Kurland, C. G. 2005. What a tangled web: barriers to rampant horizontal gene transfer. *BioEssays* 27:741–47.

———. 2010. The RNA dreamtime. *BioEssays* 32:866–71.

Kurland, C. G., Canbäck, B., and Berg, O. G. 2003. Horizontal gene transfer: a critical view. *Proceedings of the National Academy of Sciences U S A* 100:9658–62.

Kurland, C. G., Collins, L. J., and Penny, D. 2006. Genomics and the irreducible nature of eukaryotic cells. *Science* 312:1011–14.

Kutschera, U., and Niklas, K. J. 2004. The modern theory of biological evolution: an expanded synthesis. *Naturwissenschaften* 91:255–76.

Lake, J. A. 2009. Evidence for an early prokaryotic endosymbiosis. *Nature* 460: 967–71.

Lake, J. A., and Rivera, M. C. 2004. Deriving the genomic tree of life in the presence of horizontal gene transfer: conditioned reconstruction. *Molecular Biology and Evolution* 21:681–90.

Lake, J. A., Skophammer, R. G., Herbold, C. W., and Servin, J. A. 2009. Genome beginnings: rooting the tree of life. *Philosophical Transactions of the Royal Society, Series B* 364:2177–85.

Lane, C. E., and Archibald, J. M. 2008. The eukaryotic tree of life: endosymbiosis takes its TOL. *Trends in Ecology and Evolution* 23:268–75.

Lane, N. 2002. *Oxygen, the molecule that made the world.* Oxford: Oxford University Press.

———. 2005. *Power, sex, suicide: mitochondria and the meaning of life.* London: Oxford University Press.

———. 2009. *Life ascending: the ten great inventions of evolution.* New York: W. W. Norton.

———. 2011. Energetics and genetics across the prokaryote-eukaryote divide. *Biology Direct* 6:35.

Lane, N., Allen, J. F., and Martin, W. 2010. How did LUCA make a living? Chemiosmosis in the origin of life. *BioEssays* 32:271–80.

Lane, N. and Martin, W. 2010. The energetics of genome complexity. *Nature* 467:929–34.

———. 2012. The origin of membrane bioenergetics. *Cell* 151:1406–16.

Lapierre, P., Shial, R., and Gogarten, J. P. 2006. Distribution of F- and A/V-type ATPases in *Thermus scotoductus* and other closely related species. *Systematic and Applied Microbiology* 29:15–23.

Lartigue, C., Glass, J. I., Alperovich, N., Pieper, R., Parmar, P. P., Hutchison, C. A., et al. 2007. Genome transplantation in bacteria: changing one species to another. *Science* 317:632–38.

Lazcano, A., and Miller, S. L. 1996. The origin and early evolution of life: prebiotic chemistry, the pre-RNA world, and time. *Cell* 85:793–98.

Leander, B. S. 2008. A hierarchical view of convergent evolution in microbial eukaryotes. *Journal of Eukaryotic Microbiology* 55:59–68.

Lee, R. E. 1989. *Phycology.* 2nd ed. Cambridge: Cambridge University Press.

Leigh, J. A. 2002. Evolution of energy metabolism. In J. T. Staley and A.-L. Reysenbach, eds., *Biodiversity of microbial life,* 103–20. New York: Wiley-Liss.

Lenski, R. E., Ofria, C., Pennock, A. T., and Adami, C. 2003. The evolutionary origin of complex features. *Nature* 423:139–44.

Lester, L., Meade, A., and Pagel, M. 2005. The slow road to the eukaryotic genome. *BioEssays* 28:57–64.

Lewontin, R. 2001. *The triple helix: gene, organism and environment.* Cambridge, MA: Harvard University Press.

Lifson, S. 1997. On the crucial stages in the origin of animate matter. *Journal of Molecular Evolution* 44:1–8.

Lincoln, T. A., and Joyce, G. F. 2009. Self-sustained replication of an RNA enzyme. *Science* 323:1229–32.

Lindahl, P. A. 2004. Stepwise evolution of nonliving to living chemical systems. *Origins of Life and Evolution of the Biosphere* 34:371–89.

Line, M. A. 2002. The enigma of the origin of life and its timing. *Microbiology* 148:21–27.

Lithgow, T., and Schneider, A. 2009. Evolution of macromolecular import pathways in mitochondria, hydrogenosomes and mitosomes. *Philosophical Transactions of the Royal Society, Series B* 365:799–817.

Liu, R., and Ochman, H. 2007. Stepwise formation of the bacterial flagellar system. *Proceedings of the National Academy of Sciences U S A* 104:7116–21.

Lombard, J., Lòpez-Garcia, P. and Moreira, D. 2012. The early evolution of lipid membranes and the three domains of life. *Nature Reviews Microbiology* 10: 507–15.

Long, M., Betràn, E., Thornton, K., and Wang, W. 2003. The origin of new genes: glimpses from the young and old. *Nature Reviews Genetics* 4:865–75.

Lonhienne, T. A., E. Sagulenko, R. I. Webb, D.-C. Lee, J. Franke, D. P. Devos, et al. 2010. Endocytosis-like protein uptake in the bacterium *Gemmata obscuriglobus*. *Proceedings of the National Academy of Sciences U S A* 107: 12883–88.

Lòpez-Garcia, P., and Moreira, D. 1999. Metabolic symbiosis at the origin of eukaryotes. *Trends in Biochemical Sciences* 24:88–93.

———. 2006. Selective forces for the origin of the eukaryotic nucleus. *BioEssays* 28:525–33.

Luisi, P. L. 2003. Autopoiesis—a review and a reappraisal. *Naturwissenschaften* 90:49–59.

Luisi, P. L., Ferri, F., and Stano, P. 2006. Approaches to semi-synthetic minimal cells: a review. *Naturwissenschaften* 93:1–13.

Lukeš, J., Leander, B. S., and Keeling, P. J. 2009. Cascades of convergent evolution: the corresponding evolutionary histories of euglenozoans and dinoflagellates. *Proceedings of the National Academy of Sciences U S A* 106:9963–70.

Lynch, M. 2007. The frailty of adaptive hypotheses for the origins of organismal complexity. *Proceedings of the National Academy U S A* 104 (suppl 1): 8597–604.

Maezawa, K., Shigenobu, S., Taniguchi, H., Kubo, T., Aizawa, S., and Morioka, M. 2006. Hundreds of flagellar basal bodies cover the cell surface of the endosymbiotic bacterium *Buchnera aphidicola* sp strain APS. *Journal of Bacteriology* 188:6539–43.

Makarova, K. S., Yutin, N., Bell, S.D., and Koonin, E. V. 2010. Evolution of diverse cell division and vesicle formation systems in Archaea. *Nature Reviews Microbiology* 8:731–41.

Mansy, S. S., Schrum, J. P. Krishnamurthy, M., Tobè, S., Treco, D. A., and Szostak,

J. W. 2008. Template-directed synthesis of a genetic polymer in a model protocell. *Nature* 454:122–25.

Margolin, W. 2009. Sculpting the bacterial cell. *Current Biology* 19:R812–22.

Margulis, L. 1970. *Origin of eukaryotic cells.* New Haven, CT: Yale University
Press.

———. 1993. *Symbiosis in cell evolution.* 2nd ed. New York: W. H. Freeman and
Company.

Margulis, L., Chapman, M., Guerrero, R., and Hall, J. 2006. The last eukaryotic
common ancestor (LECA): acquisition of cytoskeletal motility from aerotoler-
ant spirochetes in the Proterozoic Eon. *Proceedings of the National Academy
of Sciences U S A* 103:13080–85.

Margulis, L., and Sagan, D. 2002. *Acquiring genomes: a theory of the origin of species.* New York: Basic Books.

Marin, B., Nowack, E. C., and Melkonian, M. 2005. A plastid in the making: evidence for a second primary endosymbiosis. *Protist* 156:425–32.

Martin, W. 2005. Archaebacteria (Archaea) and the origin of the eukaryotic nucleus. *Current Opinion in Microbiology* 8:630–37.

Martin, W., Baross, J., Kelley, D., and Russell, M. J. 2008. Hydrothermal vents and
the origin of life. *Nature Reviews Microbiology* 6:805–14.

Martin, W., Dagan, T., Koonin, E. V., Dipippo, J. L., Gorgarten, P. J., and Lake,
J. A. 2007. The evolution of eukaryotes. *Science* 316: 542–45.

Martin, W., and Koonin, E. V. 2006. Introns and the origin of nucleus-cytosol compartmentation. *Nature* 440:41–45.

Martin, W., and Müller, M. 1998. The hydrogen hypothesis for the first eukaryote.
Nature 342:37–41.

Martin, W., and Russell, M. J. 2002. On the origin of cells: a hypothesis for the
evolutionary transitions from abiotic geochemistry to chemoautotrophic prokaryotes, and from prokaryotes to nucleated cells. *Philosophical Transactions
of the Royal Society of London, Series B* 358:59–85.

Martin, W., and Russell, M. J. 2007. On the origin of biochemistry at an alkaline
hydrothermal vent. *Philosophical Transactions of the Royal Society of London,
Series B* 362:1887–925.

Martin, W. F. 2012. Hydrogen, metals, bifurcating electrons, and proton gradients:
the early evolution of biological energy conservation. *FEBS Letters* 586:485–
93.

Massana, R. 2011. Eukaryotic picoplancton in surface oceans. *Annual Review of
Microbiology* 65:91–110.

Matsuzaki, M., Misumi, O., Shin, I. T., Maruyama, S., Takahara, M., Miyagishima,
S. Y., et al. 2004. Genome sequence of the ultrasmall unicellular red alga
Cyanidioschzyon merolea 10 D. *Nature* 428:653–57.

Maynard Smith, J. 1986. *The problems of biology.* Oxford: Oxford University
Press.

Maynard Smith, J., and Szathmàry, E. 1995. *The major transitions in evolution.* Oxford, UK: W. H. Freeman.

Mayr, E. 1982. *The growth of biological thought: diversity, evolution and inheritance.* Cambridge, MA: Harvard University Press.

———. 1997. The objects of selection. *Proceedings of the National Academy of Sciences U S A* 94:2091–94.

———. 1998. Two empires or three? *Proceedings of the National Academy of Sciences U S A* 95:9720–23.

McInerney, J. O., Martin, W. F., Koonin, E. V., Allen, J. F., Galperin, M. J., Lane, N., et al. 2011. Planctomycetes and eukaryotes: A case of analogy, not homology. *BioEssays* 33:810–17.

McInerney, J. O., Cotton, J. A., and Pisani, D. 2008. The prokaryotic tree of life: past, present . . . and future? *Trends in Ecology and Evolution* 23:276–81.

McShea, D. S., and Brandon, R. N. 2010. *Biology's first law: the tendency for diversity and complexity to increase in evolutionary systems.* Chicago: University of Chicago Press.

Mead, W. R. 2004. *Power, terror, peace and war.* New York: Alfred A. Knopf.

Meyer, J. R., Dobias, D. T., Weitz, J. S., Barrick, J. E., Quick, R. T., and Lenski, R. E. 2012. Repeatability and contingency in the evolution of a key innovation in phage lambda. *Science* 335:428–32.

Miller, K. R. 2007. Falling over the edge. *Nature* 447:1055–56.

Miller, S. L. 1953. Production of amino acids under possible primitive earth conditions. *Science* 117:528–29.

Moorbath, S. 2005. Dating earliest life. *Nature* 434:155.

Moreira, D., and Lòpez-Garcia, P. 2009. Ten reasons to exclude viruses from the tree of life. *Nature Reviews Microbiology* 7:306–11.

Moreno, A., and Ruiz-Mirazo, K. 2009. The problem of the emergence of functional diversity in prebiotic evolution. *Biology and Philosophy* 24:585–605.

Morowitz, H. J. 1992. *Beginnings of cellular life.* New Haven, CT: Yale University Press.

Mulkidjanian, A. Bychkov, A. Y., Dibrova, D. V., Galperin, M. Y., and Koonin, E. V. 2012. Origin of first cells at terrestrial anoxic geothermal fields. *Proceedings of the National Academy of Sciences U S A* 109:821–30.

Mulkidjanian, A. Y., E. V. Koonin, K. S. Makarova, S. L. Mekhedov, A. Sorokin, Y. I. Wolf, et al. 2006. The cyanobacterial genome core and the origin of photosynthesis. *National Academy of Sciences U S A* 103:13126–31.

Mulkidjanian, A. Y., Galperin, M. Y., and Koonin, E. V. 2009. Co-evolution of primordial membranes and membrane proteins. *Trends in Biochemical Sciences* 34:206–15.

Mulkidjanian, A. Y., Makarova, K. S., Galperin, M. Y., and Koonin, E. V., 2007. Inventing the dynamo machine: the evolution of the F-type and V-type ATPases. *Nature Reviews Microbiology* 5:892–99.

Müller, M. 1993. The hydrogenosome. *Journal of General Microbiology* 139:2879–89.

Müller, M., Mentel, M., van Hellemond, J. J., Henze, K., Woehle, C., Gould, S. B., et al. 2012. Biochemistry and evolution of anaerobic energy metabolism in eukaryotes. *Microbiology and Molecular Biology Reviews* 76: 444–95.

Mullins, E. (1974) 2001. *The pilgrimage to Santiago*. Northampton, MA: Interlink Publishing Group.

Nelson-Sathi, S. N., Dagan, T., Landau, G., Janssen, A., Sted, M., McInerny, J. O., et al. 2012. Acquisition of 1000 eubacterial genes physiologically transformed a methanogen at the origin of the haloarchaea. *Proceedings of the National Academy of Sciences U S A* 109: 20537–42.

Nicholls, D. G., and Ferguson, S. J. 1992. *Bioenergetics 2*. London: Academic Press Limited.

Nilsson, D.-E., and Pelger, S. 1994. A pessimistic estimate of the time required for an eye to evolve. *Proceedings of the Royal Society of London, Series B* 256:53–58.

Nisbet, E., Bendall, D., Howe, C., and Nisbet, E. 2008. Photosynthesis and atmospheric evolution. *Philosophical Transactions of the Royal Society of London, Series B* 363.

Nitschke, W., and Russell, M. J. 2009. Hydrothermal focusing of chemical and chemiosmotic energy supported by delivery of catalytic Fe, Ni, Mo/W, Co, S and Se forced life to emerge. *Journal of Molecular Evolution* 69:481–96.

Noble, D. 2006. *The music of life: biology beyond the genome*. Oxford: Oxford University Press.

Nowack, E. C. M., and Grossman, A. R. 2012. Trafficking of protein into the recently established photosynthetic organelles of Paulinella chromatophora. *Proceedings of the National Academy of Sciences U S A* 109:5340–45.

Nowack, E. C. M., and Melkonian, M. 2010. Endosymbiotic associations within protists. *Philosophical Transactions of the Royal Society of London, Series B* 365:699–712.

Nowack, E. C. M., Melkonian, M., and Glöckner, G. 2008. Chromatophore genome sequence of *Paulinella* sheds light on the acquisition of photosynthesis by eukaryotes. *Current Biology* 18:410–18.

Olsen, G. J., and Woese, C. R. 1997. Archaeal genomics: an overview. *Cell* 89:991–94.

Olson, J. M. 2006. Photosynthesis in the Archaean era. *Photosynthesis Research* 88:109–17.

Oparin, A. I. 1938. *The origin of life*. New York: Dover.

Orgel, L. E. 1998. The origin of life: a review of facts and speculations. *Trends in Biochemical Sciences* 23:491–95.

———. 2000. Self-organizing biochemical cycles. *Proceedings of the National Academy of Sciences U S A* 97:12503–7.

————. 2004. Prebiotic chemistry and the origin of the RNA world. *Critical Reviews in Biochemistry and Molecular Biology* 399:99–123.

————. 2008. The implausibility of metabolic cycles on the prebiotic earth. *PloS Biology* 6 (1): e18.

Osteryoung K. W., and Nunnari, J. 2003. The division of endosymbiotic organelles. *Science* 302:1698–3.

Pace, N. 1991. Origin of life—facing up to the physical setting. *Cell* 65:531–33.

Pace, N. R. 1997. A molecular view of microbial diversity and the biosphere. *Science* 276:734–40.

Pace, N. R. 2001. The universal nature of biochemistry. *Proceedings of the National Academy of Sciences U S A* 98:805–8.

————. 2006. Time for a change. *Nature* 441:289.

————. 2009a. Problems with "procaryote." *Journal of Bacteriology* 191:2008–10.

————. 2009b. Mapping the tree of life: progress and prospect. *Microbiology and Molecular Biology Reviews* 73:565–76.

Pallen, M. J., and Matzke, N. 2006. From the *Origin of Species* to the origin of bacterial flagella. *Nature Reviews Microbiology* 4:784–90.

Palmer, J. D. 2003. The symbiotic birth and spread of plastids: how many times and whodunit? *Journal of Phycology* 39:4–11.

Parfrey, L. W., Lahr, D. J. G., Knoll, A. H., and Katz, L. A. 2011. Estimating the timing of early eukaryotic diversification with multigene molecular clocks. *Proceedings of the National Academy of Sciences U S A* 108:13624–29.

Patterson, D. J. 1996. *Free-living fresh water protozoa—a colour guide*. New York: Wiley.

Penny, D. 2005. An interpretive review of the origin of life research. *Biology and Philosophy* 20:633–71.

Pepper, J. W., and Herron, M. D. 2008. Does biology need an organism concept? *Biological Reviews* 83:621–27.

Peretò, J., Lòpez-Garcia, P., and Moreira, D. 2004. Ancestral lipid biosynthesis and early membrane evolution. *Trends in Biochemical Sciences* 29:469–77.

Pester, M., Schleper, C., and Wagner, M. 2011. The Thaumarchaeota: an emerging view of their phylogeny and ecophysiology. *Current Opinion in Microbiology* 14:300–6.

Pigliucci, M. 2007. Do we need an expanded evolutionary synthesis? *Evolution* 61:2743–49.

————. 2008. Is evolvability evolvable? *Nature Reviews Genetics* 9:75–82.

Pilhofer, M. Ladinsky, M. S., McDowall, A. W., Petroni, G., and Jensen, G. J. 2011. Microtubules in bacteria: ancient tubulins build a five-protofilament homolog of the eukaryotic cytoskeleton. *PLoS Biology* 9(12): e1001213. doi:10.1371/journal.pbio.1001213.

Pinker, S. 1998. *How the brain works*. London: Penguin.

Pisani, D., Cotton, J. A., and McInerney, J. O. 2007. Supertrees disentangle the chimerical origin of eukaryotic genomes. *Molecular Biology and Evolution* 24:1752–60.

Poehlein, A., Schmidt, S., Kasten, A.-K., Goenrich, M., Vollmers, J., Thurmer, A., et al. 2012. An ancient pathway combining carbon dioxide fixation with the generation and utilization of a sodium ion gradient for ATP synthesis. *PLoS One* 7(3): e33439.

Pollack, G. H. 2001. *Cells, gels and the engines of life: a new, unifying approach to cell function.* Seattle, WA: Ebner and Sons.

Poole, A. M., and Penny, D. 2006. Evaluating hypotheses for the origin of eukaryotes. *BioEssays* 29:74–84.

Popa, R. 2004. *Between necessity and probability: searching for the definition and origin of life.* New York: Springer Verlag.

Powner, M. W., Gerland, B., and Sutherland, J. D. 2009. Synthesis of activated pyrimidine ribonucleotides in prebiotically plausible conditions. *Nature* 459:239–42.

Price, D. C., Chan, C. X., Yoon, H. S., Yang, E. C., Qiu, H., Weber, A. P. M., et al. 2012. *Cyanophora paradoxa* genome elucidates origin of photosynthesis in algae and plants. *Science* 335:843–47.

Pross, A. 2003. The driving force for life's emergence: Kinetic and thermodynamic considerations. *Journal of Theoretical Biology* 220:393–406.

———. 2004. Causation and the origin of life: metabolism or replication first? *Origins of Life and Evolution of the Biosphere* 34:307–21.

———. 2005a. On the emergence of biological complexity: Life as a kinetic state of matter. *Origins of Life and Evolution of the Biosphere* 35:151–66.

———. 2005b. On the chemical nature and origin of teleonomy. *Origins of Life and Evolution of the Biosphere* 35:383–94.

———. 2012. *What is life? How chemistry becomes biology.* Oxford: Oxford University Press.

Queller, D. C., and Strassman, J. E. 2009. Beyond society: the evolution of organismality. *Philosophical Transactions of the Royal Society of London, Series B* 364:3143–55.

Rasmussen, B., Fletcher, I. R., Brocks, J. J., and Kilburn, M.R. 2008. Reassessing the first appearance of eukaryotes and cyanobacteria. *Nature* 455:1101–4.

Rasmussen, S., Bedau, M. A., Chen, L., Deamer, D., Krakauer, D.C., Parkard, N. H., and Stadler, P. F. eds. 2009. *Protocells: bridging nonliving and living matter.* Cambridge, MA: MIT Press.

Raymond, J., Zhaxybayeva, O., Gogarten, J. P., Gerdes, S. Y., and Blankenship, R. E. 2002. Whole genome analysis of photosynthetic prokaryotes. *Science* 298:1616–19.

Rees, M. 1999. *Just six numbers: the deep forces that shape the universe.* London: Weidenfeld and Nicolson.

Reyes-Prieto, A., Weber, A. P. M., and Bhattacharya, D. 2007. Origin and establishment of the plastid in algae and plants. *Annual Review of Genetics* 41:147–68.

Reynaud, E. G., and Devos, D. P., 2011. Transitional forms between the three domains of life and evolutionary implications. *Proceedings of the Royal Society of London, Series B* 278:3321–28.

Richards, T. A., and Cavalier-Smith, T. 2005. Myosin domain evolution and the primary divergence of eukaryotes. *Nature* 436:1113–18.

Richardson, D. J., 2000. Bacterial respiration, a flexible process for a changing environment. *Microbiology* 146:551–71.

Rivera, M. C., and Lake, J. A. 2004. The ring of life provides evidence for a genome fusion origin of eukaryotes. *Nature* 431:152–55.

Rodriguez-Ezpelata, N., Brinkmann, H., Burey, S. C., Roure, B., Burger, G. L., Loffelhardt, W., et al. 2005. Monophyly of primary photosynthetic eukaryotes: green plants, red algae and glaucophytes. *Current Biology* 15:1325–30.

Roger, A. J. 1999. Reconstructing early events in eukaryotic evolution. *American Naturalist* 154:5146–63.

Roger, A. J., and Hug, L. A. 2006. The origin and diversification of eukaryotes: problems with molecular phylogenetics and molecular clock estimation. *Philosophical Transactions of the Royal Society, Series B* 361:1039–54.

Roger, A. J., and Simpson, A. G. B. 2009. Evolution: revisiting the root of the eukaryote tree. *Current Biology* 19:R165–67.

Rosslenbroich, B. 2006. The notion of progress in evolutionary biology—the unresolved problem and an empirical suggestion. *Biology and Philosophy* 21:41–70.

Russell, M. J., and Hall, A. J. 1997. The emergence of life from iron monosulfide bubbles at a submarine hydrothermal redox and pH front. *Journal of the Geological Society of London* 154:377–402.

Saier, M. H. 2003. Tracing pathways of transport protein evolution. *Molecular Microbiology* 48:1145–56.

Saladino, R., Crestini, C., Pino, S., Costanzo, G., and Di Mauro, E. 2012. Formamide and the origin of life. *Physics of Life Reviews* 9:84–104.

Sapp, J. 1987. *Beyond the gene: cytoplasmic inheritance and the struggle for authority in genetics*. Oxford: Oxford University Press.

———. Cytoplasmic heretics. 1998. *Perspectives in Biology and Medicine* 41:224–42.

———. 2003. *Genesis: the evolution of biology*. Oxford: Oxford University Press.

———. 2005. The prokaryote-eukaryote dichotomy: meanings and mythology. *Microbiology and Molecular Biology Reviews* 69:292–305.

———. 2009. The new foundations of evolution: on the tree of life. Oxford: Oxford University Press.

Sapra, R., Bagramyan, K., and Adams, M. W. W. 2003. A simple energy-conserving system: proton reduction coupled to proton translocation. *Proceedings of the National Academy of Sciences U S A* 100:7545–50.

Schirrmeister, B., Antonelli, A., and Bagheri, H. C. 2011. The origin of multicellularity in cyanobacteria. *BMC Evolutionary Biology* 11:45.

Schneider, E. D., and Sagan, D. 2005. *Into the cool: energy flow, thermodynamics, and life*. Chicago: University of Chicago Press.

Schoepp-Cothenet, B., van Lis, R., Atteia, A., Baymann, F., Capowiez, L., Ducluzeau, A.-L., et al. 2013. On the universal core of bioenergetics. *Biochimica et Biophysica Acta* 1827:79–93.

Schopf, J. W. 1993. Microfossils of the early archean apex chert: new evidence of the antiquity of life. *Science* 260:640–46.

———. 2006. Fossil evidence of Archaean life. *Philosophical Transactions of the Royal Society of London, Series B* 361:869–85.

Shapiro, J. A. 2011. *Evolution: a view from the 21st century*. Upper Saddle River, NK: F T Press Science.

Shapiro, L., McAdams, H. H., and Losick, R. 2002. Generating and exploiting polarity in bacteria. *Science* 298:1942–46.

———. 2009. Why and how bacteria localize proteins. *Science* 326:1255–28.

Shapiro, R. 2006. Small molecule interactions were central to the origin of life. *Quarterly Review of Biology* 81:105–25.

———. 2007. A simpler origin for life. *Scientific American* 296:47–53.

Sheridan, P. P., Freeman, K. H., and Brenchley, J. E. 2003. Estimated minimal divergence times of the major bacterial and archaeal phyla. *Geomicrobiology Journal* 20:1–14.

Shi, T., and Falkowski, P. G. 2008. Genome evolution in cyanobacteria: the stable core and variable shell. *Proceedings of the National Academy of Sciences U S A* 105:2510–15.

Shiflett, A. M., and Johnson, P. J. 2010. Mitochondrion-related organelles in eukaryotic protists. *Annual Review of Microbiology* 64:409–29.

Simonson, A. B., Servin, J. A., Skophammer, R. G., Herbold, C. W., Rivera, M. C., and Lake, J. A. 2005. Decoding the genomic tree of life. *Proceedings of the National Academy of Sciences U S A* 102 (suppl 1): 6608–13.

Sleigh, M. 1989. *Protozoa and other protists*. 2nd ed. Cambridge: Cambridge University Press.

Smith, E., and Morowitz, H. J. 2004. Universality in intermediary metabolism. *Proceedings of the National Academy of Sciences U S A* 101:13168–73.

Smolin, L. 1997. *The life of the cosmos*. Oxford: Oxford University Press.

Snel, B., Huynen, M. A., and Dutilh, B. A. 2005. Genome trees and the nature of genome evolution. *Annual Review of Microbioloy* 59:191–209.

Snyder, L. A. S., Loman, N. J., Fütterer, K., and Pallen, M. J. 2008. Bacterial flagellar diversity and evolution: seek simplicity and distrust it? *Trends in Microbiology* 17:1–5.

Sogin, M. L. 1994. The origin of eukaryotes and evolution into major kingdoms. In Bengtson, S., ed., *Early life on earth*, 181–92. New York: Columbia University Press.

Sonea, S., and Mathieu, L. G. 2000. *Prokaryotology: a coherent view*. Quebec: Les Presses de l'Universite' de Montreal.

Speijer, D., 2011. Does constructive neutral evolution play an important role in the origin of cellular complexity? *BioEssays* 33:344–49.

Stanier, R. Y. 1970. Some aspects of the biology of cells and their possible evolutionary significance. *Symposia of the Society for General Microbiology* 20:1–38.

Stanier, R.Y., Doudoroff, M., and Adelberg, E. 1963. *The microbial world.* 2nd ed. New York: Prentice Hall.

Stanier, R. Y., and Van Niel, C. B. 1962. The concept of a bacterium. *Archiv für Mikrobiologie* 92:17–35.

Stechmann, A., and Cavalier-Smith, T. 2002. Rooting the eukaryote tree by using a derived gene fusion. *Science* 297:89–91.

Steinberg-Yfrach, G., Liddell, P. A., Hung, S.-C., Moore, A. L. Gust, D., and Moore, T. A. 1997. Conversion of light energy to proton potential in liposomes by artificial photosynthetic reaction centres. *Nature* 385:239–41.

Stetter, K. O. 2006. Hyperthermophiles in the history of life. *Philosophical Transactions of the Royal Society of London, Series B* 361:1837–43.

Stolper, D. A. Revsbech, N. P. and Canfield, D. E. 2010. Aerobic growth at nanomolar oxygen concentration. *Proceedings of the National Academy of Sciences U S A* 107:18755–58.

Sullivan, W. T. III, and Baross, J. A. eds. 2007. *Planets and life, the emerging science of astrobiology.* Cambridge: Cambridge University Press.

Summons, R. E., Bradley, A.S., Jahnke, L. L., and Waldbauer, J. R. 2006. Steroids, triterpenoids and molecular oxygen. *Philosophical Transactions of the Royal Society of London, Series B* 361:951–68.

Suttle, C. A. 2005. Viruses in the sea. *Nature* 437:356–61.

Swithers, K. S., Fournier, G. P., Green, A. G., Gogarten, J. P., and Lapierre, P. 2011. Reassessment of the lineage fusion hypothesis for the origin of double membrane bacteria. *PLoS One* 6 (8): e23774.

Szathmáry, E. 1999. The origin of the genetic code: amino acids as cofactors in an RNA world. *Trends in Genetics* 15:223–29.

Szostak, J. W., Bartel, D. P. and Luisi, P. L. 2001. Synthesizing life. *Nature* 409:387–90.

Taylor, J.S., and Raes, J. 2004. Duplication and divergence — The evolution of new genes and old ideas. *Annual Review of Genetics* 38:615–43.

Theobald, D. L. 2010. A formal test of the theory of universal common ancestry. *Nature* 465:219–22.

Therer, M. D., and Saier, M. H. 2009. Bioinformative characterization of the P-ATPases encoded within the fully sequenced genomes of 26 eukaryotes. *Journal of Membrane Biology* 229:115–30.

Thomas, L. 1974. *The lives of a cell.* Toronto: Bantam Books.

Timmis, J. N., Ayliffe, M. A., Huang, C. Y., and Martin, W. 2004. Endosymbiotic gene transfer: organelle genomes forge eukaryotic chromosomes. *Nature Reviews Genetics* 5:123–35.

Tomitani, A., Knoll, A. H., Cavanaugh, C. M., and Ohno, T. 2006. The evolutionary diversification of cyanobacteria: molecular-phylogenetic and paleontological perspectives. *Proceedings of the National Academy of Sciences U S A* 103:5442–47.

Trefil, J., Morowitz, H. J., and Smith, E. 2009. The origin of life. *American Scientist* 97:206–13.

Trevors, J. T., and Abel, D. L. 2004. Chance and necessity do not explain the origin of life. *Cell Biology International* 28:729–39.

Trevors, J. T., and Psenner, R. 2001. From self-assembly of life to present-day bacteria: a possible role for nanocells. *FEMS Microbiology Reviews* 25:573–82.

Ueno, Y., Yamada, K., Yoshida, N., Maruyama, S., and Isozaki, Y. 2006. Evidence from fluid inclusions for microbial methanogenesis in the early Archaean era. *Nature* 440:516–19.

Valas, R. E., and Bourne, P. E. 2011. The origin of a derived superkingdom: how a gram-positive bacterium crossed the desert to become an archaeon. *Biology Direct* 6:16.

Valentine, D. L. 2007. Adaptations to energy stress dictate the ecology and evolution of the Archaea. *Nature Reviews Microbiology* 5:316–23.

Vendeville, A., Larivière, D., and Fourmentin, E. 2011. An inventory of the bacterial macromolecular components and their spatial organization. *FEMS Microbiology Reviews* 35:395–414.

Vetsigian, K., Woese, C., and Goldenfeld, N. 2006. Collective evolution and the genetic code. *Proceedings of the National Academy of Sciences U S A* 103:10696–701.

Villareal, L. P. 2005. *Viruses and the evolution of life*. Washington, DC: American Society for Microbiology Press.

Wacey, D., Kilburn, M. R., Saunders, M., Cliff, J., and Brasier, M. D. 2011. Microfossils of sulphur-metabolizing cells in 3.4-billion-year-old rocks of Western Australia. *Nature Geosciences* 4:698–702.

Wächtershäuser, G. 1992. Groundworks for an evolutionary biochemistry: the iron-sulfur world. *Progress in Biophysics and Molecular Biology* 58:85–201.

———. 2003. From pre-cells to Eukarya—a tale of two lipids. *Molecular Microbiology* 47:13–22.

———. 2006. From volcanic origins of chemoautotrophic life to Bacteria, Archaea and Eukarya. *Philosophical Transactions of the Royal Society of London, Series B* 361:1787–808.

Waldbauer, J. R., Newman, D. K., and Summons, R. E. 2011. Microaerobic steroid biosynthesis and the molecular fossil record of Archean life. *Proceedings of the National Academy of Sciences U S A* 108:13409–14.

Waldbauer, J. R., Sherman, L. S., Sumner, D. Y., and Summons, R. E. 2009. Late Archean molecular fossils from the Transvaal supergroup record the antiquity of microbial diversity and aerobiosis. *Precambrian Research* 169:28–47.

Walsh, D. A., and Doolittle, W. F. 2005. The real "domains" of life. *Current Biology* 15:R237–40.

Wang, M., Yafremava, L. S., Caetano-Anollés, D., Mittenthal, J. E., and Caetano-Anollés, G. 2010. Reductive evolution of architectural repertoires in proteomes and the birth of the tripartite world. *Genome Research* 17:1572–85.

Webster, G., and Goodwin, B. 1996. *Form and transformation: generative and relational principles in biology*. Cambridge: Cambridge University Press.

Westall, F. 2005. Life on the early earth: a sedimentary view. *Science* 308:366–67.

———. 2009. Life on an anaerobic planet. *Science* 323:471–72.

Westall, F., de Ronde, C. E. J., Gordon, S., Grassineau, N., Colas, M., Cockell, C. S., and Lammer, H. 2006. Implications of a 3.472 to 3.333 gyr-old subaerial microbial mat from the Barberton greenstone belt, South Africa, for the UV environmental conditions on the early earth. *Philosophical Transactions of the Royal Society of London, Series B* 361:1857–75.

Wheelis, M. L., Kandler, O., and Woese, C. R. 1992. On the nature of global classification. *Proceedings of the National Academy of Sciences U S A* 89:2930–34.

White, D. 2007. *The physiology and biochemistry of prokaryotes*. Oxford: Oxford University Press.

White, L. 1968. *Machina ex Deo*. Cambridge, MA: MIT Press.

Whitman, W. B. 2009. The modern concept of the procaryote. *Journal of Bacteriology* 191:2000–5.

Whitman, W. B., Coleman, D. C., and Wiebe, W. J. 1998. Prokaryotes: the unseen majority. *Proceedings of the National Academy of Sciences U S A* 95:6578–83.

Whittaker, R. H. 1969. New concepts of kingdoms of organisms. *Science* 163:150–63.

Wicken, J. S. 1987. *Evolution, thermodynamics and information: extending the Darwinian program*. New York: Oxford University Press.

Williams, T. A., Foster, A. G., Nye, T. M., Cox, C. J., and Embley, T. M. 2012. A congruent phylogenomic signal places eukaryotes within the Archaea. *Proceedings of the Royal Society of London, Series B* 279:4870–79.

Wochner, A., Attwater, J., Coulson, A., and Holliger, P. 2011. Ribozyme-catalyzed transcription of an active ribozyme. *Science* 332:209–12.

Woese, C. R. 1987. Bacterial evolution. *Microbiological Reviews* 51:221–71.

———. 1998a. The universal ancestor. *Proceedings of the National Academy of Sciences U S A* 95:6854–59.

———. 1998b. Default taxonomy: Ernst Mayr's view of the microbial world. *Proceedings of the National Academy of Sciences U S A* 95:11043–46.

———. 2000. Interpreting the universal phylogenetic tree. *Proceedings of the National Academy of Sciences U S A* 97:8392–96.

———. 2002. On the evolution of cells. *Proceedings of the National Academy of Sciences U S A* 99:8742–47.

———. 2004. A new biology for a new century. *Microbiology and Molecular Biology Reviews* 68:173–86.

————. 2007. The birth of the Archaea. In Garrett, R. A., and Klenk, H.-P., eds., *Archaea: evolution, physiology and molecular biology*, 1–15. Oxford: Blackwell.

Woese, C. R., Kandler, O., and Wheelis, M. L. 1990. Towards a natural system of organisms: proposal for the domains Archaea, Bacteria and Eucarya. *Proceedings of the National Academy of Sciences U S A* 87:4576–79.

Woese, C. R., Olsen, G. J., Ibba, M. and Soll, D. 2000. Aminoacyl-tRNA synthetases, the genetic code, and the evolutionary process. *Microbiology and Molecular Biology Reviews* 64:202–36.

Wolf, Y. I., and Koonin, E. V. 2007. On the origin of the translation system and the genetic code in the RNA world by means of natural selection, exaptation, and subfunctionalization. *Biology Direct* 2:14.

Wong, T., Amidi, A., Dodds, A., Siddiqi, S., Wang, J., Yep, T., et al. 2007. Evolution of the bacterial flagellum. *Microbe* 2:335–40.

Xiong, J. and Bauer, C. E. 2002. Complex evolution of photosynthesis. *Annual Review of Plant Physiology and Plant Molecular Biology* 53:503–21.

Yarus, M. 2010. *Life from an RNA world: the ancestor within*. Cambridge, MA: Harvard University Press.

Yoon, H. S., Hackett, J. D., Ciniglia, C., Pinto, G., and Bhattacharya, D. 2004. A molecular timeline for the origin of photosynthetic eukaryotes. *Molecular Biology and Evolution* 21:809–81.

Young, K. D. 2006. The selective value of bacterial shape. *Microbiology and Molecular Biology Reviews* 70:660–703.

Yutin, N., and Koonin, E. V., 2012. Archaeal origin of tubulin. *Biology Direct* 7:10.

Yutin, N., Makarova, K. S., Mekhedov, S. L., Wolf, Y, I., and Koonin, E. V. 2008. The deep archaeal roots of eukaryotes. *Molecular Biology and Evolution* 25:1619–30.

Zhaxybayeva, O., and Gogarten, J. P. 2004. Cladogenesis, coalescence and the evolution of the three domains of life. *Trends in Genetics* 20:182–87.

Zhaxybayeva, O., Lapierre, P., and Gogarten, J. P. 2005. Ancient gene duplications and the root(s) of the tree of life. *Protoplasma* 227:53–64.

Index

A page number followed by the letter f refers to a figure or its caption.

Acanthamoeba, Mimivirus of, 50
acetogenesis, 77, 82, 237n6
acidic environments, Cavalier-Smith on, 63, 64
acidobacteria, 83
acquired characters, inheritance of: in endosymbiosis, 191; by prion state, 204
acritarchs, 155–56, 155f
actinobacteria: origin of diderms and, 202; origin of eukaryotes and, 116–17
actins: Archaean, 123; crenactin and, 97; distant homologs of, 112; of eukaryotic cytoskeleton, 118; highly conserved sequences of, 97; MreB protein and, 96, 120; in phagocytosis, 98
adaptation, 195–97, 198; evolutionary progress and, 224–26, 227; lateral gene transfer and, 199; organism as unit of, 36; at progenote level, 56. *See also* natural selection
aerobic Bacterium, energy transduction in, 72f
agency, 225, 226
algae: as eukaryotes, 6; in Haeckel's Protista, 4; with many nuclei, 7; without plasma membrane, 231n8. *See also* glaucophyte algae; green algae (*Chlorophyta*); red algae (*Rhodophyta*)
Allen, J. F., 139, 141
α-proteobacterial endosymbionts, 129, 130f; chronology and, 109, 123; genomes of, 137, 138; initially supplying hydrogen,

115, 132; origin of eukaryotic cell and, 115, 218, 219f; parallels between mitochondria and, 128, 131. *See also* endosymbiotic origin of mitochondria
amino acids, evolutionary substitutions of, 31–32
amino acid transfer RNA synthetases, 56
Amoebozoa, 107, 108f, 109; *Acanthamoeba*, Mimivirus of, 50; *Entamoeba*, mitosomes of, 144; fossils resembling, 154, 155f; testate, 2, 37f, 142, 154, 155f
amphiphiles, 176
Ancient Virus World, 186–87
Andersson, S. G. E., 132
animals: atmospheric oxygen and, 158, 159; in current eukaryote phylogeny, 107, 108f, 109; as eukaryotes, 5, 6, 37, 39; Haeckel's kingdom of, 4; in tree of life, 22, 23f, 30, 30f, 106
antibiotic resistance, 44, 65
apicomplexan parasites, 144
Archaea: in Archean eon, 150; Bacterial acquisition of genes from, 34; Bacterial features distinguished from, 44–45, 101, 102; Bacterial genes acquired from, 26; Bacterial merger with, hypothesized, 114–16, 218; bacteriorhodopsin in, 42; Cavalier-Smith on, 23, 61f, 62–64, 163, 217; cell division in, 45, 97, 240n35; with complex internal organization, 46–47; cytoskeleton of, 96–98, 112; discovery of, 21, 22, 216, 232n2; in emerging synthesis,

Archaea (*cont.*)
    218, 219f; Eukaryal affinities with,
        111–12, 114, 115, 118, 119, 122, 123, 125;
        general features of, 39, 42–45; geochemi-
        cal origin theorized for, 181; Gupta on,
        64–66; halobacteria, 21, 84; methano-
        genesis in (*see* methanogenesis); phyla
        of, 29, 30f, 43; protests against separate
        domain for, 22–23, 62, 65–66; specializa-
        tions commonly found in, 60; time of
        divergence from Bacteria, 32; time of
        Eukarya divergence from, 162; in uni-
        versal phylogenetic tree, 22, 23f, 24, 29,
        30, 30f; viruses of, 50; Wood-Ljungdahl
        pathway and, 77. *See also* prokaryotes
archaellum, 89, 90f. *See also* flagella, of
    Archaea
Archaeplastida, 107, 108f, 133, 133f, 140
Archean eon: absence of atmospheric
    oxygen in, 158; atmospheric methane
    in, 160; fossils from, 149f, 151–53;
    methanogenic Archaea in, 150, 152;
    nitrogen, sulfur, and phosphorus cycles
    in, 161; not to be confused with Ar-
    chaea, 242n1; origin of life in, 146; pro-
    karyotic cell origin in, 162; reservations
    about eukaryotes in, 156; on timeline of
    earth history, 146, 147f
Archezoa, 28f, 106–7, 239n2
Arendt, D., 121
Aristotle, 3
Ascherson, Neal, 35
ATP (adenosine triphosphate), 69–75, 70f,
    72f, 74f; Archaeal flagella powered by,
    89, 90f; in coupled reaction networks,
    173; produced in photosynthesis, 70–71,
    72f, 73, 83, 84, 85; pyrite production with
    generation of, 178
ATP/ADP antiporter, 137
ATPases: of Archaea vs. Bacteria, 44, 73,
    74f; in common ancestor LUCA, 57, 58,
    76; of Eukarya, 44, 73, 74f, 85, 112–13,
    123, 240n35; in genetic core, 27; P-type,
    113. *See also* ATP synthases
ATP synthases: A_1A_0 type, 44; energy trans-
    duction using, 70, 71, 72f, 73, 80; F_1F_0
    type, 44, 73, 74f, 138, 139, 141; lateral
    gene transfer and, 26, 233n15; of mito-
    chondria, 128, 139; phylogenetic trees
    based on sequences of, 24; of plastids,
    127, 141; primordial, 81; respiratory

redox chains and, 84; sodium-
    translocating, 77; summary of structure
    and function of, 80; V_1V_0 type, 73, 74f,
    85. *See also* ATPases; chemiosmotic
    energy transduction
autocatalytic processes, 168–69; at
    hydrothermal vents, 180. *See also* self-
    replication
autonomous agents, 189, 245n50
autonomy: evolutionary increase in, 216,
    224–25, 226, 229; of stem-group cells,
    100
autopoietic systems, 47, 86, 189, 245n50
autotrophic origins, 179–81. *See also* hydro-
    thermal vents

bacteria: early attempts to define and clas-
    sify, 4, 5, 6; as useful informal term, 24
Bacteria: Archaeal features distinguished
    from, 44–45, 101, 102; Archaeal merger
    with, hypothesized, 114–16, 218; cyto-
    skeleton of, 8, 41, 96–98, 112, 120; in
    emerging synthesis, 219f; Eukaryal affin-
    ities with, 111–12, 114, 115, 118, 119, 122,
    123; general features of, 39–42; genes
    acquired from Archaea, 26; geochemi-
    cal origin theorized for, 181; membrane
    organization in, 40f; phyla of, 29, 30f;
    with structure similar to eukaryotes, 46
    (*see also* planctomycetes); subdomains
    of, 42 (*see also* diderms; monoderms);
    time of divergence from Archaea, 32; in
    universal phylogenetic tree, 22, 23f, 24,
    29, 30, 30f. *See also* prokaryotes
bacteriophage lambda, 212
bacteriorhodopsins, 26, 42, 84. *See also*
    rhodopsins
Bamford, Jaana, 186
banded iron formations, 150, 158
Bangiomorpha, 154, 155f
Bapteste, Eric, 27
Barghoorn, Elso, 147
Barry, Rachael, 98
Bdellovibrio, 123, 131
Behe, Michael, 88, 211–12, 213
Belloc, Hilaire, 78
Bènard cells, 171
bikonts, 107, 108f, 109
bioenergetics, expansion of, 81–85, 82f. *See
    also* chemiosmotic energy transduction;
    energy; photosynthesis

biomarkers: Archaea and, 150; of cyanobacteria, 148, 153, 243n23; of eukaryotes, 113, 153, 157, 243n23; of protists, 154
Blobel, Günter, 94
blue-green algae. *See* cyanobacteria
Brandon, Robert, 10, 197, 224, 246n13
Brasier, Martin, 152
Britten, Roy, 12
Buchnera aphidicola, 89–90
Buick, Roger, 152
Burgess Shale, x, 145, 209
Burton, Richard, 229
Butterfield, Nick, 154, 155f

Cairns-Smith, A. G., 179
Cambrian explosion, x, 154, 209
carbon isotopes: in fossilized organic matter, 152; in sedimentary rock, 150
catalytic closure, 171
Caulobacter: cytoskeleton of, 98; shape of *C. crescentus*, 197
Cavalier-Smith, Thomas: alternative narrative defended by, 23, 61–64, 61f, 163, 217, 236n18; on Archezoa, 106–7; on autogenous origin of eukaryotes, 116–17; on early fossils misinterpreted as eukaryotes, 157; on last eukaryotic common ancestor, 110–11; on membrane heredity, 94–95; on mitochondrial precursor organism, 131; on origin of nucleus, 121; on plastid origin, 136; on rare events in evolution, 210; on timing of mitochondrial origin, 131
cell: continuity of, 3, 13–14, 92–98, 202–3; gene-centered view and, 12–15, 232n19; size of, 45–46; today's concept of, 7–8
cell division: in Archaea, 45, 97, 240n35; in Eukarya, 118–19; L-forms with crude form of, 98; in LUCA, 187; by mitosis, 38, 104, 111, 119, 120; plasma membrane in, 176; proteins involved in, 96, 97, 120
cell evolution, 15–17; beginning with advent of translation, 99; Behe on intelligent design and, 211–12; constraints on our knowledge of, 87–88; convergent evolution in, 195–96; emerging synthesis on, 216–20, 219f; enriched by lateral gene transfer, 34, 199; irretrievably lost record of, 34, 212–13; key questions on, 17, 216; limitations of Darwinian framework for, 100, 102–3, 210, 213–14, 246n2;

membrane heredity and, 95; prions and, 204–5; relationship to modern synthesis, 191–92; seemingly intermediate forms and, 46–47; sources of information on, x–xi, 161–62; spatial organization and, xii, 102; as subject on margins of science, xi; summary of current understanding, 98–103, 99f; timeline for, ix–x, 52, 99f, 147f, 161–63, 217. *See also* evolution; origin of cells; progenotes; protocells; RNA World; universal phylogenetic tree
cell heredity (structural heredity), 12–15, 88, 93, 203. *See also* membrane heredity
cell theory, 2–3
cell walls: in Cavalier-Smith analysis, 62; of eukaryotes, 38; MreB protein in building of, 96; of prokaryotes, 41, 62, 96
cenancestor, 53. *See also* LUCA (last universal common ancestor)
central dogma of molecular biology, 183
centrioles, 111, 120, 203
channel-forming peptides, 79. *See also* transport systems
Chargaff, Erwin, 68
Chatton, Édouard, 5
chemiosmotic energy transduction, 70–75, 72f; advantages of, 75; in LUCA, 58, 187, 217; mitochondrial proteins of, 138; plastid proteins of, 141; primordial, 177–79, 178f, 181, 182. *See also* ATP synthases
chemoautotrophic pathways, 77, 78; absent from eukaryotes, 112
chemolithotrophic organisms, 69; LUCA as, 76; none eukaryotic, 85
Chloroflexus, 62, 217
chlorophyll, 83, 84; in green algae, 135
chloroplasts, 126; defined, 127; mRNAs of, 128. *See also* plastids
cholesterol biosynthesis, 62, 63
Chromalveolata, 107, 108f
Churchill, Winston, 164
cilia, 5
ciliates, 37f; structural heredity in, 203. *See also* protists
citric acid cycle (TCA cycle), 167, 179
classification of living things: Darwin's phylogenetic image of, 18–19, 20, 28; into five kingdoms, 6–7; Linnaean, 18; into prokaryotes and eukaryotes, 5–7; replacement of traditional system, 22–24. *See also* universal phylogenetic tree

clathrin-like proteins, in planctomycetes, 46
Claverie, Jean-Michel, 50
clays, 167, 171
coenzymes, as modified nucleotides, 184
common ancestor, 17, 22, 23f, 24. *See also*
    LECA (last eukaryotic common ances-
    tor); LUCA (last universal common
    ancestor)
complexity: basic replicator leading to,
    169–70; Behe's attack on evolutionary
    theory and, 88; defined, 10, 224, 246n13;
    difficulty in discerning origin of, 87, 91;
    duplication of genes leading to, 193,
    226; energy required for, 177; eukary-
    otic, 118–21, 139, 206–8, 225; evolution
    toward increase in, 197–98, 224–25, 229;
    gratuitous, in absence of selection, 198;
    Kauffman on, 189; minimal in prokary-
    otes, 225; natural selection and, 225–26,
    246n14; neutral evolution in generation
    of, 197, 198, 235n7, 246n14; at origin of
    life, 222; of prebiotic reaction networks,
    171–73, 221; stem-group cells and, 100.
    See also organization, biological
computer simulations of evolution, 87,
    91–92, 103
conditioned reconstruction, 66–67
contingency, 2, 17, 52, 227
convergent evolution, 76, 91, 195–96, 227
Conway Morris, Simon, 188, 221, 227
creation, Conway Morris on, 227
creationists, 212, 229
crenactin, 97
Crenarchaeota, 29, 30f, 43, 44; cell division
    in, 97; origin of eukaryotes and, 116, 123,
    240n35; ring of life and, 66; viral genes
    with homologs in, 48
crescentin, 98
Crick, Francis, 183, 214
cryptic organisms, 29
Cuvier, Georges, 33
cyanelles of glaucophyte algae, 128, 134–35
Cyanidioschzyon merolea, 108
cyanobacteria: Archean biomarkers of, 153;
    evolutionary stasis of, 149–50; fossils
    of, 148, 149f; oxygenic photosynthesis
    invented by precursors of, 158; oxygen
    produced by, 123–24, 162; photosynthe-
    sis in, 40f, 42, 73, 76, 83; in process of be-
    coming plastids, 142–43; as prokaryotes,

5, 6; rebutted early Archean evidence of,
    152; resting bodies of, 148–49; ribosomal
    RNA and, 21; stromatolites constructed
    by, 148, 149f, 150–51; structural com-
    plexity of, 205; time of emergence of, 32,
    162; transformation to plastids, 109–10,
    127–28, 129, 130f, 133, 134, 137, 140–42,
    202, 219, 219f
cytidine triphosphate synthase, 98
cytochromes, 80, 84
cytoplasmic heretics, 94
cytoskeleton: as agent of continuity, 93;
    of eukaryotes, 5, 38; genes specifying
    proteins of, 14; in LUCA, 187; order im-
    posed by, 13; origin of eukaryotes and,
    118, 119–20; possible origins of, 97–98;
    of prokaryotes, 8, 41, 96–98, 112, 120.
    See also actins; microtubules; tubulins

Dacks, J. B., 120
Darwin, Charles: cell evolution and, 210; on
    common origin of living things, 52; on
    evolution of animals, 154; on evolution
    of complicated organs, 87; Haeckel and,
    4; on mechanism of evolution, 15; on
    mystery of flowering plants, 104; tree of
    life and, 18–19, 20, 28. *See also* natural
    selection
Darwinian thresholds, 55, 56
Davidov, Y., 116
Davies, Paul, 228
Davis, Rowland, 114
Dawkins, Richard, 14–15, 36, 170, 173, 195,
    232n14
Deamer, David, 177
Deinococcus, 233n15
Dennett, Daniel, 232n14
Denton, Michael, 227
Deschamps, P., 134
diderms, 40f, 42; Cavalier-Smith analysis
    and, 62, 64, 210, 217; Gupta analysis and,
    42, 61f, 65; hypothetical symbiotic origin
    of, 67, 202, 247n20
dihydrofolate reductase, 109
dioxygen reductase, 85
directed panspermia, 214
diversity, biological: extremes of, 45–47;
    mostly microbial, 35; tendency of
    increase in, 197–98
divisions. *See* phyla

DNA, 182, 185, 186–87; noncoding, 38, 39, 118. *See also* genes

DNA polymerases: Archaeal vs. Bacterial, 44, 59; in artificial vesicles, 176; of Eukarya, 59, 111; two kinds of, 186; of viruses, 59

domains, 22, 29, 33, 35; conventional doctrine of, 67, 216–17, 218; emergence of, 55–56, 100–102, 209; gene transfer between, 33; viral theory for origin of, 186. *See also* Archaea; Bacteria; Eukarya; universal phylogenetic tree

Doolittle, Ford, 27, 123, 127, 238n4

Doolittle, Russell, 32

Douzery, E. J. P., 32

duplication of DNA sequences, 192–93, 199, 226

dynamins, 140

Ediacaran fossils, 154

Eigen, Manfred, 169

Eiseley, Loren, 1, 16

Eldredge, Niles, 208

electrochemical potential gradient, 71; in photosynthesis, 83, 84; primordial, 80–81. *See also* chemiosmotic energy transduction

electron bifurcation, 237n6

Embley, T. M., 125

endocytosis: planctomycetes and, 46, 206; sterol-containing membranes and, 159

endosymbiosis, 201–2; common in eukaryotic cells, 8; in contemporary protists, 130; inheritance of acquired characters based on, 191; as mechanism of large jumps, 102; membranes acquired through, 95; phagocytosis and, 98, 111, 117; rare among prokaryotes as hosts, 66, 116, 123; secondary, 129–30, 130f, 133, 133f, 135–36, 202; tertiary, 129, 133. *See also* symbiosis

endosymbiotic origin of eukaryotic cell, 27, 33, 201–2, 203; archezoan hypothesis and, 106; cellular complexity resulting from, 206–8, 210; energetics and, 85, 124, 206–7; events involved in, 129–36, 130f; genes acquired by, 123; three modes of endosymbiosis in, 129, 130f

endosymbiotic origin of mitochondria, 33, 108f, 109, 126–27, 128, 129–32, 130f, 201–2;

in Cavalier-Smith version, 117, 131, 218; molecular basis of transformation in, 136–40; timing of, 113, 122, 123–24, 131, 162, 218. *See also* α-proteobacterial endosymbionts

endosymbiotic origin of plastids, 33, 108f, 109–10, 126–28, 129–30, 130f, 132–37, 140–42, 162, 202

energy: essential for biological activities, 68; evolvability and, 206–8; expansion of bioenergetics, 81–85, 82f; for origin of cells, 177–79, 222; prehistory of energy conservation, 79–81; prokaryotes' broad exploitation of sources for, 205; three biological sources of, 69. *See also* chemiosmotic energy transduction; metabolism; photosynthesis

energy flow, and organization, 171–73, 172f

Entamoeba, mitosomes of, 144

Enterococcus hirae, 71

enzymes. *See* DNA polymerases; metabolic enzymes; proteins; ribozymes; RNA polymerases

eocyte, 66

Ephrussi, Boris, 94

epigenetic inheritance, 13, 88, 192, 194, 205, 226

Epuloppiscium fishelsoni, 45, 207

Escherichia coli: complexity of, 8–9; experiments on evolution in, 193; flagellar direction of rotation in, 197; flagellin variants of, 88; FtsZ protein in division machinery of, 96; genes transferred from elsewhere, 26, 34; as gram-negative bacterium, 42; growth on nanomolar levels of oxygen, 151; as model organism, 4, 8; respiratory chain in, 71; ribosomes in, 20

ESCRT-III, 97

eubacteria, 21, 22. *See also* Bacteria

Euglena, 4

Eukarya: Archaeal affinities with, 111–12, 114, 115, 118, 119, 122, 123, 125; atmospheric oxygen and rise of, 159; Bacterial affinities with, 111–12, 114, 115, 118, 119, 122, 123; biomarkers of, 113, 153, 157, 243n23; Cavalier-Smith on, 61f, 62–64, 163; cell organization of, 112, 118–21; chimeric genomes of, 33, 66, 115, 116, 129, 218; current phylogeny of, 107–11, 108f; divergence time of, 32, 162;

Eukarya (*cont.*)

in emerging synthesis, 218, 219f; energy sources for, 85 (*see also* energy); evolvable compared to prokaryotes, 205–8; fossil record of, 113, 154–57, 162, 163; as four kingdoms in earlier classification, 6; general features of, 6f, 36–39; as great evolutionary leap forward, 104, 225; hypotheses about origins of, 113–18, 218; lateral gene transfer in microbes belonging to, 25; metabolism of, 38–39, 112–13; plausible narrative of origin, 121–25; prokaryotes distinguished from, 5–7, 104; ring of life and, 61f, 66–67, 115; root between unikonts and bikonts, 107, 108f, 109; signature proteins of, 117, 119, 124; as single clade, 101, 105; unicellular, 37f (*see also* protists); in universal phylogenetic tree, 22, 23f, 24, 29–30, 30f. *See also* endosymbiotic origin of eukaryotic cell; LECA (last eukaryotic common ancestor); picoeukaryotes; protists

eukaryote-prokaryote distinction, 5–7, 104; proponents of maintaining, 23

Euryarchaeota, 29, 30f, 43; cell division in, 96, 97; histones in, 44; methanogenesis in, 76; polysaccharide related to peptidoglycan, 41

evolution: cell heredity and, 12, 93; central role of genes in, 192–94; computer simulations of, 87, 91–92, 103; episodic features of, 208–10; events that underpin, 91; the modern synthesis, 190–92, 246nn1–2; neglect of microbes in understanding, 190–91; progressive, 223–26; purpose of, 226–29; tendency for increasing diversity and complexity, 197–98; transformation in nature of, 55–56. *See also* cell evolution; convergent evolution; natural selection; neutral evolution

evolutionary distance, 20, 31

evolvability, 205–8

exaptation, 89–90, 212

Excavata, 107, 108f

extraterrestrial origin of life, 165, 214

eye: gastropod, 87; vertebrate, 87

fatty acids: of plasma membranes, 44–45, 60, 95, 112; from plastids, 144; vesicles assembled from, 167, 176

fatty alcohols, vesicles assembled from, 176

fermentation: absent from earliest cells, 76; Bacterial energy transduction by, 72f; in eukaryotes, 85, 123; in hydrogenosomes, 143–44. *See also* glycolysis

ferredoxin oxidoreductase, 77

ferredoxins, 80

Ferry, James, 178

Field, M. C., 120

Finnigan, G. C., 246n14

Firmicutes, 61f, 65

First Cell, 99, 182, 221

first law of biology, 197, 224, 226

fitness, 195, 227

five kingdoms, 6–7; replacement of, 22

flagella, absent from LUCA, 57

flagella, of Archaea, 44, 63, 89, 90f

flagella, of Bacteria, 5, 8, 40f, 41, 44, 88–90, 90f; diversity of, 88, 196–97; homologs of proteins in, 89; proton circulation providing energy for, 71

flagella, of Eukarya. *See* undulipodia

flagellin, 41, 89, 90f

Forterre, Patrick, 49–50, 59, 186

fossils: calibrating molecular phylogeny with, 32; Cavalier-Smith's interpretation of, 63, 163; dating of, 161; earliest microfossils, 147–48, 149f, 151–53; eukaryotic, 113, 154–57, 162, 163; punctuated equilibrium and, 208–9

Fox, George, 21

FtsZ, 96–97, 120

fungi: as collectives in common cytoplasm, 7; as eukaryotes, 5, 6, 37, 39; molecular phylogenies of, 28, 29, 30f, 106, 108f, 109

Gardner, James, 227

Gemmata obscuriglobus, 46

gene expression: epigenetics and, 194; in LUCA, 59

genes: advent of, as greatest transition, 191; central role in evolution, 192–94; duplication of, 192–93, 199, 226; higher levels of order not specified by, 102; informational vs. operational, 27, 66, 114, 115, 118; vs. organisms as objects of selection, 195; organisms seen in terms of, 10–12, 232n14. *See also* lateral gene transfer

genetic code: communal evolution and, 56; as crucial step from protocells to true

cells, 183, 222; machinery for transcription and translation of, 182–83; universality of, 53, 183. *See also* transcription; translation
genetic core: coding for ribosomal RNA, 27; major claim about, 33; similar in Eukarya and Archaea, 111, 117; size of, 27; for transcription and translation, 111
genomes: cellular information not carried in, xi, 92 (*see also* cell heredity); expanded in eukaryotes, 206–7; malleable architecture of, 199–201; plasticity enhanced by prions, 204; as record of evolution, xi; transplantation of, 11, 12, 14
geochemical energy sources: cellular organization and, 103; for LUCA, 77–78; primordial, 79, 157, 168. *See also* hydrothermal vents; inorganic energy sources
geological time, 145–47, 147f
George, S. C., 153
geothermal fields, 245n38
Gerhart, John, 205
giant DNA viruses, 50
Giardia, 106, 119, 144
Gilbert, Walter, 184
Gitai, Zemer, 98
glaucophyte algae, 37f, 128, 129, 132, 133f, 134–35
glycolysis: absent in LUCA, 57, 76; ATP formation during, 70, 71, 72f, 73; different versions in Bacteria and Archaea, 76; in eukaryotes, 112. *See also* fermentation
glycoproteins, 62, 63
God, 164, 173, 225
Gogarten, J. P., 233n13, 233n15
Gould, Stephen Jay, 208, 212
gram-negative and gram-positive bacteria, 40f, 42, 62, 64–65. *See also* diderms; monoderms
Gray, M. W., 127, 246n14
great oxidation event, 158
green algae (*Chlorophyta*), 37f, 129, 132–33, 133f, 134, 135, 136
green sulfur bacteria, 83
growth, 11–12, 13
Grypania, 155f, 156
Gunflint cherts, 148, 149f, 150–51
Gupta, Radhey, 42, 61f, 64–66

Hadean eon, 146, 147f
Haeckel, Ernst, 4

Haldane, J. B. S., 165, 166, 229
Halfmann, R., 204
Haloarchaea, 26, 43, 84
halobacteria, 21, 84
Hedges, Blair, 32
heliobacteria, 83
hemoglobins, 20, 31–32, 196
heredity: in Darwinian theory, 15, 16; in modern evolutionary synthesis, 190, 191, 194. *See also* cell heredity; vertical inheritance
histones, 44, 63, 111
Hoffding, Harald, 104
homeostasis, 225
Hooke, Robert, 2
hopanes and hopanoids, 148, 153, 159
horizontal gene transfer. *See* lateral gene transfer
House, Christopher, 178
Hoyle, Fred, 56
Hsp60, 106
Hsp70, 64
hydrogenase, 77
hydrogen gas: at hydrothermal vents, 181; in Wood-Ljungdahl pathway of carbon fixation, 77
hydrogen hypothesis, 115, 132
hydrogenosomes, 143–44
hydrothermal vents: cellular organization and, 103; citric acid cycle and, 167; microfossils possibly associated with, 151; origin of life envisioned at, 60, 78, 81, 180–81, 222; viruses in early world of, 187

ice age 2.2 billion years ago, 160, 162
Ignicoccus hospitalis, 46–47
indels, 64–65
informational genes vs. operational genes, 27, 66, 114, 115, 118
inorganic catalysts, 80
inorganic electron acceptors, 85
inorganic energy sources, 69, 73, 77. *See also* geochemical energy sources
intelligent design, 211–12, 213, 227
intelligent life, 227
introns, 59, 118, 121. *See also* splicing introns
iron-loving bacteria, 150
iron sulfide clusters, 144
isoprenoid ether lipids, 44, 60, 63, 95, 96, 102, 112, 144

Jacob, François, 11–12
Javaux, Emmanuelle, 152, 156
Jèkely, G., 121
Joyce, G. F., 167
Jurkevitch, E., 116

Kandler, Otto, 22, 30, 56, 235n2
Kang, Y.-M., 231n8
Kant, Immanuel, 10, 247n3
Kauffman, Stuart, 189, 222, 245–46n50
kerogen, 152
Kim, G. H., 231n8
kinetic drive, 169–70, 221
Kirschner, Marc, 205
Klotchkova, T. A., 231n8
Knoll, Andrew, 146, 156
Koch, Arthur, 98, 99, 177–78, 178f, 182
Koonin, Eugene, 49, 121, 185, 186, 209, 210
Korarchaeota, 43; cell division in, 97
Kuhn, Thomas, 218
Kurland, Charles, 27, 122

Lake, James, 57, 66–67, 115, 116, 247n20
Lane, Nick, 78, 121, 122, 124, 139, 147,
    206–7
Lapierre, P., 233n13, 233n15
Lartigue, C., 12
lateral gene transfer, 25–27, 198–99, 233n13,
    233n15; in Cavalier-Smith analysis, 63,
    217; conflict with modern evolutionary
    synthesis, 191; dioxygen reductase and,
    85; in Gupta analysis, 65; not erasing
    cyanobacterial descent, 149; pervasive
    uncertainty introduced by, 57; of pho-
    tosynthetic genes, 83; in progenotes, 54,
    55–56, 100; in prokaryotes vs. eukary-
    otes, 25, 109, 199; in stem-group cells,
    100, 101, 199; tree structure undermined
    by, 28–29, 30, 31, 34; in viruses, 48. See
    also endosymbiosis
Lawrence, J. G., 26
LECA (last eukaryotic common ances-
    tor), 108f, 110–11, 120, 219f; attempts
    to attach a date to, 162; complex
    structures of, 119, 120; evolutionary
    burst following, 209; mitochondria of,
    129; observations to be accommodated
    by, 111–13; Phalansterium as model for,
    110f, 111; primordial scavengers and,
    124; undulipodia of, 120

Leeuwenhoek, Anton van, 3
Lenski, Richard, 91–92, 193
L-forms, 98
life: cosmic significance of, 226–29; defini-
    tions of, 47, 245n50; not deducible from
    laws of physics and chemistry, 2, 188;
    order manifested by, 1–2; physical
    environment shaped by, 146. See also
    origin of life
light, as primordial energy source, 178–79.
    See also photosynthesis
Lincoln, T. A., 167
Lindquist, S., 204
Line, Martin, 214
Linnaeus, 18
lipids, membrane: in Cavalier-Smith
    analysis, 63; contrast to expectations
    from universal tree, 95; of eukaryotes vs.
    prokaryotes, 112, 159; LUCA and, 58f,
    60. See also fatty acids; isoprenoid ether
    lipids; sterols
Liu, R., 238n4
long branch attraction, 107
Lòpez-Garcia, P., 114, 121
LUCA (last universal common ancestor),
    52–53; in alternate histories, 61f; an-
    aerobic world of, 76, 83; Cavalier-Smith
    on, 61f, 62, 217; cell division machinery
    in, 96, 97; DNA or RNA genome of,
    58–59; emergence of domains from, 58f,
    100–102; energy metabolism of, 75–78,
    79, 82f, 83–84; in geochemical theories,
    181; habitat of, 78; likely characteris-
    tics of, 53, 57–60, 58f, 223; with many
    of today's capabilities, 186, 187, 220;
    membrane lipids and, 96; photosynthesis
    possibly present in, 83; as progenote,
    57–58, 67; as protoeukaryote, 58f, 59–60,
    236n14; ring of life and, 61f, 66; root of
    the tree and, 57; as stage in protracted
    process, 57; standard account of, 216–17,
    218, 219f; as thermophile, 60, 76, 77,
    78, 236n17; viruses and, 48; Woese's
    rethinking of, 53–57. See also universal
    phylogenetic tree
Lyell, Charles, 208
Lynch, M., 246n14

macromolecules, abiotic generation of, 167
malaria parasites, 144, 211

Margulis, Lynn: on autopoietic systems, 47; on endosymbiosis, 66, 126; on origin of eukaryotes, 114; on protoctists, 6; on undulipodia, 5, 234n6

Martin, William: on bioenergetics, 80, 82, 124; on complexity of eukaryotes, 206–7; on genome phylogeny, 31; on origin of eukaryotes, 115, 122, 124, 125, 132, 139; on origin of nucleus, 121

Maturana, Humberto, 47, 86, 245–46n50

Mayr, Ernst, 22–23

McShea, Daniel, 10, 197, 224, 246n13

Mead, Walter Russell, 190

membrane heredity, 14, 93–96, 176, 203; of plastids and mitochondria, 93–94, 141. *See also* cell heredity

membranes: essential to cell evolution, 175–77; of eukaryotes, 5, 38, 73, 85, 120; evolution toward autonomy and, 225; of First Cell, 182; hydrothermal fields and, 78; in LUCA, 187; mitochondrial, 93, 128, 138; nuclear, 118, 119, 121; photosynthetic pigments embedded in, 83, 84; of planctomycetes, 59; plastid, 93–94, 141; of prokaryotic cells, 39, 40f, 120; puzzling origin of, 222; redox carriers in, 80; RNA World and, 186; transport carriers in, 79. *See also* lipids, membrane; membrane heredity; plasma membrane

Mereschkowski, Konstantin, 136, 180

messenger RNA, 39, 182–83

metabolic enzymes: eukaryotic, resembling those of Bacteria, 112; lateral transfer of genes for, 26–27; polymerization into filamentous structures, 98

metabolism: of eukaryotes, 38–39, 112–13; evolving from the RNA World, 186, 187; in First Cell, 221; prebiotic reaction networks leading to, 171–73, 172f, 174, 221; precellular, 179; prokaryotic versatility of, 205, 217, 225; in stem-group cells, 100. *See also* energy

meteorites: bombardment of earth by, 60, 78; organic molecules in, 166, 167, 168, 176, 179

methane, atmospheric levels of, 159–60

Methanococcus jannaschi, 26

methanogenesis: antiquity of, 64, 150, 152, 162; chemiosmotic energy transduction in, 72f, 73; discovery of archaebacteria

and, 21; in Euryarchaeota, only phylum with, 43; invented later than LUCA, 75–76; in origin of eukaryotes, 114–15; origin of mitochondria and, 132; as uniquely Archaean invention, 43, 64, 81–82, 160

Methanosarcina mazei, 26

methylation of genes, 194

microbes: early attempts to classify, 3–4; as most of living world, 22, 35; neglected in modern evolutionary synthesis, 190–91; shapes of, and neutral evolution, 197. *See also* prokaryotes; protists

microsporidia, 106, 107

microtubules, 5, 38. *See also* tubulins

Miller, Kenneth, 212

Miller, Stanley, 166

Mimivirus, 50

mind, 216, 227, 228, 229

Mitchell, Peter, 70

mitochondria: ATPases of, 24, 73, 74f; diminished derivatives of, 143–44; energy for eukaryotes and, 38, 85; genomes of, 128, 132, 137–39, 207; in last eukaryotic common ancestor, 111; limitations of genomic approach to, 139–40; origin of eukaryotic cell and, 115; perpetuation of membrane structure of, 93; possibly acquired by phagocytosis, 117, 218; ribosomes in, 128, 138, 139; viral-type DNA polymerase in, 59. *See also* endosymbiotic origin of mitochondria

mitosis, 38, 104, 111, 119, 120

mitosomes, 144

Monera, 4, 6

monoderms, 40f, 42; Cavalier-Smith analysis and, 62, 64, 210, 217, 218; Gupta analysis and, 61f, 65; Lake on fusion of, 67

Moreira, D., 114, 121

Morowitz, Harold, 171

mosaic evolution, 63

MreB, 96–97, 120

Mulkidjanian, A. Y., 80, 83, 245n38

Muller, H. J., 232n14

Müller, Miklòs, 115, 132

multicellular organisms: atmospheric oxygen and, 159; convergent evolution in, 195; time of appearance, x, 146, 154; understanding of evolution based on, 190. *See also* animals; fungi; plants

multiverse, 228
mutations, 192, 196, 198, 211–12, 213. *See also* variation
Mycoplasma, genome transplantation in, 11
myosin family genes, 109
myxobacteria, 114, 205

nanobacteria, 45
natural selection, 2, 194–98; of basic replicator, 169–70; cell organization and, 93, 102; cells required for, 16; complexity and, 87, 88, 89–90, 91, 225–26, 246n14; epigenetic changes and, 194; evolutionary progression and, 225–26; genomic rearrangements and, 201; incompleteness of explanation based on, 88; intelligent design argument against, 211–12; in modern evolutionary synthesis, 190; networks of chemical reactions and, 221; vs. neutral evolution, 196, 197; objects acted on by, 36, 87, 92, 195; opposing increase in diversity and complexity, 197–98; in origin of cells, 173–75, 181, 187, 223; origin of life not explained by, 188–89; producing flagella, 88, 89–90; producing organelles, 91; in progenotic lineages, 56, 100; requiring preexisting spontaneous order, 189; self-organization in face of, 103. *See also* adaptation; evolution
Negibacteria, 62, 210, 217
Nelson-Sathi, S. N., 26
Neomura, 62–63, 64, 210
neutral evolution, 56, 103, 196, 197, 198, 235n7, 246n14
Newton, Isaac, 126
Nilsson, D.-E., 87
nitrogen, 160–61
nitrogenase, 160
nitrogen fixation, 148, 150, 160
Noble, Dennis, 14–15
noncoding DNA, 38, 39, 118
nuclear membrane, 118, 119, 121
nucleoid, 8, 40f, 45, 46
nucleomorph, 135f, 136
nucleus, 5, 37–38, 104; origin of, 50, 114, 121; variations of, 7

obcells, 244n22
Ochman, H., 26, 238n4
Oesterhelt, Dieter, 84

oligonucleotide catalogs, 20, 21
Omnis cellula e cellula, 3, 92
oömycetes, 7
Oparin, A. I., 165, 166
operational genes vs. informational genes, 27, 66, 114, 115, 118
Ophistokonta, 107, 108f, 109
order, 1–2; beginnings of, 168–75; defined, 1, 9; growing spontaneously out of complex systems, 189. *See also* organization, biological
organelles: of eukaryotes, 5, 119; evolution of, 87, 90–91, 201; rise of complex cells depending on, 139. *See also* endosymbiosis
organism: definitions of, 10, 36; loss of appreciation for, 11, 35–36; as object of selection, 36, 92, 195
organization, biological: of bacterial cell, 8–9; cell evolution and, xii, 93; cell theory and, 2; continuity of, 3, 15, 93, 202–3; defined, 9–10; emergence of protocells and, 182; energy flow and, 171–73, 172f; epigenetic origin of, 88; higher levels not specified by genes, 193; key question on, 17; mystery of, 1, 165; three cellular themes of, 36. *See also* complexity; order; self-organization; spatial organization of cells
Orgel, Leslie, 171, 183
origin of cells, 51, 165, 190, 214, 220–23. *See also* cell evolution; protocells
origin of life: in Archean eon, 146, 151, 153; as continuing enigma, 187–89, 221, 226; cosmic significance of, 227–29; energy source for, 157; extraterrestrial, 165, 214; in gap between molecules and cells, 168; key questions on, 17; lack of coherent research framework for, 164, 189; as problem in geochemistry, 165 (*see also* geochemical energy sources); as scientific, not supernatural, problem, 164–65, 173; synonymous with emergence of protocells, 99; synonymous with origin of cells, 16. *See also* origin of cells
Ostreococcus tauri, 108
oxygen, atmospheric: last eukaryotic common ancestor and, 111; from photosynthesis, 83, 127; planetary changes due to,

83, 146; respiration enabled by, 83, 84, 85, 150–51; rise in, 123–24, 147f, 150–51, 157–60, 162

Pace, Norman, 29
Paracoccus denitrificans, 128, 131
Paulinella chromatophora, 142–43
Pelger, S., 87
peptidoglycan: of Bacteria, 40f, 41, 42; Cavalier-Smith analysis and, 62, 217; in cyanelles of glaucophyte algae, 128, 135; Gupta analysis and, 65; of Mimivirus, 50; of *Paulinella* chromatophores, 142; of planctomycetes, 46
peroxisomes, 119
phagocytosis: in acquisition of mitochondria, 117, 218; in acquisition of plastids, 115, 132; dependence on endomembranes, 118; endosymbiosis and, 98, 111, 117; by last eukaryotic common ancestor, 111; origin of, 117–18; origin of eukaryotes and, 116–17, 122, 123; rise of eukaryotes and, 208
Phalansterium, 110f, 111
Phanerozoic eon, 146, 147f, 157, 158
phospholipids, vesicles assembled from, 176
photosynthesis: anoxygenic, 82f, 152, 153, 157; Bacterial, 42, 72f, 73, 82–84, 82f; in Cavalier-Smith narrative, 217; chemiosmotic energy transduction in, 70–71, 72f; disseminated widely by symbiosis, 202; in eukaryotes, 69, 85; in eukaryotes, Proterozoic, 154, 155f; evidence from Archean pointing to, 152; invention of oxygenic type, 82f, 157–58; lateral gene transfer for apparatus of, 26; primary production of organic matter and oxygen by, 69; by two billion years ago, 148. *See also* light; plastids
photosystems, 83, 158
phycobilisomes, 135
phyla, 22; of Archaea, 29, 30f, 43; of Bacteria, 29, 30f
phylogeny, 18, 19. *See also* universal phylogenetic tree
picoeukaryotes, 37f, 104, 108, 119
picoflora, planktonic, 43, 46
Pinker, Steven, 86
pirellulosome, 40f, 46
planctomycetes, 40f, 46, 59, 113, 205, 235n14

planktonic picoflora, 43, 46
plants: in current eukaryote phylogeny, 107, 108f; as eukaryotes, 5, 6, 37, 39; Haeckel's kingdom of, 4; plastid diversification and, 133; in tree of life, 22, 23f, 30, 30f, 106
plasma membrane: as agent of cell heredity, 14; of Archaea, 43, 44; of Bacteria, 40f, 41, 42, 44–45; essential to cell evolution, 175–77; of Eukarya, 45; lipid composition of, 95, 112; of LUCA, 58; necessity of, 7, 231n8. *See also* membranes
plastids: ATPases of, 24, 73, 74f, 127; chloroplasts, 126, 127, 128; cryptic, 144; defined, 127; diminished derivatives of, 143, 144; dispersal of, 132–36, 133f; DNA of, 127–28, 134; endosymbiotic origin of, 33, 108f, 109–10, 126–28, 129–30, 130f, 132–37, 133f, 140–42, 162, 202; energy for eukaryotes and, 38, 85; genomes of, 140–41, 207; molecular basis of transformation to, 136–37, 140–42; perpetuation of membrane structure of, 93–94; possibly acquired by phagocytosis, 115, 132; ribosomes of, 127–28; secondary, 129, 130f, 133–34, 133f, 135–36, 135f; still in the making, 142–43; tertiary, 133, 136; time of origin, 162. *See also* photosynthesis
Pollack, G. H., 231n8
polyphosphates, 179
posibacteria, 210, 217, 218
prebiotic synthesis, 166–68
Precambrian, 146
prions, 203–5
progenotes, 54–55, 56, 57, 58f, 220–21; defined, 99; LUCA as, 57–58, 67; viruses helping to shape, 187; Woese on, 78, 99. *See also* stem-group cells
prokaryote-eukaryote distinction, 5–7, 104; proponents of maintaining, 23
prokaryotes: chemiosmotic energy transduction in, 71, 72f, 73; common gene pool of, 25, 27; cytoskeletons of, 8, 41, 96–98, 112, 120; eukaryotes distinguished from, 5–7, 104; evolving by gene acquisition, 199 (*see also* lateral gene transfer); general features of, 5–6, 6f, 39, 40f; as kingdom Monera, 6, 21; phylogenetic issues with, 19, 22, 28; specializations of,

prokaryotes (*cont.*)
  60; time of appearance of, 100; as useful
  though controversial term, 24, 233n12.
  See also Archaea; Bacteria
Pross, Addy, 169–70
proteasomes, in Cavalier-Smith's analysis,
  62
proteins: cellular roles of, 182–83, 222;
  chronologies based on sequences of,
  31–32; conserved insertions and deletions in, 64–65; eukaryotic signature
  proteins, 117, 119, 124; glycoproteins,
  62, 63; lateral gene transfer and, 26,
  233n13, 233n15; phylogenetic trees
  based on sequences of, 24–25; selectively
  neutral mutations in, 196; speculations
  on evolution of, 185–86; structural (*see*
  cytoskeleton)
proteobacteria, 61f, 65, 66; as endosymbionts in contemporary protists, 130;
  photosynthesis in, 83
proteorhodopsins, 84
Proterozoic eon, 146, 147f; anoxic ocean
  depths in, 159; Archaea in, 150; fossils
  from, 148, 149f, 154–55, 162–63
protists: anaerobic, 29, 106, 143–44; controversies about classification and evolution
  of, 105–11; divergence time of, 32, 162;
  as eukaryotes, 5, 6, 29, 37; in five-
  kingdom classification, 6; in fossil record, 154, 155f; Haeckel's kingdom of, 4;
  hydrogenosomes in, 143–44; as majority
  of eukaryotes, 105; neutral evolution in,
  197; number of known species, 39, 107;
  photosynthetic, 110, 162; sampler of, 37f.
  See also picoeukaryotes
protocells, 99, 175–82; defined, 175; everted,
  244n22; evolution to become LUCA,
  223; geochemical origin of, 179–81; lack
  of persuasive hypothesis for, 182, 221;
  lateral gene transfer and, 199; prebiotic
  synthesis and, 166, 168; RNA World and,
  183–86; time of origin, 99f, 100
protoctists, 6, 105
protozoa, 4, 6; convergent evolution in, 195.
  See also Amoebozoa; protists
punctuated equilibrium, 208–9
purpose of evolution, 226–29
purpose or function: kinetic drive of replication and, 170; of organisms, 36; of
  organization, 2, 9–10, 17

pyrite formation, 177–78, 178f, 180
pyrophosphate, 179

quantum evolution, 61f, 63, 117, 209, 210
quinones, from meteorites, 179

radiolaria, 37f, 197
Ramón y Cajal, Santiago, 126
random variation, 2, 213
Rasmussen, B., 153
Reclinomonas, 137
red algae (*Rhodophyta*), 37f, 129, 132, 133,
  133f, 134, 135, 136; fossils resembling,
  154, 155f
redox processes, 79–80; in photosynthesis,
  72f, 83, 127; in respiration, 71, 84. *See
  also* respiratory chain
redox protein construction kit, 80, 84
reductionism, 36, 222, 227
reductive evolution, 143–44
Rees, Martin, 228
regulatory dynamics of cellular system, 12
regulatory genes, mutations of, 102
replication. *See* self-replication
reproduction, 11, 29, 36, 173, 175, 182, 194,
  195, 217. *See also* cell division; self-
  replication
respiration: chemiosmotic energy transduction in, 70–71; photosynthesis and, 83,
  84–85
respiratory chain, 71, 72f; evolution of,
  84–85, 237n22; lateral gene transfer and,
  26; mitochondrial genes encoding, 139;
  mitochondrial resemblance to bacterial
  version, 124, 128, 131, 138. *See also*
  redox processes
resting bodies of cyanobacteria, 148–49
Rhizaria, 107, 108f, 142
Rhodobacter, flagella of, 88, 197
rhodopsins: conjectured to exist in LUCA's
  day, 77. *See also* bacteriorhodopsins
Rhodospirillum rubrum, 131
ribosomal proteins: Archaeal, 44; in genetic
  core, 27
ribosomal RNA: Bacterial phylogeny
  according to, 42; as best tracer for
  organismal lineage, 29, 32–33; databases
  of sequences of, 29; early phylogenetic
  research with, 20–22; eukaryote phylogeny according to, 106, 107; in genetic
  core, 27; as measure of evolutionary

distance, 31; of mitochondria, 128, 130, 131, 139; of *Paulinella* chromatophores, 142; of plastids, 128, 134; sequence differences between Bacteria and Archaea, 21, 22, 44; tree of life inferred from (*see* universal phylogenetic tree)

ribosomes: Archaeal vs. Bacterial, 44; basic structure and function of, 20; speculations on evolution of, 185–86; unlikely lateral transfer of genes for, 26

ribozymes, 167, 182, 184, 185, 186, 221

Rickettsia: ATP imported by, 143; as candidate mitochondrial precursor, 131

Rieske factors, 80

ring of life, 61f, 66–67, 115

Rivera, M. C., 115

RNA: catalytic properties of, 169 (*see also* ribozymes); quest for self-replicating species of, 167, 169, 173–74; in transcription and translation, 182–83. *See also* messenger RNA; ribosomal RNA; transfer RNAs

RNA polymerases: Archaeal, 44, 111; in artificial vesicles, 176; Bacterial, 111; eukaryotic, 111; function of, 182; in genetic core, 27, 111; phylogenetic trees based on sequences of, 24; of plastids, 128; primordial, DNA polymerases substituted for, 59

RNAse-P, 184

RNA World, 49f, 183–87; evolution of ATP synthase and, 80; self-replication in, 183–85, 221; viruses in, 49; Woese on, 54, 59

Rosslenbroich, Bernd, 224–25

Russell, Michael, 81, 180

Saier, Milton, 79

Salmonella, flagella of, 88

SAM (sorting and assembly machinery), 137

Sapp, Jan, 5, 210, 235n2

Schleiden, Matthias, 2–3

Schopf, William, 152

Schrödinger, Erwin, xi

Schwann, Theodor, 2

selection. *See* natural selection

self-assembly, 12, 13; abandoned theories of, 221; of compartments bounded by lipid membranes, 176

self-awareness, 227

selfish gene, 14–15

self-organization, 13, 88; built into fabric of the universe, 228–29; of complex systems, 189, 212; energy flow and, 171; in face of natural selection, 103; vs. intelligent design, 211; of metabolic cycles, 174, 221; origin of cells and, 223; spatial patterns in cells and, 102

self-replication, 168–70, 173–74; in First Cell, 182; protocells and, 175; in RNA World, 183–85, 221. *See also* autocatalytic processes

serpentinization, 181

Shapiro, James, 200–201

Shapiro, Robert, 172, 178

Shial, R., 233n15

Shuiyousphaeridium, 156

Silver, S., 99, 182

Simpson, George Gaylord, 209

Smith, John Maynard, 16, 47

Smolin, Lee, 228

Snel, Berend, 31

sodium ion current: in acetogenic Bacteria, 77, 237n6; in marine Bacteria, 72f, 73; in world before LUCA, 80

Sogin, Mitchell, 106

Sonneborn, Tracy, 94

spatial organization of cells: cell evolution and, xii; cytoskeleton and, 97; eukaryotic vs. prokaryotic, 112; not mandated by genome, 13–14, 93. *See also* organization, biological

Speijer, D., 246n14

Spirochaetes: flagella of, 88, 197; ribosomal RNA and, 21; undulipodia hypothesized to descend from, 114

splicing introns: LUCA mechanisms for, 59; spliceosomes for, 118, 119, 184, 198

split genes, 112, 118

spontaneous generation, 188

Sprat, Thomas, xii

S-proteins, 43, 65

Sputnik, 50

SSU rRNA, 20. *See also* ribosomal RNA

Stanier, Roger, 5–6, 19

stem-group cells, 99–101, 187, 199, 220–21. *See also* progenotes

steranes, 153, 159, 162

sterols: in actinobacteria, 117; cholesterol biosynthesis, 62, 63; of eukaryotic membranes, 159

Stoeckenius, Walter, 84
stromatolites, 148, 149f, 150–51
structural heredity. *See* cell heredity
structural mutations, 93
submarine vents. *See* hydrothermal vents
substrate-level energy transduction, 70, 75, 76; in hydrogenosomes, 143
Sulfolobus, 97
Sup35, 204
Swithers, K. S., 247n20
symbiosis: gene transfer associated with, 34, 109–10; in protists, 110; tree structure undermined by, 34. *See also* endosymbiosis
synthetic biology, x–xi, 29
syntrophic associations, 66; in origin of eukaryotes, 114, 115, 116
system, defined, 9
Szostak, Jack, 176

Tappania, 155f, 156
TCA cycle (citric acid cycle), 167, 179
testate amoebas, 2, 37f, 142, 154, 155f
Thaumarchaeota, 43; origin of eukaryotes and, 116, 123, 218
thermophiles: Archaean, 43, 44, 97; Bacterial, acquiring Archaean genes, 26, 33–34; Cavalier-Smith on, 63, 64, 117; glycoprotein walls and, 62; LUCA as, 60, 76, 77, 78, 236n17; membrane lipids of, 63, 96; origin of eukaryotes and, 117; of phylum Thermotoga, 26
Thermoplasma, 114
Thermoproteales, cell division in, 97
Thermotoga, 26, 95, 96
Thiomargarita namibiensis, 45
Thioploca, 156
Thomas, Lewis, 126
thymidylate synthase, 109
timeline of cell evolution, ix–x, 52, 99f, 147f, 161–63, 217. *See also* universal phylogenetic tree
TOC and TIC (translocon outer/inner chloroplasts), 140, 141
TOM and TIM (translocon outer/inner mitochondria), 137, 138
transcription: Cavalier-Smith on, 63; in genetic core, 111; invention of, 55, 100; in LUCA, 58; machinery of, 182–83
transcription factors of Archaea, 44, 111

transfer RNAs, 182–83, 185; amino acid t-RNA synthetases and, 56
translation: Cavalier-Smith on, 63; as central problem in origin of life, 185; in genetic core, 111; invention of, 54–55, 91, 99, 100; in LUCA, 58; machinery of, 182–83; RNA World and, 185, 186
transport systems, 79, 80–81, 85; eukaryotic endomembranes of, 120; in LUCA, 186, 187; required with phospholipid bilayer, 176
transposons, 192, 200, 201
tree of life, 17; Darwin on, 18–19. *See also* universal phylogenetic tree
Trichomonas vaginalis, 106
tubulins, 118; Archaean, 123; distant homologs of, 112; highly conserved sequences of, 97; in planctomycetes, 46; prokaryotic FtsZ protein related to, 96, 97, 120. *See also* microtubules
2-methylhopanes, 148, 153
Tyler, Stanley, 147
type III secretion system, 89, 90f

Ueno, Y., 160
undulipodia: Margulis on, 5, 234n6; origin of, 114, 120–21; structure of, 38; unease about Darwinian evolution of, 213; of unikonts and bikonts, 109; unlike prokaryotic flagella, 119, 234n6
Unibacteria, 62
unikonts, 107, 108f, 109
universal phylogenetic tree, 19; alternatives to conventional version of, 60–67, 61f; based on whole genomes, 31, 216; cells as proper basis of, 220; current status of simile of, 219–20; current version of, 29–30, 30f; doubts concerning, 22–24, 27–29, 30–31; eukaryotes on, 22, 23f, 29–30, 30f, 106–7; membrane lipids and, 95–96; mostly composed of microbes, 22; origin of cells and, 51; phyla of Archea on, 29, 30f, 43; protein sequences and, 24–25; research leading to, 19–24, 23f; root of, 22, 51, 57, 59, 73, 216, 233n7; two principal claims of, 33; as web with rampant gene transfer, 27, 28, 28f, 33, 34, 191, 199. *See also* LUCA (last universal common ancestor); timeline of cell evolution

Urey, Harold, 166
urkaryote, 218, 219f

Valentine, David, 44
Van Niel, C. B., 5–6, 19
Varela, Francisco, 47, 86, 245n50
variation, 192–94; in Darwinian theory, 15, 16; diversity and complexity increased by, 197–98; intelligent design argument and, 211–12; in modern evolutionary synthesis, 190, 191, 194; not always random, 191; random, 2, 13; from rearrangement of genomes, 199–201. *See also* lateral gene transfer; mutations
Venter, Craig, 11
Verrucomicrobia, 46, 113
vertical inheritance: dominant even in prokaryotes, 220; emergence of three main domains and, 55–56, 101; as pivot of evolution, 31, 199. *See also* genetic core
vesicles: abiotic assembly of, 166, 167, 176–77; laboratory experiments with, 176, 178–79
Vetsigian, K., 56
Vibrio, flagella of, 88
viral genes: few cellular homologs of, 48; hallmark genes, 49
Virchow, Rudolf, 3, 12, 92
viruses: abundance of, 47; as agents of primordial genetic exchange, 100; as challenge to modern synthesis, 191; classification systems and, 7; five great clusters of, 48, 49f; general features of, 47–50; giant, 50; between the living and the nonliving, 47–48; origin of, 48, 219f,

220; origin of DNA in world of, 186; questioning the nature of, 17

Wächtershäuser, Günter, 177, 180
Waddington, Conrad, 15, 203
Waldbauer, J. R., 153
Weinberg, Steven, 226
Wheelis, Mark, 22, 30
White, Lynn, 23–24
Whittaker, Robert, 6
wholeness, 189, 222, 247n3
Wigner, Eugene, 215
Williams, George, 232n14
Wochner, A., 167
Woese, Carl: on ancestral archaeon, 44; on archaebacteria, 21, 22, 216; Cavalier-Smith analysis and, 63; on eukaryote energy metabolism, 85; interpretation of cell origin, 98; on last universal common ancestor, 53–57, 59; on lateral gene transfer, 100, 199; origin of eukaryotes and, 33, 117, 122, 218; on progenotes, 78, 99; ribosomal RNA research of, 20–22, 23, 30; RNA World and, 54, 59, 183, 184; on root of universal tree, 51; on translation, 91
Wolf, Yuri, 185
Wood-Ljungdahl pathway of carbon fixation, 77, 179
work, biological, 69, 71

Zhaxybayeva, O., 233n13, 238n4
Zillig, Wolfram, 22, 56, 235n2
zircons, 161
Z ring, 96, 97, 98